ENERGY PRODUCTION AND RESERVOIR WATER QUALITY

A Guide to the Regulatory, Technical, and Theoretical Basis for Required Studies

SPONSORED BY
Environmental Effects Committee of the Energy Engineering Division of the American Society of Civil Engineers

EDITED BY
James Martin
John Edinger
John Higgins
John Gordon

ASCE American Society of Civil Engineers
1801 ALEXANDER BELL DRIVE
RESTON, VIRGINIA 20191–4400

Library of Congress Cataloging-in-Publication Data

Energy production and reservoir water quality : a guide to the regulatory, technical, and theoretical basis for required studies / sponsored by Environmental Effects Committee of the Energy Engineering Division of the American Society of Civil Engineers ; edited by James Martin ... [et al.].
p. cm.
Includes bibliographical references and index.
ISBN-13: 978-0-7844-0896-4
ISBN-10: 0-7844-0896-3
1. Water quality. 2. Reservoir ecology. 3. Limnology. 4. Water quality management. 5. Hydroelectric power plants--Environmental aspects. 6. Coal-fired power plants--Environmental aspects. I. Martin, James Lenial, 1947- II. American Society of Civil Engineers. Environmental Effects Committee.
TD370.E54 2007
628.1'683--dc22

2007030020

American Society of Civil Engineers
1801 Alexander Bell Drive
Reston, Virginia, 20191-4400

www.pubs.asce.org

ISBN 13: 978-0-7844-0896-4
ISBN 10: 0-7844-0896-3
Manufactured in the United States of America.

Foreword

The ASCE Energy Engineering Division held a Workshop 2000 attended by all committee chairs to discuss problems of mutual interest for which task committees might be formed. The Environmental Effects Committee, the Fossil Generation and the Hydropower Committee brought up the problem of the effect of water quality regulations on power generation operations. This was truly a joint problem since power generation at both fossil and hydropower operations could affect the water quality within and downstream of impoundments and water quality conditions within a reservoir could often limit hydropower generation. In addition, many steam electric power plants are located along hydropower reservoirs; the thermal discharges from these plants can affect reservoir water quality, and the plants' withdrawals can entrain resident or migratory fish and other aquatic organisms.

A result of Workshop 2000 was to form a small group from the two committees to more specifically define the objectives of a task committee and to seek funding to support it. The group decided to form a task committee on "Energy Production and Reservoir Water Quality" within the Environmental Effects Committee. The objective of the task committee was to produce a guideline document on study methods and procedures for investigating the interrelationship between energy production and water quality in reservoirs.

The group applied for and obtained support through an ASCE Opportunity Fund award. The ASCE Opportunity Fund was created by the ASCE Board of Direction, and administered through the Opportunity Fund Committee, to encourage development of innovative programs that promote civil engineering and its goals, enhance value to members, or advance the public image of the profession. Additional funds for final manuscript preparation were made available through the ASCE TAC budget to the Energy Engineering Division. The task committee assembled to undertake the study and prepare the guidelines is shown in the list of contributors. Each of these individuals volunteered their own time to the task committee for meetings, correspondence, document writing and assembly. Some of their employees funded portions of the expenses of meetings, made facilities available for meetings, and provided editorial and library assistance. The volunteer contributions of time and their own employer resources leveraged the ASCE funding contributions by about 10:1.

There are a large number of steps in bringing a working group and document like this together. A detailed outline of the document was prepared and distributed to the contributors in early 2001. Working sessions were held in the Spring and Fall of 2001 with different groups of contributors to further revise the outline and to make writing assignments. Drafts of chapters prepared following these sessions underwent extensive review, re-write, and reorganization at a working session in the Spring of 2002. Contributors to individual chapters met and corresponded among themselves during the Fall and Winter of 2002, and chose a lead author for each chapter to bring the ii contributions together. The chapter lead authors met in the Spring of 2003 to review an assembled draft of the document, do further editing, and to prepare remaining material for copyediting and formatting. A small internal editorial group met to review the copyedited document, select and contact final outside reviewers, and prepare any required revisions in response to outside reviewer comments. The final document was submitted to ASCE Publications for printing and distribution in 2006.

At the time the Environmental Effects Task Committee was beginning this project, the ASCE TAC Engineering Mechanics Division (EMD) was completing a ASCE Publications report on Environmental Fluid Mechanics. The EMD made available draft summaries of its text chapters to the Task Committee for review, to serve as a reference and to minimize duplication of effort. In addition, the Task Committee held a session on its work at the EMD annual conference in 2002. As a result, this publication and the EMD report are highly complementary; the EMD report deals mostly with the theoretical aspects of environmental fluid mechanics, while this document is intended primarily to assist with resolving actual situations.

J. Russell Manson
John Eric Edinger
Task Committee Co-chairs
Environmental Effects Committee
ASCE Energy Engineering Division

Contributors

Jerad D. Bales
Water Resources Engineer
United States Geological Survey
3916 Sunset Ridge Road
Raleigh NC 27607

Jeffery A. Ballweber
Assoc, Dir. for Governmental Affairs
Water Resource GeoResources Inst.
Mississippi State University Box 9652
Mississippi State, MS 39762

Edward M. Buchak
Manager
Surfacewater Modeling Group
ERM Inc.
350 Eagleview Boulevard, Suite 200
Exton, PA 19341-1180

Sharon G. Campbell
Aquatic Ecologist
U.S. Geological Survey
2150 Center Avenue
Ft Collins, CO 80526-8118

John Eric Edinger
Consultant
63 Crestline Rd.
Wayne, PA 19087

John A. Gordon
Professor Emeritus
Tennessee Tech University
1318 Dave Huddleston Road
Cookeville, TN 38501

Craig Hesterlee
USEPA REGION 4
61 Forsyth Street, S.W.
Atlanta, GA 30303-8960

John M. Higgins
Program Manager
Tennessee Valley Authority
1101 Market Street (LP 3D-C)
Chattanooga, TN 37402

Mary Ann Hosko
Hydropower Consultant
250 North 24th St.
Camp Hill, PA 17011

Rajeev Jain
Principal
QUALMOD, LLC
2300 Placid Way
Ann Arbor, MI 48105

G. Lynn Jarrett
Cumberland Environ. Group, LLC
P.O. Box 446
Henryville, IN 47126

George Krallis
Surfacewater Modeling Group
ERM Inc.
350 Eagleview Boulevard, Suite 200
Exton, PA 19341-1180

Katherine F. Lindquist, PE
(Deceased)
Former Manager, Hydrothermal Team
Tennessee Valley Authority
PO Box 1649
Norris, TN 37828

J. Russell Manson
Computational Scientist
Macaulay Land Use Research Institute
Craigiebuckler
Aberdeen
Scotland

James L. Martin
Prof. and Kelly Gene Cook, Sr. Chair
Department of Civil Engineering
Mississippi State University
Box 9546
Mississippi State, MS 39762

Jerry Miller
Water Quality Scientist
U.S Bureau of Reclamation
Upper Colorado Region
125 South State Street
(UC-242)
Salt Lake City, Utah 84138-1147

Greg Pelletier
Department of Ecology
P.O. Box 47600
300 Desmond Drive
Olympia, WA 98504-7600

Scott Wells
Professor and Chair
Department of Civil Engineering
Portland State University
1719 S.W. 10th
PO Box 751
Portland OR 97207-0751

Table of Contents

1. INTRODUCTION

Mary Ann Hosko

John Eric Edinger

Reservoirs and impoundments are used in a number of energy production processes ranging from hydroelectric generation to providing cooling water for steam electric generation. Reservoir water quality both within the reservoir and in its downstream releases is continually being studied when changes in operations are made, new facilities added, or changes in water quality regulations come about. Energy production and reservoir water quality is an ongoing challenge, where energy production processes can influence reservoir water quality, and reservoir water quality can limit energy production.

This book on energy production and reservoir water quality brings together the regulatory framework driving the need for technical evaluations of the issues being raised and details of the tools presently available for assessing those issues. The regulatory framework presented in Chapter 2 shows that there are over 100 rules and regulations alone along with various combinations of these related to energy production and reservoir water quality. They are under the purview of numerous state and federal agencies and administered by a range of commissions and boards as well as ad hoc groups set up by the agencies for a specific project. The water quality review and regulatory process can take longer than facility construction required for changes in energy facility operations. Because of the long time frame, many projects are allowed to begin or continue operations as the final permitting, licensing and re- licensing procedures are underway.

Reservoir water quality studies require sound quantitative analyses based on hydrodynamic and water quality modeling. Chapter 3 presents background descriptions of the kinds of processes that can take place for different classes of reservoirs. Chapter 4 presents details of the hydrodynamic and transport relationships used in the analyses to quantitatively describe their circulation processes. Chapter 5 presents the quantitative relationships used to describe the water quality processes taking place within reservoirs as included in nutrient cycling eutrophication modeling, water sediment exchange formulations, and toxics substances processes.

Modeling is the name given to the quantitative analysis of reservoirs that incorporates the Chapter 4 hydrodynamic relationships and the Chapter 5 water quality relationships. A description of models and of the steps required to apply models are presented in Chapter 6. There are a number of publicly available models, but the steps in using these models are common to all model applications. Model applications are based on an assessment of available data and of processes as key elements that support problem definition and modeling objectives. The specialized modeling steps identified in Chapter 6 flow from these elements: identifying real time simulation periods worthy of study; mapping the prototype bathymetry to a finite difference grid; and acquiring, formatting, and reviewing the time-series boundary condition data. Once formal modeling has begun, confirmation becomes the focus, following by management simulations (problem solving) using the now-credible model application. Chapter 6 also addresses the historical background of modeling as a foundation for classifying models and provides guidance for model selection based on the technical and institutional properties of models.

Six case studies have been prepared that illustrate different aspects of applying the principles and procedures presented in the first six chapters. The cases include the study of a whole river system operating model to optimize power production yet meet water quality objectives; the interaction of a smaller hydroelectric reservoir with a larger storage impoundment affected by nutrient loadings with water quality consequences; analysis of a run of the river reservoir to determine effects of hydroelectric releases on downstream water quality; a systems analysis of a steam electric station condenser cooling water supply to meet once through cooling lake discharge temperature objectives; analysis of overflow weirs behind a hydroelectric dam to improve downstream water quality; and, a study of partial cooling tower operations to maintain cool water zones of refuge and passages for fishes in a reservoir arm and its influence on dissolved oxygen.

1.1. Reservoirs and Energy Production

In 2000, the existing capacity of U.S. electric utilities totaled 604,514 megawatts. Of the existing capacity, conventional and nuclear steam-electric units accounted for 73 percent (441,235 megawatts). Nuclear units accounted for 14 percent; hydroelectric (conventional), 12 percent; gas turbine, 8 percent; pumped storage hydroelectric, 3 percent; combined cycle, 3 percent; and internal combustion, geothermal, solar, wind and other, 1 percent.

Reservoirs support both hydroelectric generation and thermal generation. Reservoirs support once through cooling for thermal generation and evaporative make up water when thermal plants are operated on cooling towers. The US hydroelectric and thermal generation capacity is examined here along with the benefits, consequences, and future development of each source of electrical power.

Hydroelectric Generation

In a recent Electric Power Research Institute (EPRI) presentation of water and a sustainable future (Ref: 1) it was noted that there is a fast growing demand for clean, fresh water for a variety of purposes. All regions of United States are vulnerable to water shortages and there are many significant concerns associated with water use, and reuse. These include:

- Environmental protection and enhancement are significant cost items.
- Dependency of electricity supply and demand on water availability, for thermal cooling water and hydroelectric production, may result in shortages of electricity supply
- Despite the multi-purpose values of reservoirs, there remains a negative perception that may limit future reservoir development.

Indeed the multipurpose value of dams and reservoirs is dramatically illustrated in Figure 1-1. (Ref: 4)

With recreation reported as the primary purpose of dams in the U.S. according to the USACE National Inventory of Dams (Ref:4) and the reluctance to develop new reservoirs, there is an increasing realization that multipurpose use of dams and reservoirs must be studied to develop energy uses from existing facilities. With hydroelectric generation representing only 2 % of the dam primary uses, and reservoir uses for thermal generating requirements represented in the other 7%, much of the reservoir capacity is already allocated to other socio-economic sources. Preserving and increasing reservoir capacity for energy purposes will be key to national energy security.

As renewable resources, hydroelectric facilities utilize water directly and are not consumptive users of water. The 160,000 MW of hydroelectric generation in North America; representing the greatest renewable source of energy today, relies on reservoirs and rivers for operation.

Like wind, solar, geothermal and biomass, hydropower is a renewable source of electricity. Water, its "fuel," is essentially infinite, replenished by the hydrologic cycle, which is powered by the sun. By generating carbon-free electricity, hydropower avoids burning fossil fuels and releasing an amount of carbon dioxide that equals the annual exhaust of 62.2 million passenger cars, or half of the cars on U.S. roads.

Today's hydropower turbines are capable of converting 90 percent of available energy into electricity - that is more efficient than any other form of generation. Even the best fossil fuel power plant is only about 50 percent efficient.

Hydropower's operational flexibility - its unique ability to change output quickly - is highly valued, and will become even more so in a competitive market. Its unique voltage control, load-following and peaking capabilities help maintain the stability of the electric grid ensuring economic growth and a high quality of life.

The Future of Hydroelectric Production on Reservoirs

To provide a more accurate assessment of the domestic undeveloped hydropower capacity, the DOE Hydropower Program developed a computer model, Hydropower Evaluation Software (HES). HES allows the personal computer user to assign environmental attributes to potential hydropower sites, calculate development suitability factors for each site based on the environmental, legal, and institutional attributes present, and generate reports based on these suitability factors. Modeling of the undeveloped hydropower resources in the United States, based on environmental, legal, and institutional constraints, has identified 5,677 sites that have a total undeveloped capacity of about 30,000 megawatts. Table 1-1 shows the developed and potential hydropower resources at existing dams, as modeled by the DOE HES program.

There are 4,316 MW of "incremental" hydropower available at sites with existing hydroelectric facilities. "Incremental" hydropower is defined as capacity additions or improved efficiency at existing hydro projects.

According to river basin analyses, there are nearly 70,000 Megawatts (MW) of potential hydropower generation in the U.S. when only engineering and economic factors are considered. When screening for environmental, legal and institutional factors at potential sites, there are 29,780 MW of hydro generation-most of which can be developed without the construction of a single, new dam. There are 16,998 MW available at dams without hydroelectric capacity.

Additionally, the U.S. Department of Energy (DOE) has an ongoing interest in assessing the hydropower potential of the United States. (Ref:5) In FY 2000, DOE initiated planning for an assessment of hydropower potential for low head (less than 30 ft) and low power (less than 1 MW) resources. The assessment estimated that the total hydropower potential of the conterminous United States is approximately 120,000 MW. Low head/low power potential makes up about 19,000 MW of the total available potential, in conventional, unconventional, and micro hydro systems. In addition to the low head/low power potential, it is estimated that there is a total of 20,000 MW of high head (30 ft or greater)/low power potential available in the 48 states. Most of this development would require unconventional systems or micro hydro technology requiring significant turbine and system configuration research and development. The fact that this source of distributed power could be realized without the need for water impoundments is a positive attribute, but conducting the R&D necessary to develop these resources is uncertain.

Thermal Generation

Most large-scale electric power plants require water to operate. Daily utilization figures indicate that nuclear, fossil fuel, and geothermal power plants require over 195 billion gallons of water per day. However, since 98 percent of the water used in power plants is returned to its source, distinctions are made between use and consumption. Figure 1-2 illustrates the usages.

Water use is a measure of the amount of water that is withdrawn from an adjacent water body (reservoirs, streams, rivers, estuaries, etc.), passes through various components of a power plant, and is then ultimately discharged back into the original water body. Environmental concerns surrounding water use center around any chemical or physical alteration of the water body and any impacts these changes may have on the plants, fish, and animals that reside in the ecosystem. Water consumption refers to water evaporation lost during the cooling cycle. The primary concerns surrounding water consumption is how best to utilize this essential resource, especially in areas, such as deserts in the West, where water is in short supply.

Thermal electric generating facilities make electricity by converting water into high-pressure steam that drives turbines. Once water has gone through this cycle, it is cooled and condensed back to water and then reheated to drive the turbines again. The process of condensation requires a separate cooling water body to absorb the heat of the steam. These condenser systems typically consist of banks of thousands of one-inch diameter tubes, through which cooling water is run, and over which the hot steam and water is circulated. There are two cooling technologies are in use today:

Closed-cycle systems discharge heat through evaporation in cooling towers and recycle water within the power plant. The water required to do this is comparatively small since it is limited to the amount lost through the evaporative process. Because of the expense associated with closed-cycle cooling, once-through systems are far more common.

Once-through systems require the intake of a continual flow of cooling water. The water demand for the once-through system is 30 to 50 times that of a closed cycle system, however, most of the once-through cooling water flow is not consumed. The amount of water used for power plant cooling also varies by each specific power plant's electricity generating technology and size. Table 1-2 summarizes these uses by cooling type.

The Future of Thermal Energy Production on Reservoirs

A large portion of thermal generation is sited on hydroelectric reservoirs. Existing run of river hydroelectric impoundments in the Southeastern US presently support 0.03 MWe/Acre to 0.07 MWe/Acre of steam electric capacity on once-through cooling. FERC estimates that there is a total of 3 million acres of hydroelectric reservoir surface area in the US. If all of this could be loaded at the presently supported capacity, then existing hydroelectric reservoirs could potentially support an additional 90,000 MWe to 210,000 MWe of steam electric capacity. Much of the hydroelectric reservoir surface area is on large impoundments in the Western and Northwestern US that would not be loaded to the indicated levels. More stringent siting requirements being imposed on steam-electric stations would further reduce this estimate of additional capacity. It is estimated that only about one-forth of the potential or about 20,000 MWe to 50,000 MWe additional steam electric generation will be sited for once through cooling on hydroelectric reservoirs. It is expected that the new capacity will be located at existing steam-plant sites to take advantage of supporting infrastructure for fuel handling, fly ash disposal, and site security.

1.2. Uses of Report

The major purpose of this report on Energy Production and Reservoir Water Quality is to serve as a guideline for how to develop studies related to this problem, to provide the background to the available analysis tools, and to provide examples of different types of studies. Areas of required research and investigation are indicated throughout the report. Eventually the required water quality studies need to be related project economic decisions. It should be a goal of the ASCE Energy Engineering Division to foster ASCE's interest in becoming a Standards setting professional group by expanding these guidelines to standards for the hydrodynamic and water quality modeling required for the study of energy production and reservoir water quality.

1.3. References

1. EPRI, Smith, B, Goldstein, B., and Carns, K., *Water and Sustainability –The EPRI Research Plan* Water & Sustainability Workshop, Washington, DC, July 25, 2002.

2. National Hydropower Association (NHA) www.hydro.org.

3. U. S. Geological Survey (USGS). Estimated use of water in the United States in 1990 Thermoelectric Water Use http://water.usgs.gov/watuse/wupt.html

4. U.S. Army Corps of Engineers (USACE), National Inventory of Dams (NID), www.tec.army.mil/nid.

5. U.S. Department of Energy, Energy Efficiency and Renewable Energy, Wind and Hydropower Technologies, Idaho Operations Office, Low Head/Low Power Hydropower Resource Assessment Project, Idaho National Engineering and Environmental Laboratory, Idaho Falls, ID, October 2003.

6. U.S. Department of Energy, U.S. Hydropower Resource Assessment, Idaho National Engineering and Environmental Laboratory, Idaho Falls, ID. December 1998

Table 1-1 US Hydropower Potential at Existing Dams Ref:(6)

Category	Number Of Projects	Capacity (MW)	Adjusted Capacity
With Power	389	7,820	4,316
W/O Power	2,527	29,625	16,998
Undeveloped	2,761	32,452	8,466
Total	5,677	69,897	29,780

Table 1-2 Cooling Water Withdrawal and Consumption (Evaporation) rates for Common Thermal power plant and Cooling System Types (Ref:1)

Plant and Cooling System Type	**Water Withdrawal (gal/MWh)**	**Typical Water Consumption (gal/MWh)**
Fossil/biomass/waste-fueled steam, once-through cooling	20,000 to 50,000	~300
Fossil/biomass/waste-fueled steam, pond cooling	300 to 600	300-480
Fossil/biomass/waste-fueled steam, cooling towers	500 to 600	~480
Nuclear steam, once-through cooling	25,000 to 60,000	~400
Nuclear steam, pond cooling	500 to 1100	400-720
Nuclear steam, cooling towers	800 to 1100	~720
Natural gas/oil combined-cycle, once-through cooling	7500 to 20,000	~100
Natural gas/oil combined-cycle, cooling towers	~230	~180
Natural gas/oil combined-cycle, dry cooling	~0	~0
Coal/petroleum residuum–fueled combined-cycle, cooling towers	~380* includes gasification process water	~200

Figure 1-1 Dam and Reservoir Primary Purposes Ref: (4)

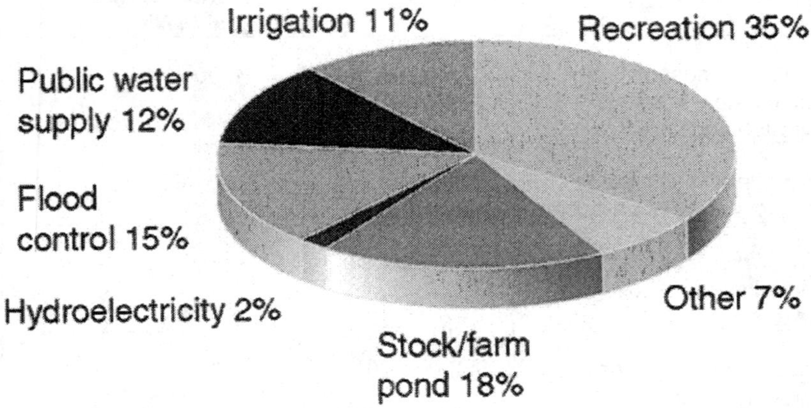

Source: U.S. Army Corps of Engineers, National Inventory of Dams

Figure 1-2 Thermoelectric Water Power Source and Disposition, 1990, in percent

195,000 gallons per day, Ref:(3)

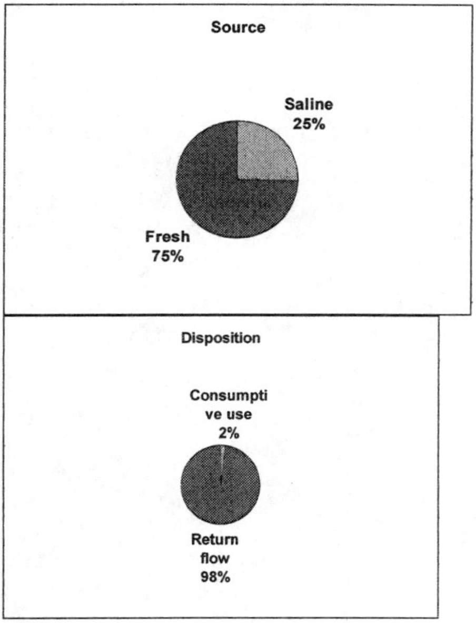

2. REGULATORY FRAMEWORK

James L. Martin

Craig Hesterlee

Jeffery A. Ballweber

The regulatory framework affecting power generation is complex and in a state of flux. There are a variety of regulations affecting, and agencies involved in, the permitting or licensing and operation of energy production facilities. While not intended to be exhaustive, the purpose of this Chapter is to review some of these major issues within the context of energy production and reservoir water quality. In the design of studies related to this topic it is necessary to first identify the studies' goals and objectives. Often those goals are driven to some extent to address management decisions related to regulatory issues.

Like all water resources development projects, there are a variety of federal and state laws, regulations, executive orders and case law impacting energy production facilities, some of the more relevant federal provisions are tabulated below. By some count, for example, there are over 25 federal statutes that impact some part of the hydropower licensing process (FERC 2000b), many of which are related to natural resource issues. Individual state's laws and compacts between states also impact the licensing and permitting process and the operation of power generation facilities. The legal environment governing the planning, construction, operation and maintenance and re-licensing of these types of facilities have evolved over time and continue to evolve. States may vary in the manner in which laws and regulations are implemented. For example, states take various approaches to administer and enforce their own equivalent of the federal Clean Water Act (CWA), such as in the listing of impaired waterbodies [under Sections 305(b) and 303(d)], and the implementation of water quality certifications or wavers (under Section 401). In addition, the CWA includes Sections 316(a) and 316(b) that influences the joint operations of steam electric power plants and hydroelectric projects as they affect water temperatures and intake entrainment.

The interpretation of these laws by the Federal Court system, and in some cases their requirements for the enforcement of existing laws (such as provisions of the CWA) is also impacting the present regulatory environment. Thus, the regulatory issues impacting the licensing or permitting and operation of power generation facilities, and the litigious nature of those issues, make the regulatory environment in which power generation facilities must operate complex. To understand these issues it is helpful to categorize legal authority as 1) authorization; 2) procedural, and 3) regulatory. This is an inexact categorization and many individual statutes elements of all three categories. Still,

2.1. <u>Authorizations</u>

1902: Reclamation Act: Authorized Secretary of the Interior to develop irrigation and hydropower projects in the 17 Western States

1906: Town Sites and Power Development Act: Authorized Secretary of the Interior to lease surplus power or power privileges

1920: Federal Water Power Act: Regulated hydroelectric development of navigable waterways

1933: Tennessee Valley Authority Act: Authorized establishment of the Tennessee Valley Authority with a fairly broad mission including the development of hydro-electric power

1935: Rivers and Harbors Act: Authorized Grand Coulee Dam for construction

1939: Reclamation Project Act: Extended the contract term to 40 years for sale of power or lease of power privileges, giving preference to public entities

1944: Flood Control Act: Authorized Secretary of the Interior to market power from Army Corps of Engineers projects and authorized the Pick-Sloan Missouri Basin Program

1977: Department of Energy Organization Act: Transferred existing Power Marketing Administrations to the Department of Energy and created Western Area Power Administration

1978: Public Utility Regulatory Policies Act: Encourages small-scale power production facilities; exempted certain hydroelectric projects from Federal licensing requirements, and required utilities to purchase - at "avoided cost" rates - power from small production facilities that use renewable resources

1980: Energy Security Act: Exempted small-scale hydroelectric power from some licensing requirements

2.2. Regulatory Emphasis

1972: Coastal Zone Management Act (CZMA): Requires state certification that all federally conducted or supported activities, including development projects directly affecting the coastal zone, are consistent with approved state coastal management programs. The CZMA was reauthorized in 1990 with the requirement (§ 6217) for states with federally approved coastal zone management programs develop and implement Coastal Nonpoint Control Programs.

1973: Endangered Species Act: Protects listed threatened and endangered species and their critical habitat

1977: Clean Water Act (CWA): Aims to improve environmental water quality to meet national goal of fishable/swimmable waters. For the most part, the CWA is administered by individual states which have the choice of either setting their own water quality standards or accepting national standards set by EPA. The CWA specifically includes Section 316(a) and 316(b) related to steam plant cooling water discharges and intake entrainment. State agencies also issue permits under the National Pollutant Discharge Elimination System, and usually administer the Total Maximum Daily Load program. Finally, states monitor compliance and enforce the act subject to EPA oversight and preemption. Only very rarely have states sought delegation to issue "dredge and fill" (wetlands) permits.

1996: Sustainable Fisheries Act (Amendments to the Magnuson-Stevens Fishery Conservation and Management Act): Requires [Section 303(a)(7)] identification of Essential Fish Habitat (EFH) for managed species and identification of measures to conserve and enhance the habitat necessary for fish to carry out their life cycles

2.3. Procedural Emphasis

1980: Pacific Northwest Electric Power Planning and Conservation Act: Authorized Bonneville Power Authority to plan for and acquire additional power resources

1980: Crude Oil Windfall Profit Tax: Provided tax incentives to small-scale hydropower producers

1974: Fish and Wildlife Coordination Act: Ensures equal consideration of fish and wildlife protection in the activities of Federal agencies

1968: Wild and Scenic Rivers Act: Protects rivers in their natural state by excluding them from consideration as hydroelectric power sites

1969: National Environmental Policy Act: Ensures that environmental considerations are systematically taken into account by Federal agencies

1986: Electric Consumers Protection Act: Amended the Federal Power Act to remove public preference for relicensing actions; gives equal consideration to non-power values e.g., energy conservation, fish, wildlife, recreation, etc.) as well as to power values when making license decisions

1992: Energy Policy Act of 1992: An act to provide for improved energy efficiency. Includes provisions to allow for greater competition in energy sales and amendments to Section 211 of the Federal Power Act.

There are a variety of federal agencies which have some role in the regulation of power generation facilities. The U.S. Environmental Protection Agency is the primary federal agency responsible for implementing the Clean Water Act (CWA) provisions which require, for example, National Discharge Elimination System (NPDES) permits for any discharge of "pollutants (which will be defined further later) and include thermal and storm water discharges) into waters of the United States. NPDES permits are also required for cooling water intakes (under Section 316(b) of the CWA).

Hydroelectric power projects built by the federal government are authorized by Congress and constructed primarily by the U.S. Department of the Interior (Bureau of Reclamation), the U.S. Army Corps of Engineers, and the Tennessee Valley Authority. For most non-federal hydroelectric power projects, the Federal Energy Regulatory Commission must issue a license authorizing construction, or in the case of an existing project, continued project operation. Most hydroelectric projects serve other purposes such as navigation, flood damage reduction, recreation, and irrigation, and flow augmentation.

Federal Agencies that may be involved in a relicensing may include the Fish and Wildlife Service, National Marine Fisheries Service, Forest Service, Bureau of Land Management, National Park Service, Bureau of Indian Affairs, Bureau of Reclamation, Army Corps of Engineers, and the Environmental Protection Agency. The federal resource agencies may exercise their authority under Sections 4(e), 10(j), and 18 of the Federal Power Act, the Fish and Wildlife Coordination Act and the Endangered Species Act, and other federal statutes.

State agencies implement the CWA provisions, including NPDES permitting and 401 water quality certifications, and participate in relicensing through Section 10(j) of the Federal Power Act, the Fish and Wildlife Coordination Act, the Clean Water Act, the Coastal Zone Management Act, and the National Historic Preservation Act. Specific state laws and regulations may also impact the permitting and licensing processes. In *S.D. Warren Company vs. Main Board of Environmental Protection*, the Supreme Court upheld the 401 certification process as part of the hydropower licensing procedure (Landers 2006)

There are a number of initiatives designed to simplify, clarify and coordinate some of the regulatory issues, particularly as related to coordination among federal resource agencies involved in the licensing or permitting process, some of which will be discussed in this Chapter.

2.4. Water Quantity and Use Issues

Water supply issues are becoming increasingly important in the design and operation of water projects. Water quantity and use issues directly impact the quantity of water available for power generation. In addition, water quantity and quality issues are inexorably linked, and water quality issues also directly impact the quantity of water available for power generation. Examples of regulatory issues impacting water quantity and use include state's permitting authorities, state compacts, and federal water resource and land management. Examples of regulatory issues related to water quality impacting water quantity and use include water quality criteria, permitting and standards issues, instream flow issues, and effects of flows on fish and wildlife habitat.

States and local agencies have the primary responsibility for developing and managing water supplies and permitting water usage. State laws concerning the appropriation of water, conditions and preference for use, and administration of water rights and distribution vary widely and a discussion of state regulations is beyond the scope of this text. The applicability of state and local law also varies. For example, drought-induced water issues, such as minimum instream flows, are not within the purview of local government. States typically permit the use of state's waters, where the permit requires review of the use effects on other water rights, bays and estuaries, conservation, water availability, public welfare, and other considerations. Permits and authorizations may also be required related to water quality issues. In 1994, the United States Supreme Court (Jefferson Vs Washington) ruled that flow is an essential element of water quality standards. The Court held that if there is not sufficient flow to support the uses designated in water quality standards, water quality standards are violated. As such, the state's role in implementing provisions of the Clean Water Act impacts other water usage issues, such as listing and permitting discharges as discussed in the following sections. For example, Section 401 of the Clean Water Act requires an applicant for a federal license or permit to apply for certification from the state that the potential discharge from the activity will not violate state standards. The state must deny certification if compliance cannot be ensured. Any such federal license, such as a FERC hydropower license, must include these conditions. The certification may include water quality and instream flow requirements to protect the designated beneficial use of the impacted waters. In *S.D. Warren Company vs. Main Board of Environmental Protection*, the Supreme Court upheld the 401 certification process as part of the hydropower licensing procedure (Landers 2006)

Interstate compacts governing allocation and flow of water impact the quantity of water available for power generation and other uses. Interstate compacts are agreements between states that contractually bind them to the agreement, and in which compliance can be enforced by Congress and the courts. Commonly, compacts involving water use issues require the consent of Congress. Also commonly, interstate water commissions are established to implement the compacts in conjunction with federal and state agencies operating water resource management facilities. The Council of State Governments (CSG, http://ssl.csg.org/compactlaws/comlistlinks.html) lists nearly 200 compacts between states believed to be in effect in 2001, of which the majority deals with apportionment of interstate waters between affected states. Additional information on individual compacts is provided by the CSG (1997, 1998). Examples include a variety of compacts for the Colorado River Basin (including the Colorado River Compact of 1922 and Upper Colorado River Compact of 1948), Rio Grande (1938), Costilla Creek (1963), Arkansas

River (1963) and Pecos River involving the states of Kansas, Wyoming, Utah, Arizona, Colorado, New Mexico and Texas. Examples of new and pending compacts include those between Georgia and Alabama for the Alabama-Coosa-Tallapoosa river system (ACT), and between Georgia, Alabama, and Florida for the Apalachicola-Chattahoochee-Flint river system (ACF). Under these compacts, the three states will develop agreements for allocating and managing the waters of the interstate river systems to meet future water needs. One of the issues involved in the ACT and ACF compacts is the management of federal reservoirs such as lakes Lanier, West Point, and Allatoona, all of which have power generation authorizations. An issue, for example, is whether reservoir operating levels should preserve sufficient water supplies to meet human water needs at all times, including severe droughts, which would impact how the federal facilities have traditionally operated. State compacts, and the agencies involved in the implementation of those compacts, which may include the Army Corps of Engineers, Bureau of Reclamation, Compact Commissions, and state water quality agencies, are also the subject of litigation regarding the affect of those compacts on environmental issues.

The federal role in water quantity and use issues is complex and involves a wide variety of agencies that have a function related to water in the United States. The Federal Regulatory Commission must license all non-federal hydropower facilities, and FERC licensing issues are discussed later in this Chapter. Under the Federal Powers Act [Section 4(e), Equal Consideration] , FERC is required to give equal consideration to development and environmental values, including irrigation, flood control and water supply in issuing a hydropower license.

Three agencies primarily involved in the construction and operation of water projects and the management of federal lands are the U.S. Army Corps of Engineers, U.S. Bureau of Reclamation and the U.S. Bureau of Land Management. Sections 6 and 8 of the 1944 flood control act, as amended, allow the Corps of Engineers and Bureau of Land Management to allocate water for storage and agricultural water supply. The 1958 Water Supply Act provided for cities and industries to tap into Corps of Engineers reservoir storage to meet municipal and industrial water supply needs. Under U.S. Code (Title 43 Public Lands, Chapter 12, Subchapter I), storage may be included in any reservoir project planned, constructed, or operated by the Corps of Engineers or the Bureau of Reclamation in order to impound water for present or anticipated future demand or needs for municipal or industrial water. Such use can be considered in estimating the economic value of the project. Typically, projects can be authorized for such uses as flood-protection and fish and wildlife enhancement, whose costs are borne by the federal government. However, costs for local uses such as hydropower or water supply must be borne by the local sponsor. Changes in the operations of existing or authorized projects, such as to include water supply storage, which would involve major changes in the structure or operation of the facility, can only be made with the approval of Congress.

Federal agencies, such as the U.S. Corps of Engineers, Bureau of Land Management, Forest Service, Bureau of Reclamation, National Park Service and other federal agencies are also involved in land management, which involves water rights and use issues. Under the doctrine of reserved federal water rights, the courts assume that the United States government, in dedicating property from the public domain for a specific public purpose intended to reserve sufficient water rights to accomplish that purpose. The doctrine of federal reserved water rights has been a source of controversy in western states as water supplies diminish and water required for such federal purposes increases. Many of the regulatory issues impacting water quantity and use are related to water quality issues. Under the Western Water Policy Review Act of 1992, Congress directed the President to undertake a comprehensive review of Federal activities in the nineteen Western States affecting the allocation and use of water resources, and to submit a report of findings to the President and Congress. The President appointed the Western Water Policy Review Advisory Commission (Commission).

Federal activities impacting water quantity and use issues as related to water quality were addressed in the National Environmental Policy Act (NEPA). In enacting the National Environmental Policy Act (NEPA), Congress recognized that nearly all federal activities affect the environment in some way and mandated that before federal agencies make decisions, they must consider the effects of their actions on the quality of the human environment. As part of NEPA, Congress assigned the Council for Environmental Quality (CEQ) within the Executive Office of the President with the task of ensuring that federal agencies meet their obligations under the Act. Additional responsibilities were provided to the CEQ by the Environmental Quality Improvement Act of 1970. The Federal Water Pollution Control Act Amendments of 1961 also stipulated that Federal Agencies consider storage to regulate streamflow for the purpose of water quality control during the planning of any reservoir.

An additional regulatory issue has been the authority of the U.S. Environmental Protection Agency (EPA) to address water quantity as it relates to water quality. Under Sections 101(g; the "Wallup Amendment") and 510(2) of the Clean Water Act (CWA), the EPA is proscribed against any unnecessary Federal interference with state's water rights. However, Section 510(2) further states that state's water rights are not to be impaired "except as expressly provided in the Act," and issues impacting water quality may frequently impact water quantity and use issues. For example the Water Quality Standards Handbook (USEPA 1994), indicates that: Section 101(a) of the CWA recognizes the importance of preserving the physical integrity of the Nation's water bodies, and physical parameters such as flow, temperature and other factors identify any physical limitations that may preclude the attainment of a designated use. A number of EPA's water quality criteria are flow based, and flows often directly impact water quality as well as loads as permitted under the National Pollutant Discharge Elimination System (NPDES). The CWA requires that state's designated uses to meet the CWA's fishable/swimmable goals including designated uses to provide for the protection and propagation of fish, shellfish and wildlife. In Jefferson v. Washington (1994) the Supreme Court held that minimum flows may be defined and implemented to ensure that the uses (designated and existing uses) of a water quality standard are not impaired.

An additional issue related to implementation of the CWA [Section 303(d)] is that of effluent trading, which is encouraged by the USEPA (USEPA 1999a,b, 2004) where a discharger may be requested to, or choose to, mitigate their discharge by reducing the load from another source. This allows, for example, a point source to arrange for control of non-point source discharge(s) in a watershed in lieu of upgrading its own treatment, as long as the Total Maximum Daily Load (TMDL, see following section for definition). An additional issue in effluent trading is related to increased flows. For example, Crossman et al. (2000) described a TMDL study on the South Fork Holston River near Kingsport, TN, where it was concluded that water quality standards could be achieved with an increased flow (from 750 to 1000 cfs in a pulsing versus a peaking operation), coupled with flow aeration, at a cost two orders of magnitude less than the cost of an advanced wastewater treatment facility.

2.5. Water Quality Impairment Lists and Pollution Allocation Issues

State water quality programs have primary responsible for over-seeing the development of water quality impairment lists and subsequent pollution allocation. EPA reviews and approves these lists and backstops the states' pollution allocation responsibilities. Fundamental to the Clean Water Act of 1977 is the feature requiring states to develop water quality standards for waterbodies, monitor for compliance with these standards, and report impairments. States then are required to develop watershed-based pollutant load reduction goals and allocations for waterbodies deemed impaired. These watershed-based pollutant allocations are collectively known as total maximum daily loads (TMDLs).

According to EPA over 218 million Americans live within 10 miles of a water quality impaired waterbody. Over 5 million lake acres are considered impaired. Excess sediments, nutrients, and pathogens are considered leading reasons for impairments.

From 1972 to 1985, EPA focused the majority of its resources on upgrading treatment of wastewater from direct regulated point sources such as municipal and industrial facilities. The Clean Water Act has called for the development of TMDLs since 1977, and in 1985, EPA first issued TMDL regulations with provisions for addressing non-point sources of pollution. In 1992, the regulations were revised with provisions for states to list impaired waterbodies every two years. Additional revisions to TMDL regulations were proposed in 1999 (largely to address state concerns regarding listing requirements) and adopted in 2000, but a congressional rider adopted the day after the rule was finalized prohibits EPA from implementing the 1999 regulations. Currently, most state programs and the regulated community are pushing for major modifications to the existing TMDL rule, whereas most environmental organizations prefer to see the current rule unchanged.

2.6. Standards and Criteria

Often cited as a "cornerstone" of the Clean Water Act, the water quality standards program requires states to set standards for their waters (CWA (Section 303)). Water quality standards must have three components: 1) beneficial designated uses, 2) numerical criteria protective of designated uses and 3) anti-degradation policies. Standards have the impact of setting water quality goals for waterbodies as well as establishing water quality based treatment practices for dischargers. Standards are adopted for each waterbody within a state, and states may segment waterbodies so that different standards may exist for the same waterbody where appropriate. The states are responsible for adopting standards, and EPA may establish standards if the states fail to adopt suitable standards. States are required to review their adopted standards every three years and modify them as necessary. EPA reviews new or revised state standards and has issued guidance to the states regarding minimum methodologies that should be used to develop standards acceptable under the Clean Water Act.

Dischargers must target water-quality based effluent limits when technology-based effluent limits fail to result in the obtainment of the receiving waterbody's water quality standards. Standards can be waived if a discharger proves that the social and economic cost of meeting the ambient water quality standards outweighs the environmental benefit.

2.7. Impairment Determination

State are required to report waters that are impaired in a report known as the 303(d) report after the requirement's citation in the regulations (CWA Section 303(d)). Reporting methodologies vary from state to state as do impairment thresholds. States should consider all existing and readily available water quality-related information for making impairment determinations. States are required to provide "good cause" for not including a waterbody on the 303(d) list at EPA's discretion. The 303(d) list has certain categories whereby states determine the necessity for development of a TMDL based on the nature of the listed waterbody's impairment and remediation status. Because of increasing controversy between state regulators, environmentalists, and industry and farming advocates, Congress authorized the National Research Council to review the effectiveness of the Clean Water Act's current TMDL requirements. Particular emphasis in the subsequent report to Congress indicated that the TMDL program is hampered by a pervasive lack of data to support impairment listing and subsequent TMDL development (NRC, 2001). Many state water quality programs, as well as EPA, are considering modifications to their impaired waters rules that will allow for waterbodies to be placed into a planning subcategory in their 303(d) list, whereby there exists some data that indicates impairment might exist, but not sufficient data to determine impairment exists and develop pollution allocations. Regulators hope this new category will help to focus limited monitoring resources on data gaps and improve the scientific defensibility (and implementation potential) of TMDLs that are developed. Environmental groups are concerned that a "planning" category will unnecessarily result in a delay of the development of TMDLs.

2.8. Total Maximum Daily Loads

TMDLs represent the amount of pollutant a waterbody can receive and still maintain water quality standards. Throughout the 1970s and 1980s, the Clean Water Act requirements to develop TMDLs to address point and non-point sources of pollution under CWA Section 303 were largely ignored by both the states as well as EPA. Regulators focused Clean Water Act resources on the NPDES (national Pollution Discharge Elimination System) permitting program, controlling point sources and establishing waste load allocations for point sources. Tremendous progress was made during this period in terms of point source control. Emphasis shifted in the 1990s towards non-point source pollution abatement, largely driven by citizens' and environmental group lawsuits against EPA for not exercising regulatory authority over state programs to require development of TMDLs as stipulated by the Clean Water Act since the 1970s. Nationwide there are currently over 40,000 TMDLs scheduled for development within an 8 to 13 year time frame depending on the state.

Elements of A TMDL and Schedules

TMDLs must include a total wasteload allocation and a total load allocation. The wasteload allocation represents loading from regulated (NPDES-permitted) point sources and the load allocation (all other loads) is the sum of background (sometimes called natural) loads and non-point source loads. TMDLs should contain a margin of safety and consider seasonal variation. States are to identify waters targeted for TMDL development every two years. Waters scheduled for TMDL development should be prioritized by the states based on the waters' designated uses and magnitude of degradation.

TMDL Issues Related to Reservoir Water Quality

TMDLs are developed to address waterbody impairments due to pollutants as opposed to pollution. This represents a significant distinction. “Pollutants” are substances that degrade water quality or impair beneficial uses when loaded to a waterbody. Some examples of pollutants include sewage, garbage, heat, biological materials, toxic chemicals, sediment, and agricultural runoff. The Clean Water Act specifically defines “pollution” as the man-made or man-induced alteration of the physical, biological, chemical, or radiological integrity of water. Pollution can exist without pollutants such as biological impairment caused by dredging, widening, or impounding waterbodies. Larger peaking hydrographs attributed to increasing impervious surfaces in urbanizing areas can result in the alteration of benthic biological integrity, and this also is considered “pollution” where there exists no specific “pollutant”.

Much regulatory attention over the past decade has focused on dissolved oxygen (DO) and temperature in reservoirs. Reservoirs often become thermally and chemically stratified, and regulators can consider temperature profiles and oxygen depletion rates in both the epilimnion (upper layers) as well as the hypolimnion (lower layers). Many states list reservoirs as “impaired” when epilimnetic dissolved oxygen concentrations fall below numerical criteria determined to protect the beneficial uses of the reservoir. Many states have both daily average as well as absolute minimum criteria for dissolved oxygen. In productive waters large diurnal (daily) swings in dissolved oxygen can occur, whereby continuous monitoring might indicate a violation of a state’s absolute criteria, yet maintenance of a daily average criteria. Dissolved oxygen deficit in epilimnetic waters of reservoirs can be attributable to both pollution and pollutants, whereby conditions for large diurnal swings in DO would not exist without physical impoundment (pollution), but this condition is often exacerbated by anthropogenic loading of oxygen demanding substances (pollutants). It is appropriate to develop TMDLs to address pollutant loading to reservoirs to maintain in-pool DO criteria, and state programs should design monitoring plans that help differentiate between the impacts of upstream sources of pollutants, internal sources of pollutants, and the effects on DO caused by pollution (DO depletion caused by the physical effects of impoundment alone). Often regulators must use scenario modeling to achieve this distinction. In-pool temperature, like DO, can be impacted by both pollutants (point and non-point source heat loading above natural conditions) as well as pollution (elevated temperatures due to thermal stratification). Regulators must exercise judgment when determining if a TMDL is necessary to address in-pool violations of state temperature criteria. Some states have flexible temperature criteria whereby the physical impacts of impoundment on water temperature is considered to be a background condition and so specific numeric criteria do not apply.

Many reservoirs are listed on state 303(d) lists as being impaired for nutrient enrichment. Traditionally many states have adopted narrative nutrient criteria to allow for flexibility and adaptive management in assessing reservoir water quality. An EPA initiative is underway to encourage states to establish numerical nutrient criteria for waterbodies including reservoirs. Often states develop monitoring criteria to detect the impacts of increasing nutrient loadings on the trophic state of a reservoir. These indicators might be algal biomass, chlorophyll-a as an indicator of algal biomass, Secchi depth and light penetration, or even hypolimnetic oxygen depletion rates. TMDL developers must translate narrative standards into numerical targets and the aforementioned indicators are examples of targets that might be adopted as TMDL targets. As states develop nutrient concentrations and loads that are protective of individual waterbodies' designated uses, it is a regulatory goal that some of the uncertainties associated with TMDLs based on narrative criteria will be removed.

Often states list tailwater release as not meeting state water quality standards due to relatively elevated release temperatures when compared to non-impounded free-flowing reference water temperatures. Hypolimnetic dissolved oxygen concentrations in release waters often do not meet state DO criteria for streams, and many tailrace waters are 303(d) listed for this reason. Many reservoirs thermally stratify and then mix seasonally. The sudden mixing resulting from turnover can result in increased nutrient and oxygen demanding substance loading to the tailrace as well as iron and manganese loading. These substances, if causing a violation of state water quality standards, could be construed as pollutants and thereby require a TMDL under the Clean Water Act. As discussed in a later section, water quality can also be considered in the state's certifications under section 401 of the CWA, as recently upheld by the Supreme Court (Landers 2006), impacting hydropower licensing. Usually low dissolved oxygen or elevated temperatures in tailwater release can be considered a near-field short-term problem. Downstream from the immediate influence of the tailrace, a stream's temperature and dissolved oxygen can be more impacted by local environmental and meteorological factors than by the impoundment. Tailrace DO concentrations and temperatures can result from the physical impacts of stratification due to impoundment and also from the loading of oxygen-demanding pollutants or heat to a reservoir. TMDLs can address tailrace DO when a load reduction on an oxygen demanding pollutant results in the obtainment of State DO criteria in the tailrace. Models are often used by the regulatory community to quantify the impacts of pollutants versus pollution.

Pollutant Allocations and TMDL Implementation

As previously discussed, TMDLs can be expressed in terms other than "total maximum daily loads". To address reservoir and tailrace impairments, it is important to consider a waterbody's response time to pollutant load abatement. Due to retention time dynamics many reservoirs respond biologically and chemically to seasonal or even annual loads as opposed to daily loads. Many TMDLs are developed to address nutrient enrichment with seasonal or annual loads. States are developing criteria to measure the trophic status of reservoirs at seasonal scales and it is appropriate for allocation scenarios to also be developed at seasonal scales, particularly when wet-weather loading and non-point source loading are a significant portion of the pollutant budget. Transient and pulse loading events can be more readily quantified over a longer time scale than loads from point sources such as wastewater treatment plants.

Reservoir operators can impact the extent of "pollution" on a waterbody through reservoir operation, and they may receive TMDL pollutant allocations for tailrace pollutant loads. It is important that reservoir operators participate in the TMDL development process by finding ways to optimize water quality in release waters. Otherwise the TMDL regulation might be construed as placing an undue burden on upstream sources of pollutants when changes in reservoir operation might lead to more cost efficient and implementable environmental results. A recent Oregon TMDL, the Western Hood Subbasin Temperature TMDL (http://www.deq.state.or.us/wq/tmdls/Hood/HoodTMDLFinal.pdf) addresses the impacts that a dam structure has on river temperature. This TMDL assigns responsibility (in the form of load allocation) to the dam for changes in temperature that can be directly attributable to its presence and operation. This includes holding it accountable for some of the changes related to changes in flow that are caused by its operation. When point sources are present, EPA and the state programs enforce compliance with NPDES requirements including TMDL requirements. Sources without NPDES permits, such as dams, still receive load allocations whereby implementation responsibilities lie with other designated agencies or fall under different sections of the Clean Water Act outside of Section 303. The Clean Water Act does not give EPA or the states authority to issue NPDES permits to dams, but in receiving a load allocation as part of a TMDL, the pollutant loading associated with impoundment is accounted for in determining the capacity of a waterbody to assimilate pollutants. Although dam operators do not receive a wasteload allocation, CWA Section 401 certification provides the pathway by which the dam operators are required to meet their allocation by addressing "pollution" caused by impoundment.[1]

---

[1] Since this text was written, EPA has developed a draft document "National Management Measures To Control Nonpoint Source Pollution From Hydromodification," [Federal Register: July 17, 2006 (Volume 71, Number 136)] which continues to support voluntary compliance

2.9. Fish and Wildlife Issues

There are numerous procedural and substantive regulatory issues related to the actual or potential impact of existing and proposed power production facilities on fish and wildlife. Issues include minimum instream flow requirements, the impact of altered flows, the impact of thermal discharges, inundation of aquatic and or terrestrial habitat and reservoir water level fluctuations, fish mortality in turbines and cooling water intakes and passage, endangered species, fish and wildlife recreation, and habitat alterations such as due to scour below peaking hydropower operations. Some of the impacts of hydropower facilities on downstream water quality are discussed by Ashby et al. (1999).

The federal agencies primarily charged to conserve and manage fish and wildlife resources are the Department of the Interior, acting through the Fish and Wildlife Service (FWS), and the Department of Commerce, National Oceanic and Atmospheric Administration, acting through the National Marine Fisheries Service (NMFS). Primary procedural authorities to consider fish and wildlife issues are found in the Fish and Wildlife Act of 1956, the Fish and Wildlife Coordination Act (FWCA), and the Federal Power Act (FPA). The FWS and NMFS are also involved in implementation of the National Environmental Policy Act (NEPA), which requires Federal agencies to seriously consider potential environmental impacts in the decision making processes, and submit alternative proposed project plans to other Federal agencies and the public for review and comment. Other federal agencies involved in fish and wildlife issues include the U.S. Environmental Protection Agency (EPA), in their implementation of the Clean Water Act (CWA), which requires that state's designated uses meet the fishable/swimmable goals of that Act, and to meet the fishable goal, the designated uses must provide for the protection and propagation of fish, shellfish and wildlife. The Endangered Species Act (ESA) is the preeminent fish and wildlife regulatory law. The NMFS has federal responsibilities for marine, estuarine and anadromous fish under the FPA, the ESA, the Magnusun-Stevens Fishery Conservation and Management Act (Magnuson-Stevens Act). Important additional provisions are found in:

Clean Water Act- dredged and fill material into Waters of the United States-Section 404 requires permits to discharge dredged or fill material into water of the United States, including wetlands. Section 404 also includes specific reference to the FWS in various aspects including review of applications for permits to discharge.[2]

---

[2] Note that since this Chapter was written there have been several Supreme Court Decisions impacting the definition of wetlands requiring permitting under Section 404 of the CWA, including (2001) *Solid Waste Agency of*

Estuary Protection Act- requires Federal agencies to assess the impacts of commercial and industrial developments on estuaries.

Federal Water Project Recreation Act- Declares the intent of Congress that recreation and fish and wildlife enhancement be given full consideration as purposes of Federal water development projects if non-Federal sponsors are available.

Sustainable Fisheries Act (1996 amendments to the Magnusson-Stevens Act)- requires Regional Fisheries Management Councils to identify essential fish habitat (EFH) for each federally managed species and draft fisheries management plans. The eight Regional Fisheries Management Councils are the Gulf of Mexico Fishery Management Council; New England Fishery Management Council; Mid-Atlantic Fishery Management Council; South Atlantic Fishery Management Council; Caribbean Fishery Management Council; Pacific Fishery Management Council; North Pacific Fishery Management Council; and Western Pacific Fishery Management Council.

As is discussed below, state agencies also have explicit procedural and regulatory roles relevant to fish and wildlife issues. State's involvement includes implementation of CWA provisions, including the 401 certification process, and through their participation in relicensing through Section 10(j) of the Federal Power Act, the Fish and Wildlife Coordination Act, the Coastal Zone Management Act, and the National Historic Preservation Act.

2.10. Effects of Operations on Wildlife And Fisheries

Fish Passage Issues

One direct impact of structures such as dams is that they impede or block fish movement. The blockage may be of particular importance to the viability of species that may move to and between different habitats as part of their life cycle, such as anadromous species. The impact of dams on fish movement can be partially mitigated by the construction of fishways. The Federal Power Act (FPA) grants the Departments of Interior and Commerce (acting through the FWS and NMFS) exclusive authority to prescribe fishways, and it is mandatory that FERC include the prescribed fishways in license conditions. Similarly, the CWA 316(a) regulations as written by numerous States often require a zone of passage by specifying the percentage of a stream width or cross-section that must be within specified temperature and/or temperature rise limits. NPDES procedures are being used to require zones of passage

---

Northern Cook County (SWANCC) vs. United States Army Corps of Engineers;" (2006) *Rapanos v. United States*;" and (2006) *Carabell vs. Army Corps of Engineers.*

under cooling water discharges to tributary arms of reservoirs, and within the reservoirs themselves (GADNR, 2000)

Entrainment and Impingement Issues

One issue regarding the operation of hydroelectric power facilities is turbine mortality. In turbine mortality, fish are injured or killed when they are entrained in intake structures and pass through turbines. Physical stresses associated with turbine passage, such as pressure changes, shear and turbulence, and striking the turbine structure, are known to cause injury or mortality. The issue is of particular importance and consequence for anadromous species, such as salmon. Entrainment during pump back operations has also been an issue of importance and litigation at a number of facilities. In pump storage facilities, water is pumped from a lower reservoir to a higher reservoir when low cost pumping is available, and released during periods of high power demand. If the difference between the off-peak and on-peak energy cost values is large, the process can result in cost savings. As fish often congregate below reservoirs, pumpback operations can often entrain large numbers of fish, causing injury or mortality.

The operation of cooling water intake structures can also entrain small aquatic organisms and impinge upon larger organisms on traveling screens. Section 316(b) of the CWA requires that the location, design and capacity of a "cooling water intake structure" reflect the best available technology for minimizing adverse environmental impacts. The regulations were suspended in 1977 and since then EPA has used only draft general guidelines in the implementation of this Section. In 1995, under a consent decree, the EPA undertook rulemaking to develop Section 316(b) implementation rules (EPRI 1999, 2000a). In 2000, EPA and plaintiffs in Riverkeeper, Inc. v. Whitman (93 Civ. 0314 AGS, U.S.D.C., Southern District of New York, November 21, 2000) jointly submitted an Amended Consent Decree to the U.S. District Court, Southern District of New York, which the court approved to revise an existing court order that set deadlines for EPA to propose and take final action on Section 316(b) regulations. The amended consent decree called for the EPA to: develop rulemaking governing new facilities that employ a cooling water intake structure by 2000; governing, at a minimum, existing utilities and non-utility power producers that employ a cooling water intake structure, and whose flow levels exceed a minimum threshold to be determined by EPA by 2003; and, governing, at a minimum, existing facilities that employ a cooling water intake structure, that are not covered by the Phase II rule, and whose intake flow levels exceed a minimum threshold to be determined by EPA by 2004. Presently, EPA is also developing biocriteria which may be used to help assess the impacts of cooling water intake structures under CWA Section 316(b) (EPRI 2000b).

2.11. Water Quality and Flow

There are a variety of issues impacting power generation related to water quality and flow, a number of which are discussed in other sections of this Chapter. One common, but difficult and contentious, regulatory issue is the impact of flows on fish and wildlife below and within impoundments. Widely varying reservoir levels and downstream flows may impact fish and wildlife. For example, scour below peaking hydropower operations may have a negative impact. While seasonal variability in water levels may be beneficial, optimal reservoir level variations for fisheries management may be incompatible. Issues such as impoundment fluctuations, fish protection and downstream passage, minimum and base instream flows are often included in state 401 water quality certifications, as well as federal resource agencies reviews of FERC license applications. Relatively high instream flow release requirements can potentially mitigate the impact on fisheries, but a high instream flow release may reduce the economic viability of a project. Development at undeveloped sites could have a major impact on aquatic habitat through inundation and altered flows during and following construction, some of which can only be mitigated at considerable expense. These impacts are also often considered in state 401 water quality certifications.

Thermal releases are regulated and permitted under the National Pollutant Discharge Elimination system (NPDES). Intake structures for cooling water are also regulated for both flow and potential entrainment and impingement (under Section 316(b) of the CWA). A number of state's regulatory agencies focus on temperature and dissolved oxygen, as well as other water quality conditions affecting fish and wildlife in both the impoundment and releases, under leakage and generation flow conditions. This consideration may be included as part of the 401 water quality certification process as well as in the identification of impairments under the 305(d) and 303(d) listing processes. However, EPA does not believe that TMDLs should be the solution to problems substantially caused by hydromodification (Federal Register, Vol. 64, No. 162).

Although not regulated as point source discharges, impoundment and reservoir releases have been identified as a cause of impairment in a number of cases. Examples include impairment due to release of waters with low dissolved oxygen and high concentrations of reduced materials, habitat impairment, and impairment due to water temperatures. Emergency spills during flood conditions and turbine outages, or intentional spills for fish passage, are also an issue under conditions where they may lead to gas supersaturation and consequently gas bubble disease. Another water quality concern in many impoundments is related to excess nutrients and eutrophication, which can impact both the water quality of the impoundment as well as release water quality. Many of the potential impacts of water quantity and water quality variations on fish and wildlife will be reflected in the establishment of biocriteria, which are discussed in a following section of this Chapter.

Where impairments occur modifications in operation and maintenance, and/or structural modifications are potential options to address the impairment. Challenges may occur, for example, where it is necessary to maintain optimal conditions for downstream fisheries simultaneously with high impoundment water quality. Changes in release quantity may also be suggested as an effluent trading alternative to improve downstream quality such as by increasing flows to increase waste dilution. Changes in operations to improve quality may also conflict with other designated uses and the economic viability of the project.

2.12. Endangered Species Act Issues (ESA)

The ESA (http://endangered.fws.gov/esa.html) is a powerful environmental law designed to protect endangered and threatened species and their habitat. The ESA is administered by the FWS and NMFS. The FWS has primary responsibility for terrestrial and freshwater organisms, while the NMFS is responsible for marine species such as salmon and whales. As of December 2001, 1,254 U.S. species are listed as endangered (FWS 2002).

In the famous Snail Darter Case, TVA v. Hill 437 US 153, (1978), the Supreme Count concluded that in enacting the ESA Congress intended "to halt and reverse the trend toward species extinction, whatever the cost" and gave "endangered species priority over the 'primary missions' of federal agencies (FERC 2001c). As more species and fisheries stocks become listed, potential conflicts between endangered species and their critical habitat and power development are likely to increase. Recently, issued salmon listings in particular have great potential to significantly affect relicensing throughout California and the Pacific Northwest (FERC 2001c). For example, from 1973-1978 the FERC initiated only 31 formal ESA consultations related to hydropower projects. However, between January 1, 1999 and November 30, 2001, the FERC initiated 29 formal ESA consultations. Details of the endangered species act and licensing implications are described in FERC (2001c). The details of the licensing process are described in detail in the Hydropower Licensing Handbook (FERC 2001a).

2.13. Biocriteria

The Clean Water Act's (CWA) preeminent goal is to restore and maintain the physical, chemical and biological integrity of the nation's surface waters [Section 101(1)(2)]. Section 304 (a)(1) of the CWA requires EPA to develop water quality criteria that accurately reflects the latest scientific knowledge. Section 304(a) also provides guidance to states and tribes to set water quality standards that reflect these criteria which are developed to protect aquatic life and human health (USEPA 2000).

Historically, regulatory agencies relied on narrative criteria, toxicity testing (such as chemical specific or whole effluent toxicity, WET, tests), physical criteria, or chemical specific concentration criteria as indicators of water quality and environmental or human health effects. However, the presumption that improvements in chemical water quality will restore biological integrity has come into question during the last 20 years (Yoder and Rankin 1999). Nevertheless, standards and methods to assess biological integrity have not been established or standardized, even though biological or habitat impairment is recommended for CWA 305(b) reporting and is a leading cause for listing in many states' current CWA 303(d) lists, although criteria used have often been qualitative rather than quantitative. A key element of EPA's draft strategy for water quality standards and criteria is to provide additional guidance and assistance in implementing biocriteria (USEPA 2002).

A number of States (e.g. Maine, Ohio, North Carolina, and Florida) have adapted biological criteria designed to measure the cumulative biotic responses to stressors, rather than a concentration or level of a chemical, and EPA is in the process of developing national recommended biocriteria. Forty-two states now use multimetric assessments of biological condition, and an additional six states are developing biocriteria programs (EPRI 2000b). Some agencies, such as TVA have long established programs for monitoring the ecological integrity of streams (Carriker 1999). Guidance for bioassessments and the development of biocriteria for lakes and reservoirs is provided in "Lake and Reservoir Bioassessment and Biocriteria" (USEPA 1998a). Guidance for estuaries and coastal waters may be found in USEPA (2000). Evaluation guidelines for ecological indicators are discussed by Jackson et al. (2000). EPA is also considering future criteria initiatives for excessive sedimentation, flow and wildlife (USEPA 1998b).

As biocriteria are adapted, it may be expected that the number of systems listed [as on 303(d)] as biologically impaired will increase (Martin and Kennedy 2000). For example, of 645 streams in Ohio, biological indicators revealed impairment in 49.8 percent of the segments where chemical indicators detected none (Rankin and Yoder 1990, Yoder and Rankin 1999), which may further indicate that relying on water-quality information alone is apt to result in underestimates of environmental degradation (Yoder and Rankin 1999).

It may also be expected that the implementation of biocriteria will impact the licensing and operations of power generation facilities, and important challenges will occur upon the implementation biocriteria. Biocriteria are also central to a proposed draft regulatory framework that the EPA is developing under a consent decree to implement Section 316(b) of the CWA, which can be used to regulate cooling water intake structures (EPRI 2000a). As indicated above, biocriteria will provide a quantitative basis for identifying impairments. Once adapted, biological results also provide a legal basis for enforcements against entities discharging chemicals for which there are no existing standards, or regulating non-chemical degradation (Yoder and Rankin 1999). Challenges will also occur in the design of methods and models necessary to establish and evaluate the quantitative cause-and-effect relationships necessary to relate management alternatives to expected changes in the biological condition of the impacted system.

2.14. FERC Licensing Issues

Under the Federal Power Act (FPA, as amended), the Federal Energy Regulatory Commission has the exclusive authority to license non-federal hydropower projects on navigable waterways, federal lands, and under certain other criteria (FERC 2001a). The commission issues an original license for up to 50 years to construct, operate, and maintain non-federally owned hydropower projects. When a license expires, the federal government can take over the project, the FERC can issue a new license, or a project may be decommissioned. The FERC can also exempt small conduit hydroelectric facilities and small hydroelectric projects of 5 MW or less from licensing subject to the nature of the project and whether federal lands are involved. FERC may also issue preliminary permits and amendments to existing licenses.

A major regulatory issue involving non-federal hydropower facilities licensed by FERC is the relicensing of existing facilities. As of April 2000, FERC oversees 1005 licenses and 597 exemptions. Until 1993, relicensing was a relatively infrequent procedure that received little, if any, public attention. In 1993, 160 licenses affecting 237 dams on 105 rivers expired, representing over ten percent of all FERC-licensed dams (Reference). Between January 1993 and December 31, 2000, the licenses for 395 of these projects expired. According to FERC, over the next 15 years the licenses for an additional 238 will expires which combine to generate over half of the nation's nonfederal hydropower (GAO 2001).

The regulatory issues involving relicensing of non-federal hydropower facilities under FERC are a complex and a subject of considerable controversy. By some counts, there are in excess of 25 federal statutes that may impact the licensing of a hydropower project by disseminating various parts of the licensing process to various state and federal agencies (FERC 2000). The relicensing process itself is also the subject of reviews by a number of agencies and the subject of a number of pending or proposed legislative acts. Some of the issues that may be addressed in relicensing include: tribal rights, water quality, visual quality, land and water use, fisheries, wildlife and botanical resources, energy resources, socioeconomic factors, recreational resources, compatibility with comprehensive plans, cumulative impact assessment, flow regimes and cultural resources (EPRI 2000c). Much of the controversy involves the implementation of provisions of the FPA, as amended, and other Federal regulations and statutes related to natural resource issues, as discussed in the following sections. Some of the issues and recommendations for improving the licensing as by a National Review Group, convened by the Electric Power Research Institute and consisting of more than 30 organizations are described in EPRI (2000c).

Recommendations from an Interagency Task Force (ITF) of resource agencies for ways to clarify and coordinate procedures for incorporating resource agency recommendations, conditions, and prescriptions, in the licensing process under the FPA provisions are described by ITF (2000a), and under NEPA provisions by ITF (2000b). A review of some of the licensing issues was prepared by the General Accounting Office (GAO 2001). The recommendations of the Hydropower Reform Coalition, a consortium of national, state, and local conservation and recreation organizations are described in HRC (1997). Recommendations for license reform by the National Hydropower Association, a trade association, are presented in NHA (1999).

Guidance for license and exemption procedures can be found in FERC (2001a).[3] FERC can also issue temporary licenses and modifications to existing licenses as described by FERC (2001b). A forecast of the hydropower licenses expirations is provided by the U.S. Department of Energy's hydropower program. The FERC also provides informational databases formerly including the Commission Issuance Posting System (CIPS) and Records Information Management System (RIMS). The Commission Issuance Posting System (CIPS) provides access to documents issued by the Commission, such as: orders, notices, rulemakings, decisions, news releases, commission agenda and actions taken. The Records and Information Management System (RIMS) is a database containing documents issues or received by the commission, including an index to all documents since November, 1981; scanned images of documents since November, 1994 and documents in electronic format filed through the Commission's web enabled filing mechanism since November 2000. The CIPS and RIMS systems are presently being replaced by the FERRIS (Federal Energy Regulatory Records Information System) integrated system.

2.15. <u>Federal Power Act Provisions</u>

Many of the regulatory issues in licensing, relicensing, granting exemptions from licensing, or modification of an existing license or exemption involve natural resources, and the interactions between FERC and other agencies and organizations. The FPA provisions vary in the degree to which other

[3] Note that since this text was originally developed, an Integrated Licensing Process, as specified under Section 241 of the Energy Policy Act of 2005, has been developed to aid in streamlining the FERC licensing process, See § 5.18 18 CFR Ch. I (4–1–05 Edition).

agencies may impact the licensing process. In general, while the FPA provisions require FERC to take recommendations from state and federal wildlife agencies, these agencies can only advise FERC. They cannot mandate any environmental protections such as minimum stream flows or other measures to protect environmental values. For example, 10(a)(1) only requires that FERC consider comprehensive plans. Under 10(j) the FERC must also consider recommendations from fish and wildlife agencies, but can reject those under certain circumstances. Under 30(c), however, fish and wildlife agencies may impose mandatory conditions on exempt projects such as small facilities (defined as generating less than five megawatts of power), or new hydroelectric projects. Under Section 4(e), federal agencies that manage public lands (such as the U.S. Forest Service) may impose mandatory operating conditions in the licensing process to protect public resources. This authority only extends to federal land management agencies and applies only for projects located on some federally managed lands. Moreover, this authority cannot be used to reject relicensing of a project, but it can impose strict environmental conditions.

Specific sections of the FPA requiring that these issues be considered include (FERC 2001a):

FPA Section 4(E), Equal Consideration Standard

Under this provision (FERC 2001a), in deciding whether to issue a license, the FERC must give equal consideration to development and environmental values, including: hydroelectric development; fish and wildlife resources (including spawning grounds and habitat); visual resources; cultural resources; recreational opportunities and other aspects of environmental quality; irrigation; flood control; and, water supply (FERC 2001a). In cases where the proposed licensed project would be located on a federal reservation, the federal agency responsible for managing that land can file terms and conditions to protect the reservation that become, upon filing, mandatory upon the FERC to include in any license issued. Section 4(e) further requires that no license will be issued until the plans of the dam or other structure affecting navigation have been approved by the Commander, U.S. Army Corps of Engineers and the Secretary of the Army (USACE 1983).

FPA Section 10(A)(1) Comprehensive Development Standard

The FERC must insure that the project to be licensed is best adapted to a comprehensive plan for developing the waterway for beneficial public purposes. In making this judgment, the FERC considers comprehensive plans prepared by federal and state entities and the recommendations of federal and state resource agencies, Indian tribes, and the public, affected by the proposed project (FERC 2001a).

FPA Section 10(J)

In issuing a license, the FERC must include conditions to adequately protect, mitigate damage to, and enhance fish and wildlife and their habitats, based on recommendations of state and federal fish and wildlife agencies including the NMFS, FWS, and state fish and wildlife agencies.

FPA Section 30(C)

Exemptions from licensing are subject to mandatory terms and conditions from the FWS, the NMFS, and the state fish and wildlife agency.

FPA Section 18

The FPA authorizes the Secretaries of Commerce and the Interior to prescribe fishways at licensed projects. This Section applies to both upstream and downstream passage and any project that may impact the passage of fish present in the project area. It may also impact a project where there is a planned introduction of fish in the area. The federal agency may prescribe fishways "necessary to maintain all life stages of such fish" impacted by the project, including physical structures or devices, and project operations.

FPA Section 31(A)

After a license or exemption from licensing is issued, the FERC monitors for compliance with the license or exemption. Failure to comply can result in civil penalties or even rescission of the license or exemption.

2.16. Other Applicable Laws and Statutes

A variety of other laws and statutes impact the licensing, relicensing, exemption, or modification of a license. A number of these laws and statutes were discussed at the beginning and elsewhere in this Chapter and include: the National Environmental Policy Act, Fish and Wildlife Coordination Act, National Historic Preservation Act, Endangered Species Act, Clean Water Act, Wild and Scenic Rivers Act, Energy Policy Act, Americans with Disabilities Act, and others. The impact of these regulations and statutes as related to hydropower licensing are briefly discussed below.

National Environmental Policy Act (NEPA)

The NEPA of 1969 identified environmental protection as a major national policy objective. The NEPA requires all federal agencies involved in permitting of activities affecting the environment to evaluate environmental impacts and the significance of those impacts. In the case of non-federal hydro projects, FERC's issuance of a license constitutes a federal action, requiring a NEPA environmental review. Typically, FERC will conduct a "scoping" process to identify project issues and alternatives and then prepares an environmental assessment (EA). The EA describes the existing environment and the applicant's proposal to operate the project and to provide environmental enhancements. FERC will analyze the project's effect on environmental resources as well as power generation and will determine how the project should be operated to "balance" power and non-power values. If the analysis in the EA indicates that relicensing will significantly affect the quality of the human environment, or if the project scope or impact otherwise dictates, FERC (or their contractor) will prepare an environmental impact statement (EIS), (FERC 2001a). An Interagency Task Force (ITF 2000a,b) provided recommended NEPA procedures for resource agencies in the licensing process. The Department of Energy maintains a NEPA web site with additional information (http://www.eh.doe.gov/nepa/).

Fish and Wildlife Coordination Act (FWCA)

The FWCA requires FERC to first consult with the U.S. Department of Interior's Fish and Wildlife Service and appropriate state fish and wildlife agencies prior to granting a license to a hydro owner to control, impound or modify a stream or water body (FERC 2001a). Federal land managers also may impose conditions on hydro projects through the issuance of a special use permit (which is required for all commercial uses of federally-managed public lands).

National Historic Preservation Act (NHPA)

The NHPA requires FERC to take into account the effect of issuing a license for a project on any district, site, building, structure or object that is included in or eligible for inclusion in the National Register of Historic Places (NRHP, FERC 2001a). In such cases where there would be an effect, FERC must provide the Advisory Council on Historic Preservation the opportunity to comment on the relicensing of a project.

Endangered Species Act (ESA)

The ESA requires that FERC consult with the FWS or National Marine Fisheries Service (NMFS) to determine whether the relicensing of a project is likely to jeopardize the continued existence of any endangered or threatened species or result in destruction of critical habitat.

Wild and Scenic Rivers Act (WSRA) and the CEQ procedures

Section 7(a) of the WSR Act prohibits FERC from issuing a license for the construction of any project "on or directly affecting" a wild and scenic river and limits the power of any Federal agency to assist in the construction of any "water resources project having a direct and diverse effect on the values for which the river was established". Section 7(b) provides the same protection for rivers being studied for inclusion (FERC 2001a). The National Park Service maintains a list of all designated rivers under the WSR Act and those that are being studied for inclusion, located at http://www.nps.gov/rivers/wildriverslist.html. The National Park Service also maintains the Nationwide Rivers Inventory (NRI, http://www.ncrc.nps.gov/programs/rtca/nri/), a listing of more than 3,400 free-flowing river segments in the United States that are believed to possess one or more "outstandingly remarkable" natural or cultural values judged to be of more than local or regional significance. These segments are protected under a 1979 Presidential directive and related Council on Environmental Quality procedures, which stipulate that all federal agencies must seek to avoid or mitigate actions that would adversely affect one or more NRI segments.

Coastal Zone Management Act (CZMA) – 1972

The CZMA, as amended [Section 307(c)(3)(A)] requires that any applicant for a required Federal license or permit to conduct an activity, in or outside of the coastal zone, affecting any land or water use or natural resource of the coastal zone of that state shall provide in the application to the licensing or permitting agency a certification that the proposed activity complies with the enforceable policies of the state's approved program and that such activity will be conducted in a manner consistent with the program. The Coastal Zone Act Reauthorization Amendments of 1990 explicitly stressed coastal states' consideration of nonpoint source issues and may also impact hydropower licensing. See NOAA and USEPA (1993) for a discussion of issues.

Americans with Disabilities Act (ADA)

This act was created to protect the civil rights of persons with disabilities. Titles II and III of the ADA apply to licensee's recreation facilities. This law requires public and private entities with "public accommodations", such as hydro projects' recreation facilities, to be accessible to persons with disabilities (FERC 2001a).

Clean Water Act (CWA)

Several sections of the CWA impact, or potentially impact, the licensing process. Some of the relevant sections were discussed earlier in this Chapter. One of the more controversial sections relating to permitting is Section 401, and Section 401 permitting authority which is usually delegated to the states.

The CWA requires an applicant for a federal license or permit to conduct any activity that may result in a discharge to navigable waters to provide the federal permitting agency with a Section 401 certification or permit from the appropriate state agency in which the discharge originates. The certification must state that that the proposed discharge will comply with applicable provisions of the CWA, including water quality standards established by the state (Copeland 1998). The state must deny certification if compliance cannot be ensured. If necessary, the state can condition these activities to ensure compliance with the standard. Any such federal license, such as a FERC hydropower license, must include these conditions. . In *S.D. Warren Company vs. Main Board of Environmental Protection*, the Supreme Court upheld the 401 certification process as part of the hydropower licensing procedure (Landers 2006)

Section 401 certification has been commonly applied to hydroelectric projects seeking a FERC license. It is also applied to dredge and fill operations, under Sections 404 of the CWA and Sections 9 and 10 of the River and Harbors Act, by the U.S. Army Corps of Engineers, and permit requirements for industrial and municipal point discharges (under Section 402 of the CWA).

Section 401 of the CWA provides authority to the states to deny a permit to any new or existing hydropower project that is inconsistent with water quality standards. Section 401 provides states with two distinct powers: one, the indirect power to prevent issuance of federal permits or licenses by withholding certification; and two, the power to impose conditions upon federal permits by placing limitations on certification (Copeland 1998). This is one of the few areas where a state agency can overrule FERC's preemptive authority. Section 401 provides the state's mandate to require hydroelectric projects to meet state "basin plan" standards for water quality, including water temperature, dissolved oxygen, and turbidity. The 401 permit also requires a project to support specific beneficial uses listed in a basin plan, including power generation, consumptive use (meaning water supply), fish and wildlife habitat, and recreation.

Considerable controversy has revolved around different state's implementation of Section 401. Participation in Section 401 certification by the states is optional, and the implementation of Section 401 has been varied among states (Copeland 1998). States' ability to allocate waste loads among desired development activities has been challenged in the courts by both the hydropower industry and FERC, setting a number of legal precedents. Following several industry challenges to state authority under Section 401, in 1994 the Supreme Court held that a state may use Section 401 to impose minimum stream flow requirements in a certification to protect salmon and steelhead for a proposed hydropower facility (Public Utility District No. 1 of Jefferson County v. Washington Department of Ecology, 511 U.S. 700, 114 S.Ct. 1900 (1994). In additional litigation (Oregon Natural Desert Association V. Thomas, 940 F. Supp. 1534, D.Or. 1996) the federal district court in Oregon held that Section 401 "applies to federally permitted activities that may result in a discharge, including discharges from non-point sources," which if upheld provides state's with additional authority over non-point source discharges. Additional litigation has held that under Section 303(d), the EPA and states have authority to list and establish TMDLs for impaired waterbodies. The decisions impacting non-point sources may impact, directly or indirectly, the licensing's of facilities and land management operations.

Following the Supreme Court's PUD No. 1 decision above, states' authority under Section 401 has become a legislative issue (Copeland 1998). Federal legislation has been proposed to exempt hydropower projects from Section 401 if FERC determines that the state's certification is inconsistent with the FPA.

2.17. Other Regulatory Issues

Permitting, licensing, and operation of power generation facilities is subject to a wide variety of regulatory and natural resource management issues that may often result in competing and contradictory resource demands. Economic considerations are often not an integral part of, for example, resource agency recommendations affecting hydropower licenses. However, those recommendations or prescriptions may impact the economic viability of projects, which is an issue. Quality of life considerations are another significant regulatory issue. Quality of life considerations may impact, for example, the construction of reservoir projects for power generation purposes.

Under the present watershed paradigm for water quality management, the cumulative impact of all sources (point and non-point) on water quality is considered, such as in the determination of a TMDL. For example, the impacts of thermal discharges were the subject of TMDLs in the recent draft temperature TMDL on the Chattahoochee River, Georgia, impacting Georgia Power Company facilities (Georgia DNR 2002). An issue impacting the operation of reservoirs for power production is the implementation of those regulations. For example, although not regulated as point source discharges, reservoir releases have been identified as a cause of impairment in a number of cases. Examples include impairment due to release of waters with low dissolved oxygen and high concentrations of reduced materials, habitat impairment, and impairment due to water temperatures (Martin and Kennedy 2000).

At issue is how these impairments will be removed, and the timing and coordination of that implementation. The NPDES permit cycle, for example is typically five years, and permits within any given watershed may vary in the timing of their renewals. For federally owned facilities, operational strategies for water quality management are bounded by (1) current project authorization, (2) water quality objectives, and (3) reservoir attributes (Kennedy et al. 2000). Federal reservoirs are typically authorized for multiple purposes. Changes in the operations of existing or authorized federal projects, which would involve major changes in the structure, or operation of the facility, can only be made with the approval of Congress. For non-federal hydropower projects, the licensing process may take five or more years, and the life of the license is typically 30 to 50 years, to allow recovery of the capital costs. A licensed project frequently includes multiple structures including power-generating dams, tunnels and other facilities, such as a licensed project on the southern part of the Sierra Nevada which includes five different dams producing electricity. Each of the structures in a licensed project may have associated water quality impacts. FERC is precluded from granting a license solely for the purpose of water quality management. In addition, amendments are typically required when there is a proposed change to a project previously licensed by FERC. An amendment to a license typically requires many of the same environmental and socioeconomic considerations and coordination activities as an original license, and the amendment process is described by FERC (2001b). Therefore, the timing and coordination of changes in authorized, permitted, or licensed power generation activities as they may impact water quality is problematic.

There are a wide variety of procedural and regulatory issues which were not, or were only briefly, discussed in this Chapter. A complete discussion of these issues is beyond what could be accomplished in this format. However, it is hoped that the information provided highlights some of general procedural and regulatory issues and challenges, and further indicates sources of more detailed information. An early awareness of these issues in a power generation and reservoir water quality context is often critical to design studies and identify alternatives to respond to regulatory issues.

2.18. References

Hannan, H.H., I.R. Fuchs, and D.C. Witenberg. 1979. Spatial and temporal patterns of temperature, alkalinity, dissolved oxygen and conductivity in an oligo-mesotrophic, deep-storage reservoir in Central Texas. Hydrobiologia. Vol. 66, No 3. Pp. 209-221.

Ashby, S.L.; Myers, J.L.; Laney, E.; Honnell, D.; and Owens, C. 1999. "The Effects of Hydropower Releases from Lake Texoma on Downstream Water Quality," Journal of Freshwater Ecology, 14(1):103-112.

Carriker, N.E. 1999. "TVA's Approach to Ecological Health Assessment in Streams and Rivers," In Measures of Environmental Performance and Ecosystem Conditions, National Academy Press, Washington, D.C., pp. 260-283.

Copeland, C. 1998. "Clean Water Act Section 401: Background and Issues", National Council for Science and the Environment, Congressional Research Service Report 97-488 ENR, updated October 4, 1998.

Crossman, J.S., R. Ruane and M. Belefski. 2000. "TMDLs, Watershed-Based Pollutant Trading, and Hydropower Projects- A New Paradigm for Water Resources Management?", Water 2000

CSG. 1997. "Interstate Compacts 1997," Council of State Governments, Lexington, KY, 127 pp.

CSG. 1998. "Interstate Compacts and Agencies, 1998," Council of State Governments (Ed. W.K. Voight), Lexington, KY., 200 pp.

Department of the Interior. 2000. "Environmental Review Memorandum No. ERM00-2." U.S. Department of the Interior, Office of the Secretary, Washington, D.C. March, 2000.

EPRI. 1999. "Proceedings: 1998 EPRI Clean Water Act Section 316(b) Technical Workshop: Coolfront Conference Center," TR-112613, Electric Power Research Institute, Palo Alto, CA.

EPRI. 2000a. "Procedural Guideline for Evaluating Alternative Fish Protection Technologies to Meet Section 316(b) Requirements of the Clean Water Act Report," Report 1000551, Electric Power Research Institute, Palo Alto, CA.

EPRI. 2000b. "Evaluation of Biocriteria as a Concept, Approach and Tool for Assessing Impacts of Entrainment and Impingement under Section 316(b) of the Clean Water Act," TR-114007, Electric Power Research Institute, Palo Also, CA.

EPRI. 2000c. "Hydro Relicensing Forum: Relicensing Strategies," National Review Group Interim Publication 1000737, Electric Power Research Institute, Palo Alto, CA.

FERC. 2000. "Hydropower Licensing Policies, Procedures and Regulations Comprehensive Review under Section 603 of the Energy Act of 2000: Comments of the National Hydropower Association," Docket No. PL01-1-000," Federal Energy Regulatory Commission, Washington, D.C.

FERC. 2001a. "Hydroelectric Project Licensing Handbook," Federal Energy Regulatory Commission, Washington, DC.

FERC. 2001b. "Hydroelectric Project Handbook for Filings Other Than Licenses and Exemptions. Federal Energy Regulatory Commission, Washington, DC.

FERC. 2001c. Hydropower Licensing and Endangered Species. A guide for Applicants, Contractors and Staff. Federal Energy Regulatory Commission, Office of Energy Projects, Washington, DC. December 2001.

FWS. 2002. "ESA Basics, Over 25 Years Protecting Endangered Species," U.S. Fish and Wildlife Service, Endangered Species Program, Arlington, VA (http://endangered.fws.gov), March 2002.

GAO. 2001. "Licensing Hydropower Projects: Better Time and Cost Data Needed to Reach Informed Decisions About Process Reforms, GAO-01-499, General Accounting Office, Washington, D.C.

Georgia DNR. 2000. Special Requirements for GADNR EPD Permit Number GA 00260051. Georgia Department of Natural Resources, Atlanta, GA, November 2000.

Georgia DNR. 2002. "Draft: Chattahoochee River Temperature TMDL," Submitted to the USEPA Region 4 by the Georgia Department of Natural Resources, Atlanta, GA, June 2002.

HRC. 1997. "Relicensing Tool Kit: Guidelines for Effective Participation in The FERC Relicensing Process," by the Hydropower Reform Coalition, July 1997.

ITF. 2000a. "Interagency Task Force Report on Agency Recommendations, Conditions, and Prescriptions Under Part I of the Federal Power Act," Prepared by the Work Group on the Coordination of Federal Mandates, December 8, 2000.

ITF. 2000b. "Interagency Task Force Report on NEPA Procedures in FEFC Hydroelectric Licensing," Prepared by the Work Group on the Coordination of Federal Mandates, December 8, 2000.

Jackson, L., J. Kurtz and W. Fisher (Editors). 2000. "Evaluation Guidelines for Ecological Indicators," EPA/620/R-99/005. May 2000.

Kennedy, R. J. Clarke, W. Boyd and T. Cole. 2000. "Characterization of U.S. Army Corps of Engineers Reservoirs: Design and Operational Considerations, ERDC WQTN-MS-05, Engineer Research and Development Center, Vicksburg, MS.

Landers, J. 2006. "Supreme Court Upholds State Certifications in Hydropower Licensing," ASCE Civil Engineering News, 76(7), pp. 20-22

DAMS: Lazarus, E. 1994. "PUD No. 1 of Jefferson County v. Washington Department of Ecology: An Expansive Interpretation of State Authority under the Clean Water Act, " National Council for Science and the Environment, Congressional Research Service Report 94-601 A, July 1994.

Martin, J. and R. Kennedy. 2000. "Total Maximum Daily Loads (TMDLs): A Perspective, " ERDC/TN EEDP-01-46, U.S. Army Corps of Engineers Engineer Research and Development Center, Vicksburg, MS.

NOAA and USEPA. 1993. Coastal Nonpoint Pollution Control Program: Program Development and Approval Guidance. U.S. Environmental Protection Agency, Washington, DC.

NHA. 1999. "Draft Report: Relicensing Hydroelectric Power Projects: A Handbook for People Involved In Relicensing Hydropower Projects." National Hydropower Association, Washington, DC (available from www.hydro.org).

NRC, 2001. "Assessing the TMDL Approach to Water Quality Management." National Research Council, Water Sciences and Technology Board. National Academy Press, Washington, D.C.

Rankin, E.T. and C. Yoder. 1990. "A Comparison of Aquatic Life Impairment Protocols for Use in Rivers and Streams: Benthic Macroinvertebrates and Fish. RPA/444/4-89-001, U.S. Environmental Protection Agency, Washington, D.C.

USACE. 1983. "Engineering and Design, Corps Responsibilities for Non-Federal Hydropower Development under the Federal Power Act," Engineer Regulation 1110-2-1454, U.S. Army Corps of Engineer, Washington, D.C.

USEPA. 1989. "Wetlands and 401 Certification," U.S. Environmental Protection Agency, Office of Water, Washington, D.C.

USEPA. 1994. "Water Quality Standards Handbook, 2nd Edition," EPA-823-8-94-005, Office of Water, Washington, D.C.

USEPA. 1998a. Lake and Reservoir Bioassessment and Biocriteria Technical Guidance Document," EPA 841-B-98-007

USEPA. 1998b. "Water Quality Criteria and Standards Plan -- Priorities for the Future. EPA 822-R-98-003

USEPA. 1999a. "Protocol for Developing Sediment TMDLs (First Edition)", EPA 841-B-99-004.

USEPA. 1999b. "Draft Guidance for Water Quality-based Decisions: The TMDL Process (Second Edition)," EPA 841-D-99-001.

USEPA. 2000. "Estuarine & Coastal Marine Waters: Bioassessment & Biocriteria Guidance, " EPA-822-B-00-024, December 2000.

USEPA. 2002. "Draft Strategy for Water Quality Standards and Criteria: Strengthening the Foundation of Programs to Protect and Restore the Nation's Waters," EPA-823-R-02-001, United States Environmental Protection Agency, Office of Water, Washington, D.C. May 2002.

USEPA. 2004. "Water Quality Trading Assessment Handbook," EPA 841-B-04-001, United States Environmental Protection Agency, Office of Water, Washington, D.C, Nov. 2004

3. FUNDAMENTAL WATER QUALITY PROCESSES

J. A. Gordon

J. M. Higgins

Over the past century, the construction of dams and reservoirs has provided a variety of social benefits (e.g., navigation, flood control, power generation, water supply, recreation, and aquatic habitat). Man-made reservoirs differ from natural lakes in purpose and in operation. Reservoirs provide a control point in the river system and are subject to specific regulatory review and authorization. Reservoirs have inherent use conflicts; are viewed in terms of economic success and failures; and are subject to political influences. They may be single purpose or multipurpose, but society almost always demands that they be multi-purpose and serve a variety of needs.

Reservoirs create new hydrologic and water quality regimes that are at least partially controlled by design parameters and operational procedures. For example, reservoir water quality is influenced by reservoir depth and retention time. Downstream water quality depends on the depth from which water is withdrawn from the reservoir. Important unregulated factors that influence water quality include meteorology, inflow rates, and inflow water quality. Understanding and predicting how these factors affect water quality and beneficial uses is necessary for effective reservoir and river system management.

This chapter provides an overview of key water quality processes. Basic principles are discussed rather than the mathematical formulations that vary from model to model and are given in user manuals and supporting references. The chapter focuses on water quality processes in reservoir because 1) the basic chemical and biological processes are similar in rivers and reservoirs, 2) the major differences between rivers and reservoirs are determined more by hydrodynamic characteristics than by water quality processes, and 3) water quality processes in reservoirs are usually more significant than in rivers due to a longer reaction time (i.e., longer water storage times in reservoirs result in seasonal changes versus hourly or daily time periods that are usually considered in rivers).

3.1 Types of Reservoirs

Reservoirs vary greatly in geometric and hydraulic characteristics. Physical characteristics can influence a reservoir's purpose, design, method of operation, water quality, and the type of model most appropriate for a simulation analysis. For example, a one-dimensional model may suffice for a long, narrow reservoir with high flows relative to volume. A two-dimensional model may be suitable for a wider and deeper impoundment that experiences thermal stratification. A three-dimensional model may be required to simulate a thermal discharge into a large, deep reservoir with highly varying flow rates.

Mainstem Reservoirs

Reservoirs may be divided into three types: 1) run-of-the-river or mainstem reservoirs, 2) storage or tributary reservoirs, and 3) transition reservoirs. Run-of-the-river reservoirs tend to have short hydraulic detention times (days or weeks), little if any flood control storage, and are used primarily for navigation and hydro-electric power production. Other uses include recreation, water supply, shoreline enhancement, cooling water, and flow regulation. The reservoirs tend to be located on the mainstem of rivers where large flows, lowland topography, and populated communities required lower dams. Low heads, wide dams, high flow rates, high velocities, low to moderate water retention times, and riverine topography characterize mainstem reservoirs. Very often, the main water quality changes occur as functions of length as the impoundments are usually well mixed vertically and laterally (Figure 1).

Storage Reservoirs

Storage impoundments have a longer hydraulic detention time, often storing water for months or years. Storage reservoirs are most often found in steep valleys on tributary streams with high narrow dams. The impounded stream usually has low flows in dry periods and high flows during the runoff season. Typical uses for storage impoundments are flood control, water supply, hydropower, recreation, and water quality control. Water velocities are slow with the longer water storage times allowing significant water quality changes with length, depth, and lateral extent. One of the dominant processes in storage reservoirs is vertical stratification by water density, a process usually caused by vertical differences in water temperature (Figure 2). Once a reservoir becomes stratified, the hydrodynamics are complicated and driven by density flows, wind, and other energy sources such as waves and seiches. Reservoir stratification results in water quality changes with length, depth, and cross-section. Temperature and water storage times also vary with depth.

Transition Reservoirs

Transition reservoirs are a combination of the previous types. They are essentially run-of-the-river reservoirs which occasionally stratify due to low inflows. During stratification, the density layers are usually tilted upstream by the momentum of water inflows (Figure 3). Simulating the extent and duration of stratification, and subsequent water quality conditions, is often difficult in transition reservoirs.

3.2 RESERVOIR ZONES

The multi-zone reservoir concept originated with Hannan (1979) and has been utilized in reservoir ecology literature for over 20 years (Thornton, et al.1981, Kimmel and Groeger 1984, Miller 1984, Thornton 1990, Worth 1995). Many reservoirs in the West are located in mountainous terrain, occupy steep-sided, narrow canyons, and have a relatively narrow dendritic shape. Because they often impound large, high-order rivers, reservoirs receive high flows during spates or other runoff events laden with sediment, organic matter, and dissolved constituents. This results in distinct zones in the reservoir that reflect a riverine to lacustrine longitudinal gradient imposed by the incoming river flow into a basin obstructed by a physical structure (Kimmel et al. 1990).

Riverine Zone

The riverine zone of a reservoir is usually narrow, shallow, has relatively high velocity currents, may have a particulate load of both sediment and organic material and production may be light-limited. It has a short hydraulic retention time, usually ranging from days to several weeks. It is the zone of greatest mechanical sediment interaction with deposition during filling or rising stages of reservoir storage, and scouring or resuspension (referred to as headcutting) during drawdown. The release of nutrients, organic matter, oxygen demand, and taste and odor producing compounds is still one of the least understood and modeled functions. In larger reservoirs with annual drawdown exceeding 10 meters, thousands of tons of sediment may be moved down reservoir. High turbidity density currents and nutrient release during these periods of headcutting may provide greater internal nutrient loading than the annual external input . Understanding these mechanisms is very important in designing Total Maximum Daily Loadings (TMDL's) with expectations of measurable end-points within the reservoir or tailwater.

Transition Zone

When the reservoir basin broadens and deepens, the incoming current slows and the finer particulates begin to settle in a transition zone between the incoming river and the lake-like environment near the dam. Algae and aquatic plant productivity is usually greater than at any other location in a reservoir because nutrients are more readily available and light penetration is greater. This is usually the zone of greatest autochthonous organic production, settling, and oxygen demand. Inflows begin seeking deeps of common density, and underflow, interflow, or overflow density currents begin in the transition zone (Figure 4). The most notable and dramatic change is the plunge line, usually formed in the fall when inflowing water is cooler and more dense than the reservoir. The "plunge line" is often seen as a line with turbid water on the upstream side and warmer much clearer water on the downstream side. This marks the beginning of the transition zone. The length of the riverine and transitional zone changes with reservoir elevation, magnitude of inflow, and season.

Lacustrine Zone

The lacustrine zone in a reservoir may be deeper, wider, and is usually furthest down the longitudinal gradient in the impoundment. There is little particulate load, riverine type current velocities are usually greatly reduced; as are available nutrients, oxygen demand, algae productivity, and maximum light penetration. Seasonally strong over flow, interflow, or underflow density currents may still be present. Lacustrine Reservoirs generally have more than 1 year hydraulic retention time, retain more than 80 % of the total inflowing phosphorus and sediment budgets, and are generally much more than 40 meters deep.

All reservoirs do not have all three zones; thus a reservoir may be predominantly riverine (hydraulic retention time generally less than 120 days), transitional (hydraulic retention times generally from 120- 270 days), and lacustrine (hydraulic retention times in years). Long term storage reservoirs can change depending on stage and inflow in any particular year.

3.3 Reservoir Hydrodynamics

Hydrodynamics is one of the most important factors affecting water quality, due to the impact on hydraulic residence time, water temperature, aeration, and mixing. Flow in rivers is primarily driven by gravity as water moves from upstream to downstream with significant horizontal and vertical mixing. In reservoirs, water density and momentum become more important as the volume increases and the velocities decrease. In a stratified reservoir, inflows are distributed to levels of similar density. The momentum of a density confined inflow can create reverse flows that extend over a much larger area than eddies in a river. Reservoir discharges are similarly influenced by density and momentum.

Reservoir hydrodynamic regimes may be characterized as overflow, underflow, or interflow (Figure 4). All are common occurrences in reservoirs. Overflow occurs when the inflowing water is lighter than the water in the reservoir, usually due to a warmer temperature. The inflowing water spreads over the surface and usually increases stratification. Water quality can change rapidly within an overflow due to the warm temperatures and the availability of nutrients, light, oxygen, and mixing. Phytoplankton blooms often occur in an overflow.

Underflow occurs when the inflow is heavier (e.g., colder) than the water in the reservoir. The inflow moves along the bottom in the pre-existing river channel and lower overbank areas. An underflow may move directly through the reservoir if the dam has a low-level discharge. The water quality of an underflow may not change much due to the lower temperatures and the isolation from light and air. Cold water underflows may move through a series of impoundments in this fashion.

Interflow occurs when the density or temperature of the inflow is intermediate to that of a stratified reservoir. In this case the inflow is distributed to an intermediate layer of similar density. An interflow may also move through a reservoir quickly if a mid-level discharge is used. A low level discharge may effectively trap the interflow within the impoundment. Water quality changes will be related to the depth of the interflow, temperature, and residence time. The presence or absence of light (which is a function of depth and water clarity) is vitally important to water quality within an interflow.

Overflows, underflows, and interflows are influenced by entrance mixing, bathymetry, wind, and outflow regimes. Other mixing along the length of the reservoir can also affect hydrodynamics, (e.g., cooling water withdrawals and returns, skimmer walls, and submerged weirs). More detailed descriptions of hydrodynamics are given by Wunderlich (1971) and Krenkel and Novotny (1980).

3.4 Water Properties

Water is a unique compound that forms the basic building block for living organisms. It is a good solvent that can dissolve and transport many elements and compounds important to reservoir water quality. Water can hold gasses due to its open structure, allowing dissolved oxygen concentrations suitable for fish and other aquatic life. Water has a high surface tension, attenuates light exponentially with depth, and has great wetting properties (Reid and Wood, 1976). Many of these water properties and their underlying mechanisms determine reservoir conditions and are simulated in reservoir water quality models.

The solvent, evaporative, and flow properties of water are influenced by the structure of the water molecule. The two hydrogen atoms share an electron with the oxygen atom (H_2O) at an angle of 105 degrees. This gives the water molecule a distinct positive and negative polarity that it a good solvent. Water molecules also experience hydrogen bonding. This occurs when an atom of hydrogen is attracted to two oxygen atoms instead of one (i.e., one bond with the oxygen atom in its own water molecule and a second bond with the oxygen atom of another water molecule). Because of the hydrogen bonding, considerable energy is needed to evaporate water. Water has the highest heat of vaporization of all liquids (500-600 calories per gram). This property allows water to be effectively stored in reservoirs. If water had the heat of vaporization of similar fluids, it would quickly evaporate from a reservoir.

Overflow, interflow, and underflow in a reservoir often occurs without significant mixing. This is possible because one layer of water can move adjacent to another layer by breaking a series of hydrogen bonds between the two layers. This results in water layers being able to move as units with only a few molecules between the two layers being disturbed.

Reservoirs resist changes in water temperatures due to heat addition. This property, specific heat, makes water ideal for supporting aquatic life and can result in reservoir layers that maintain the same temperature for many months.

Another important property of water for reservoirs is the density relationship with temperature. From 4 to 100 °C, water density decreases as temperature increases. This change in water density with temperature causes reservoirs to stratify. Thus, the warmest water in a reservoir rises to the surface. In contrast and unlike most other substances, however, water also becomes less dense below 4 °C. Thus, water has two density relationships with temperature, one from 0 to 4 °C and one from 4 to 100 °C. Since water has two density relationships with temperature, reservoirs can stratify in the spring or summer at temperatures above 4 °C and again in the winter between 0 and 4 °C (Figure 5). Other factors can also effect water density (e.g., dissolved solids and suspended solids). While it may be necessary to consider these other factors, temperature is often the primary determinant of density in fresh water reservoirs.

Ice at 0 °C is less dense than liquid water. Consequently, ice floats. If it were not for this property, ice would sink and most if not all of the world's water would lie frozen at the bottom of rivers, lakes and oceans. This unique property also makes ice an insulating layer when covering an impoundment, protecting the liquid below from freezing and evaporation.

3.5 Reservoir Stratification

To illustrate reservoir stratification, consider a dimictic impoundment (stratifies twice a year) that experiences winter ice cover. In January, the ice effectively prevents wind from mixing the reservoir. The reservoir is stratified with the densest water, 4 °C, at the bottom and the warmest water, 0 °C, just under the ice (Figure 5). If mixing from inflows and outflows is small, the impoundment will stay stratified until the ice melts (ice out). After ice out in perhaps late February, the reservoir will become isothermal with depth due to solar warming and mixing from wind, inflows, and outflows (Figure 6).

Spring typically brings more sunshine and warmer inflows that promote summer stratification. The process accelerates with increasing temperatures because the density-temperature relationship is not linear (i.e., the density difference per degree increases as the temperature rises). Thus a warm reservoir stratifies more easily than a cold one. The warm inflows spread out over the surface and promote stratification. The radiant heat from the sun further warms the surface and lowers the density. The low thermal conductivity of water stores the heat and prevents it from being distributed downward to the colder layers. Reservoir operations often store spring runoff and raise the water elevation to summer pool. This reduces mixing and facilitates stratification. The forces resisting stratification include mixing due to wind and flow and surface cooling from evaporation and cool nights. As summer approaches the stratifying forces intensify. The result is a warm surface zone, the epilimnion, which floats on the colder bottom layers, the hypolimnion. In between is a transition zone, the metalimnion, where temperature changes rapidly with depth. The stratified reservoir tends to be stable as the cold hypolimnion moves the stratified center of gravity below the isothermal center of gravity.

A stratified reservoir is like a deck of cards with each card having a higher temperature than the one below it. Any cards added to the deck (i.e., inflows) are inserted at a layer of equal density or temperature. Cards warmer than the whole deck go on top (overflow) while cards colder than the deck go on the bottom (underflow). Outflows from the dam or from lateral withdrawals are taken from the outlet elevation. Depending on the amount of water withdrawn, the overlying layers will drop down to occupy the withdrawn volume.

In the fall, when the amount of daylight and solar heating decreases, night time temperatures and reservoir inflows become colder. The surface water becomes cooler, denser, and settles downward, mixing with underlying water to a more uniform temperature. The epilimnion steadily becomes thicker, colder, and denser. The hypolimnion diminishes in thickness and volume while the stratified center of gravity rises above the isothermal center of gravity. At this point, the reservoir becomes unstable and the water in the reservoir turns over and mixes from top to bottom. The turnover process typically takes a few days and is noticeable due to dark hypolimnetic waters rising to the surface.

After turnover, the reservoir remains isothermal with depth and steadily cools down until ice forms at the surface and the annual process begins again. In latitudes where ice cover is not formed, the reservoir will only experience the summer stratification (monomitic) and fall turnover.

3.6 Reservoir Heat Sources and Sinks

Heat is the vibration of molecules at a temperature above absolute zero. Heat can be observed as the temperature of a unique mass of material. Heat is defined as the amount of molecular vibration that will raise the temperature of a specific mass of water a specific amount (i.e., one Btu will raise one pound of water one degree F; one calorie will raise the temperature of one gram of water one degree C; and 4.187 Joules will raise the temperature of one gram of water one degree C).

Heat is added to a reservoir directly by industrial discharges that increase water temperature and cause thermal wedges that move across the reservoir surface. The greatest sources and sinks of heat entering a reservoir usually come from the atmosphere, however (Edinger, Brady and Geyer 1974). Essentially all of earth's heat and energy comes from the sun. As the energy moves through the atmosphere, it is attenuated, reflected and absorbed by dust and moisture. At the surface of a reservoir, incoming sources of heat include net short wave solar radiation, net long wave atmospheric radiation, and conductive heat (if the air is warmer than the water). Heat losses at the reservoir surface include back radiation, evaporation, and conduction (if the air is cooler than the water).

There can be, and often is, a balance between atmospheric input and back radiation at the reservoir surface. In this case the solar and evaporation components become more important. If the energy into the water is greater than energy out of the water, heating occurs and decreases exponentially with depth. Similarly, if the heat out of the water is greater than the heat into the water, cooling occurs and the cooler water falls, promoting more uniform temperatures and potentially leading to lake turnover.

Of course, an accurate heat assessment is important because the degree of reservoir stratification drives hydrodynamic and water quality conditions. The modeling process must be evaluated at the temperature prediction stage and must be considered verifiable before proceeding. Models will be able to compute the net temperature effect of the heat balance using meteorological input, impoundment bathymetry, and water quality data. Most models use approaches developed by Edinger, Brady and Geyer (1974).

3.7 Particle Settling and Transport

Particles in a reservoir will settle or float depending on their buoyancy. Some particles tend to agglomerate and form larger, denser particles while others, such as negatively charged clays, are self-repelling. Stoke's law for spherical particles settling in a viscous fluid includes the force of gravity pulling the particle down, an upward buoyant force, and an upward drag force if the particle is moving. The law assumes that a particle quickly reaches a terminal settling velocity where all of the forces are equal.

The settling velocity can be used to estimate the time required for a particle to settle to the bottom. Problems with using Stoke's Law arise when particles are not spherical, when they tend to agglomerate, or when the density of the particle tends to change with depth due to pressure. In a stratified reservoir, changes in water temperature, density, and viscosity occur with depth. Thus a particle may settle more slowly as it goes down and overall settling may be slower in a stratified impoundment. Some particles may reach a layer where they become neutrally buoyant and concentrate. If this happens with living organisms like phytoplankton, other water quality changes may be observed.

3.8 Light and Heat Penetration

Light and heat are forms of short wave radiation whose ability to penetrate water decreases exponentially with depth. Specific formulations can be found in Wetzel and Likens (1991), Williams, et al (1981) and French, et al (1982). The formulations allow one to calculate the effect of net solar radiation on temperature increases and to determine the light available for photosynthesis at various reservoir depths. The photic zone is typically assumed to be the depth at which 99 percent of light has been absorbed.

3.9 Dissolved Oxygen in Stratified Reservoirs

Dissolved Oxygen (DO) is vital to aquatic life and natural processes. Most regulatory agencies use DO as an indicator of aquatic health and as a surrogate enforcement parameter. Typical minimum DO criteria are 5 mg/L for warm water ecosystems and 6 mg/L for cold water ecosystems. The saturation values for water are a function of temperature and are not much higher than the typical criteria (e.g., 11.3 mg/L at 10 °C; 10.2 mg/L at 15 °C; 9.2 mg/L at 20 °C; and 8.4 mg/L at 25°C).

Surface aeration and photosynthesis are the primary sources of dissolved oxygen in a reservoir. During stratification, the processes are limited to the epilimnion where mixing and light are available. The metalimnion serves as a barrier to mixing and light penetration. Thus a large portion of a stratified reservoir is excluded from reaeration and depends primarily on the DO available at the onset of spring stratification. Many times there is not enough DO to last until the fall turnover and the hypolimnion becomes DO deficient or completely devoid of oxygen (Figure 7). In this case, reservoir discharges that are withdrawn from the hypolimnion may not meet downstream water quality standards unless some sort of artificial reaeration is provided.

DO depletion has been observed in several regions of a stratified impoundment. Typically, depletion starts in the upstream portion of the hypolimnion where inflows and bottom sediments tend to be more organic. The depletion often follows the bottom contours and can eventually extend to the dam. This zone of DO depletion is termed the benthic zone and typically extends 25 to 50 feet above the bottom contours. A second zone is just below the thermocline and extends for 25 to 50 feet below the epilimnion. This is the zone of metalimnetic DO depletion or metalimnetic minimum. The metalimnetic DO minimum often extends upstream from the dam to 75 percent of the reservoir length. In a deep stratified reservoir, there may be a considerable volume of higher DO water between these two zones. A mid-level or multi-level outlet can often selectively withdraw from this zone of higher DO. This zone can also provide a fish refuge that may result in fish kills if the zone is isolated and then depleted.

The processes that deplete DO from a stratified impoundment are primarily biological, rather than physical or chemical, as described by Cole and Hannan (1990), EPRI (1990), TVA (1990) and Morris (1978). Most DO depletion is from bacterial and phytoplankton respiration. Bacterial respiration takes the form of long term Biochemical Oxygen Demand (BOD) wherein chemosynthetic aerobic bacteria consume organics in the water column and bottom sediments. The process oxidizes organics to CO_2, reduces O_2 to water, and allows bacteria to grow. The organics may have been present in the hypolimnion at the time of stratification or brought in by an interflow, underflow, or settling plankton and weeds.

A second mechanism of depletion is DO uptake by autotrophic bacteria that oxidize organic nitrogen and ammonia to nitrate. The bacteria require up to 5 mg/L of DO per mg/L of NH_3-N oxidized. The sources of organic and ammonia nitrogen are typically municipal and industrial wastewater and agricultural discharges.

Oxygen depletion in the metalimnion is also caused by the respiration of phytoplankton that have settled out of the photic zone and become trapped and concentrated in the metalimnion. Additional DO demands are imposed by zooplankton that migrate into the metalimnion to feed on phytoplankton, and by bacteria that consume the organic material generated by the zooplankton and phytoplankton. Some metalimnetic depletion may be due to organics brought in via interflows.

Oxygen depletion in the benthic zone can be substantial due to the settling of organic material that flows into or is produced by the reservoir. Plants and aquatic life produce organic matter and eventually die and settle to the bottom. Wastewater discharges and nonpoint sources of pollution from the watershed introduce oxygen demanding materials that is deposited in reservoir sediments. Similarly, the metalimnetic minimum will exist as long as phytoplankton grow and settle in the reservoir. The richer the ecosystem in a stratified reservoir, the greater the oxygen depletion.

Dissolved oxygen is strongly related to the eutrophic state of a waterbody. Eutrophication is a natural phenomena whereby waterbodies age and are ultimately transformed into a marsh or meadow. While a natural process, eutrophication is often accelerated by inputs of excess nutrients resulting from human activities, often referred to as cultural eutrophication (Chapra 1997) or anthropogenic eutrophication.

Water bodies are often classified according to their trophic state, with the general terms being

- Oligotrophic (poorly nourished)
- Mesotrophic (moderately nourished)
- Eutrophic (well-nourished), and
- Hypereutrophic (overnourished)

While such terms are most commonly applied to lakes and reservoirs, they are also appropriate descriptions of rivers, streams and estuaries (Chapra 1997).

3.10 Anoxic Reservoir Processes

Water quality in a stratified reservoir depends on time, temperature, environmental conditions, and terminal electron acceptors. Time is required for physical, chemical, and biological processes. Temperature regulates stratification and the rate of each process including settling, chemical reactions and biochemical/biological growth rates. Other environmental conditions such as physical mixing or light intensity influence the availability of food and energy that drive biological processes like photosynthesis. Terminal electron acceptors are necessary in the metabolic process of oxidizing organic matter to provide food for aquatic life. When DO is present, O_2 serves as the terminal electron acceptor. Under anoxic conditions, new microorganisms emerge that can use other forms oxygen (e.g., NO_3, MnO_2, SO_4).

Organics in a reservoir are usually classified as: (1) allochthonous -- those external to the reservoir such as point and non-point sources from the watershed, and 2) autochthonous--those generated within the reservoir such as bacteria, algae, zooplankton, benthic organisms, and fish. The organics, together with nitrogen, phosphorus and other micro-elements, stimulate the growth of heterotrophic and autotrophic microorganisms.

Autotrophic microorganisms are those that get essential 'biologically useful energy' (BUE) from light or inorganic chemical reactions. This BUE allows autotrophs to synthesize mass from inorganic nutrients such as CO_2, N, P, K, and Si. The process can be represented by the following equation:

$CO_2 + H_2X + BUE + NH_3 \leftrightarrow C_5H_7O_2N$ (cells) + X (unbalanced).

When X is oxygen, photosynthesis results in DO production. When X is sulfur, non-oxygenic growth results in sulfide production.

Heterotrophic bacterial growth uses organics for both BUE and cell synthesis. Since there is an oxidation of organics for BUE, something must be reduced so that there are equal amounts of oxidation and reduction. The materials being reduced are termed terminal electron acceptors, TEA. Terminal electron acceptors must be present in the reservoir to oxidize organic matter and facilitate the bacterial growth. The equation representing the process is:

Organics + TEA $\leftrightarrow CO_2$ + EP (End Products)

The following table shows the common terminal electron acceptors in an aquatic environment and their end products and BUE yields as reported by Bouwer, 1992. Similar values are given by Fenchel, et al, 1998.

Terminal Electron Acceptors and their End Products and BUE (Bouwer, 1992)

Desirability	Terminal e^- Acceptor	End products	BUE Yield
Most favorable	O_2	H_2O	29.9 kcal/eq
Less favorable	NO_3	N_2	28.4
Undesirable	MnO_2	Mn^{++}	23.3
Undesirable	FeO(OH)	Fe^{++}	10.1
Undesirable	SO_4	H_2S	5.9
Unfavorable	Organics	Reduced Organics	na
Least favorable	CO_2	CH_4	5.6

Deoxygenation

In an aerated reservoir oxygen is the preferred TEA by microorganisms oxidizing organic matter for food. The loss of DO in the benthic zone, the metalimnetic zone, and often the entire hypolimnion creates an anaerobic environment in which the bacteria seek out other TEAs. The following sections describe anaerobic water quality processes that occur in a reservoir zones that experience a complete loss of DO. The resulting water quality is significantly influenced by the availability of TEAs. The depletion of DO usually lowers the ORP (Oxidation/Reduction Potential) from about 400 mV to 300 mV and the pH from neutral to 6.5.

Denitrification

Denitrification is the biological process that generates BUE by reducing nitrate to nitrogen gas. In the absence of DO nitrate becomes the TEA which allows microbes to gain BUE for synthesis. Denitrification of organics will occur in a microbial environmental that is essentially anoxic and within the temperature range of 2 to 65 °C. Since nitrate is freely soluble, the reaction takes place in the water column. Sediments play a minor role. The end product, N gas, is not a problem in most cases and denitrification is considered an asset to a stratified reservoir ecosystem. Nitrate has been added to reservoirs to facilitate this process, particularly in the benthic zone. Denitrification poises the ORP at about 200 mV and pH may drop to 6.

Ammonification

Ammonification refers to the release of ammonia from organics that are being decomposed. Bacteria first strip off ammonia from nitrogenous organics before further oxidation. In an anaerobic environment, the ammonia builds up because the only sink is cell synthesis which is minimal.

Manganese Reduction

Under anaerobic conditions, and after nitrate has been used up, manganese becomes soluble as the solid phase of pyrolucite, MnO_2, is used as a TEA. The reduction of MnO_2 to Mn^{++} is a complicated biological process that is usually associated with bottom sediments. The soluble Mn^{++} diffuses upward through the water column but is not considered toxic therein. However, withdrawals of water containing manganese concentrations above 50 µg/L will cause staining of laundry and fixtures. Sediment release rates are in the range of 1 to 10 $mg/m^2/day$. The ORP drops from 200 to 50 mV and pH remains at about 6 to 6.5.

Iron Reduction

Iron, like manganese, is reduced from sediments as solid phase hydrous oxides of iron are used for TEA's. The soluble form released is Fe++, which accumulates under anaerobic conditions. Ferrous iron is not considered toxic in a reservoir but withdrawals containing 50 µg/L will cause staining upon exposure to oxygen. The release rates in reservoirs are in the range of 10 to 350 $mg/m^2/day$. The ORP must have been at about 50 mV for two weeks before reduction begins. The pH will be 6 to 6.5.

Sulfate Reduction

When the ORP has become negative at about –150mV due to prolonged anaerobic activity, sulfate will begin to serve as a TEA with the release of sulfides and H_2S. H_2S is toxic and odorous. Withdrawals of reservoir waters containing sulfides are toxic and objectionable. Sulfate reduction reduces pH and ORP to a point where fermentations begin to occur. Release rates of 0.2 to 5 g $S^{=}/m^2/day$ have been reported.

Acidification of Organics

Fermentations occur after anaerobic conditions have become established. Organics are reduced to organic acids and alcohols which accumulate as end products. The end products produce an immediate oxygen demand when released to an aerobic environment. They are also precursors to methane formation. The pH declines as organic acids are produced. Odors are quite apparent.

Methane Formation

When anaerobic conditions and organic acids are present, CO_2 will serve as a TEA and form methane. The methane will accumulate and effervesce to the surface. The pH rises due to the destruction of organic acids. The ORP remains at about –200 mV.

Anaerobic Summary

Anaerobic conditions lead to the accumulation of ammonia, nitrogen gas, manganous ions, ferrous ions, sulfides, organic acids and methane. Some of the end products react with phosphorus and precipitate, but most accumulate. Manganese frequently rises to 2 mg/L, iron to 15 mg/L, ammonia to 2 mg/L, and BOD to 5 mg/L. The resulting hypolimnetic waters are corrosive, reactive, odorous, staining, and have an immediate and long term DO demand. Drawing down a reservoir before fall turnover can potentially release poor quality water to the detriment of downstream users.

3.11 Reservoir Discharge and Operational Effects

When and where water is withdrawn from a reservoir can significantly affect water quality, both within the reservoir and downstream. This is especially true for a stratified reservoir with a long residence time, where the quality varies greatly with time and depth. For example, discharging water from the bottom of a stratified reservoir with low dissolved oxygen concentrations will release cold, low DO water downstream. It will also reduce the amount of low DO, possibly anoxic, water within the reservoir and replace it with warmer water from the levels above the withdrawal zone. In contrast, a surface discharge will release the warmest and highest DO water downstream, preserving the cold hypolimnion with potentially low DO and anoxic products.

Hydro-power dams typically have several means for discharging water. These include a bottom sluice, a turbine intake often located near the bottom, wicket gates near the surface for releasing flood flows, and a spillway for extremely high flows or emergency use. Some dams include a selective withdrawal facility that allows water to be withdrawn from a range of lake elevations. During normal operations, water is discharged through the turbine intake. The bottom sluice may be used to draw the lake down for inspection or construction activities. The surface outlets and turbine intakes are typically used during flood control operations. Selective withdrawal is usually conducted to maintain specified downstream water quality conditions. For most applications, effective modeling requires an accurate representation of not only the withdrawal location, but also, the withdrawal zone and how it varies with discharge rate.

Reservoir operations significantly affect water quality by determining reservoir elevations, flow rates, residence times, and the timing of releases during the year. For example, a reservoir may begin the calendar year at a low elevation that provides storage capacity for spring floods. As the threat of flooding passes, the reservoir may be filled to support summer recreation. In the late summer and fall, the pool elevation may be drawn down to generate power and to recover flood storage for the next spring. Any variation in the operational pattern can potentially affect reservoir and tailwater quality (Hauser, G. E., et al. 1990). If the fall drawdown were delayed several months to provide a longer recreational season, there would be lower summer flows through the reservoir, deeper lake elevations, and a longer summer residence time. This could result in a cooler hypolimnion, greater DO depletion, and increased anoxic products.

3.12 Tailwater Quality

A tailwater is the stream reach immediately below a dam that receives the reservoir discharge. The discharges can be withdrawn from anywhere in an impoundment, but most often comes from an intake located near the dam. Sluice releases are typically taken from near the reservoir bottom. Hydropower releases may come from anywhere in the depth profile. Gate releases come from the epilimnion. The quality of the release is a function of the withdrawal location, the size of the outlet structure, and the duration and the magnitude of the release. A model must successfully simulate the withdrawal to predict the water quality effects on both the tailwater and the reservoir. The model by Dortch, et al (1992) discusses tailwater quality processes in more detail.

Tailwater Temperature

A tailwater is typically well mixed with depth. If a low level intake is used, the tailwater temperature during periods of stratification will be colder than the ambient river. The water will warm up as it travels downstream with heat exchange taking place at the water surface. The ambient conditions and water volume determine the rate of temperature increase. The length of tailwater determines whether the tailwater reaches an equilibrium temperature. Water quality standards applied to tailwaters usually include a rate of change temperature criterion for the water released.

Tailwater Dissolved Oxygen

The dissolved oxygen concentration of tailwaters created by a low level reservoir intake is typically below saturation during stratification and may be below the water quality criterion. In some cases reservoir releases may be reaerated to improve low DO (Higgins and Brock, 1999). However, since reservoirs are not generally been treated as point sources, the DO may be allowed to replenish naturally. Natural tailwater reaeration takes place at the surface. The driving force is the difference between DO saturation concentration at the tailwater temperature and the actual DO. The rate at which reaeration occurs is related to the degree of turbulence and slope of the tailwater. Tailwaters that are low volume, shallow, and turbulent will reaerate quickly, while high volume, low turbulence, deep tailwaters can flow for many miles before reaching DO saturation. Some tailwaters have been modified with aerating weirs or other devices to increase the DO and maintain minimum flows.

Tailwater Iron and Manganese Oxidation

Reservoir discharges from the hypolimnion may contain iron and manganese. Both begin to oxidize when exposed to light, DO, higher pH and ORP, bottom surfaces and turbulence. Iron oxidizes rapidly at tailwater pH levels in the presence of oxygen. Iron oxidation is a chemical reaction that forms a flocculant red precipitate. The precipitate coats substrate as it settles. Water supplies will find it difficult to remove.

Manganese is slow to chemically oxidize at pH values near neutral. The process is not sufficient to remove much, if any, dissolved manganese. Some manganese will sorb onto the flocculant iron and appear to be reduced in concentration but is easily desorbed. The most prominent mechanism for manganese removal is bacterial oxidation. Autotrophic manganese-oxidizing bacteria are attached bacteria which gain BUE from the process. The bacteria are ineffective in light and tend to grow best under rocks or other dark places, like conduits, where a black slime of cells and precipitated manganese is deposited. The bacteria are easily smothered out by iron oxide precipitates and many times all iron must be removed prior to manganese oxidation. Removing manganese from water supplies requires a strong chemical oxidizer like chlorine, chlorine dioxide, or potassium permanganate followed by filtration. Manganese in clear, cold tailwaters may move many miles downstream.

Tailwater Nutrients and Other Materials

Tailwaters usually contain lower concentrations of nutrients than the reservoir inflow because reservoirs serve as a sink for nitrogen and phosphorus. However, fall drawdowns prior to fall turnover may release considerable quantities of ammonia and phosphorus if the hypolimnion is anaerobic and the release is from a deep outlet. Phosphorus entering a reservoir will usually either settle or be converted to phytoplankton and enter the food chain. If the hypolimnion becomes anaerobic some phosphorus may re-dissolved. The dissolved phosphorus can react with iron to form precipitates. Any remaining phosphorus in the hypolimnion may be removed by reservoir withdrawals. Nitrogen in the form of ammonia will be oxidized to nitrate by nitrifying bacteria in the aerobic tailwater. Nitrogen and phosphorus are also used by phytoplankton as essential nutrients.

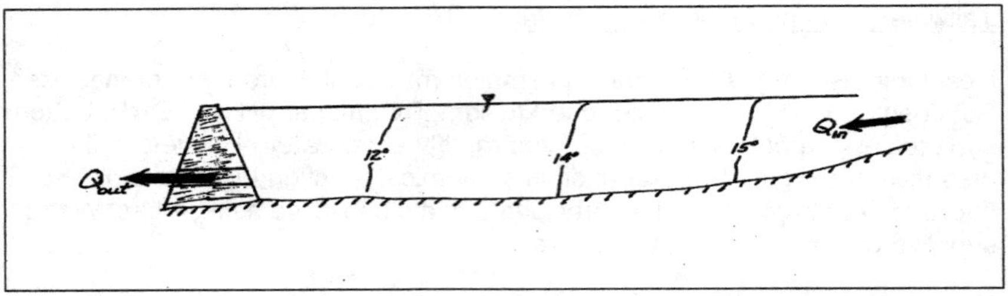

Figure 1. Mainstem Reservoir

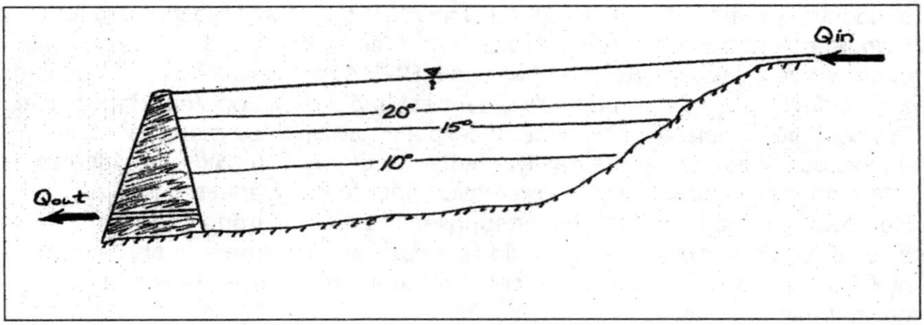

Figure 2. Storage Reservoir

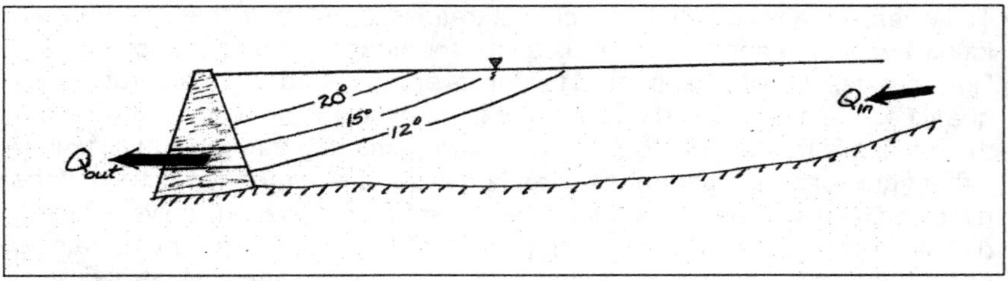

Figure 3. Transition Reservoir

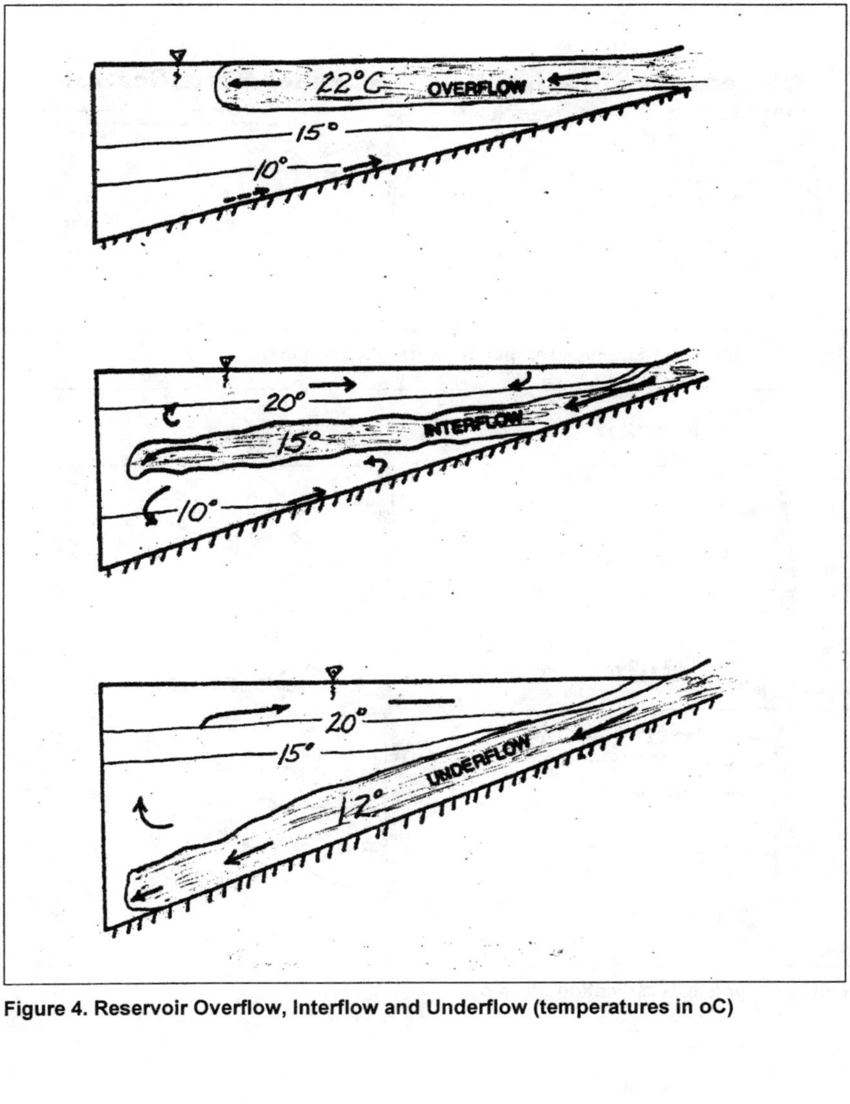

Figure 4. Reservoir Overflow, Interflow and Underflow (temperatures in oC)

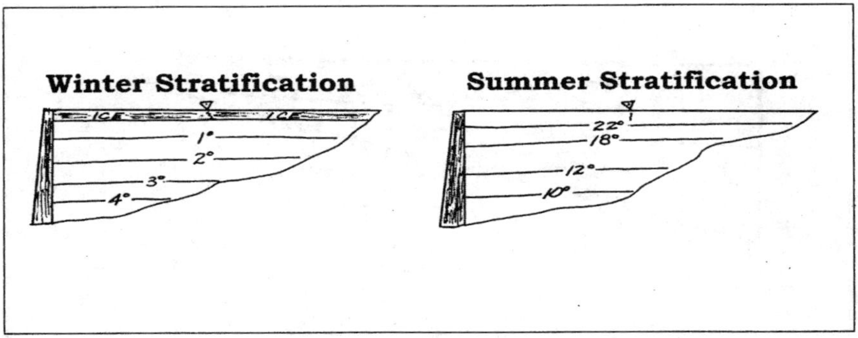

Figure 5. Winter and Summer Temperature Stratification in oC

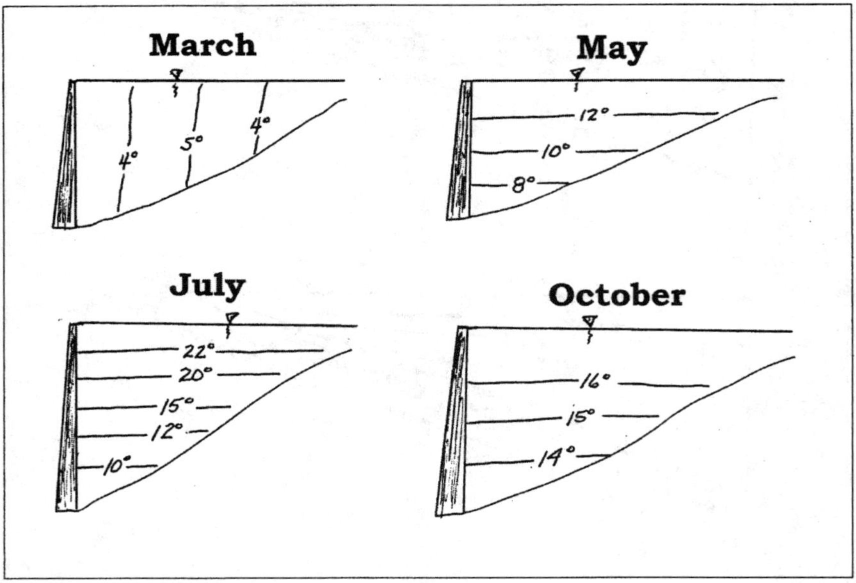

Figure 6. Typical Stratification Patterns

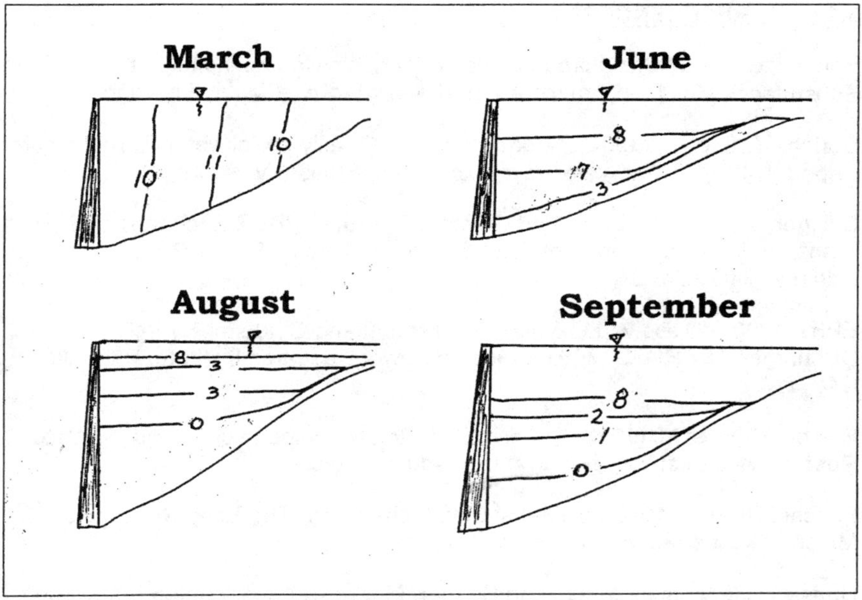

Figure 7. Typical Dissolved Oxygen Concentrations in a Stratified Reservoir (mg/L)

3.13 REFERENCES

Bouwer, E. J. (1992). "Bioremediation of Organic Contaminants in the Subsurface", Ch. 11 in **Environmental Microbiology**, Wiley-Liss, Inc.

Dortch, M. S. et al, (1992). "Modeling Water Quality of Tailwaters", Technical Report W-92-1, Army Waterways Experiment Station, Vicksburg, MS.

Edinger, J.E., Brady, D. K., and Geyer, J.C. (1974). "Heat Exchange and Transport in the Environment", Report No. 14, Electric Power Research Institute, Palo Alto, CA.

EPRI (1990). "Dissolved Oxygen Assessment and Guide for Hydro Discharges", EPRI GS- 7001s, Electric Power Research Institute, Palo Alto, CA.

French, R. H. et al, (1982). "Secchi Disk Relationships", vol. 18, no. 1, **Water Resources Bulletin**, Amer. Water Resources Assn.

Fenchel, T. et al (1998). Bacterial Biogeochemistry, The Ecophysiology of Mineral Cycling, Academic Press.

Hauser, G. E., Shiao, M. C., and Bender, M. D. (1990). "Modeled Effects od Extended Pool Level Operations on Water Quality," Report No. WR28-2-590-148, Tennessee Valley Authority, January.

Higgins, J. M. and Brock, W. G. (1999). "Overview of Reservoir Release Improvements at 20 TVA Dams," **Journal of Energy Engineering**, ASCE. April.

Krenkel, P.A. and Novotny, V. (1980). **Water Quality Management**, Academic Press.

Morris, J. W. (1978). "Dissolved Oxygen Depletion Mechanisms Operating in the Metalimnion of a Deep Impoundment", An MSCE Thesis, Tennessee Technological University, Cookeville, TN.

Reid and Wood (1976). "The Physical and Chemical Properties of Water", Ch. 2 in **Ecology of Inland Waters and Estuaries**, D. Van Nostrand Co.

Cole, T.M. and Hannan, H. H. (1990). "Dissolved Oxygen Dynamics", Ch. 4 in **Reservoir Limnology, Ecological Perspectives** by Thornton, Kimmel and Payne, Wiley Interscience.

TVA, 1990. "Tennessee River and Reservoir System Operation and Planning Review", Tennessee Valley Authority, Knoxville, TN.

Wetzel and Likens, (1991), "Light and Temperature", Exer. 2 in **Limnological Analyses**, Springer-Verlag.

Williams, D. T. et al (1981), "Determination of Light Extinction Coefficients in Lake and Reservoirs", **Surface Water Impoundments**, ASCE. (ISBN 0-87262-271-1).

Wunderlich, W. O. (1971). "The Dynamics of Density-Stratified Reservoirs", In **Reservoir Fisheries and Limnology**, G. E. Hall, Editor, Special Publication No. 8, American Fisheries Society.

Thornton, K.W., B.L. Kimmel, and F.E. Payne, eds. 1990. Reservoir Limnology:

Ecological Perspectives. John Wiley & Sons, Inc. New York, NY. 246 pp.

Thornton, K.W., R.H. Kennedy, J.H. Carroll, W.W. Walker, R.C. Gunkel, and S. Ashby. 1981. Reservoir sedimentation and water quality B a heuristic model. Proceedings of the Symposium on Surface Water Impoundments, June 2-5, 1980. American Society of Civil Engineers. Pp 654-661.

Worth, D.F. 1995. Gradient changes in water quality during low flows in run-of-the-river and reservoir impoundments, Lower Snake River, Idaho. Lake and Reservoir Management. Vol. 11, No. 3. Pp.217-224.

Miller, J. 1984. Intermountain West reservoir limnology and management options. Lake andReservoir Management. EPA 440/5/84-001. U.S. Environmental Protection Agency. Washington, DC. Pp, 272-276.

Kimmel, B.L., and A.W. Groeger. 1984. Factors controlling primary production in lakes and reservoirs: a perspective. Lake and Reservoir Management. EPA 440/5/84-001. U.S. Environmental Protection Agency. Washington, DC. Pp, 277-281.

4. NUMERICAL HYDRODYNAMIC AND TRANSPORT MODELS FOR RESERVOIRS

Scott A. Wells

J. Russell Manson

James L. Martin

Hydrodynamics and mass transport are fundamental to the study of water quality in all waterbodies. Besides the complex chemistry, physics and biology of the water quality constituent under consideration, the transport as a result of the carrier fluid is an integral part of temperature and water quality analyses. The distribution of temperature and water quality constituents within a water body can be modeled only to the detail with which the time varying flow field is known. In modern studies of water quality, it is increasingly common to include the application of hydrodynamic models as an integral component of the model application. The hydrodynamic models are then used to "drive" models of water quality, providing information used to predict the movement of water quality constituents through the constituent transport equation.

The hydrodynamic equations are based upon conservation of water mass and momentum. The water quality transport equations are based upon conservation of constituent mass and heat. An additional equation of state is used to relate the water density to salinity, temperature, and suspended solids that can affect momentum transfer. Typical assumptions of the flow field are related to the dimensionality of the system in space (one, two or three-dimensions), whether the flow field is dynamic or steady-state, and the particular turbulence closure approximation. The model must then be mapped onto a grid representing the actual water body either in 1-D, 2-D, or 3-D, and numerical methods then are used to solve the governing hydrodynamic equations.

This chapter will provide the foundation for the theoretical development of hydrodynamic, temperature and water quality numerical models, as well as turbulence closure approximations. Using simplifications from this development, sections will be developed focusing on applications of these governing equations to

- 1-D longitudinal riverine models
- 1-D vertical reservoir/lake models
- 2-D longitudinal-vertical riverine/lake/reservoir models
- 2-D horizontal models
- 3-D models

This chapter will also provide an overview of how the governing equations are solved numerically using finite difference and finite element methods.

4.1. Governing Equations for Mass, Momentum, Constituent Mass and Heat Conservation

Theoretical Basis for Flow Modeling

The fundamental equations for fluid motion are described below from fundamental principles of mass and momentum conservation. The development of the governing equations is based on a control volume of homogeneous properties as in Figure 1.

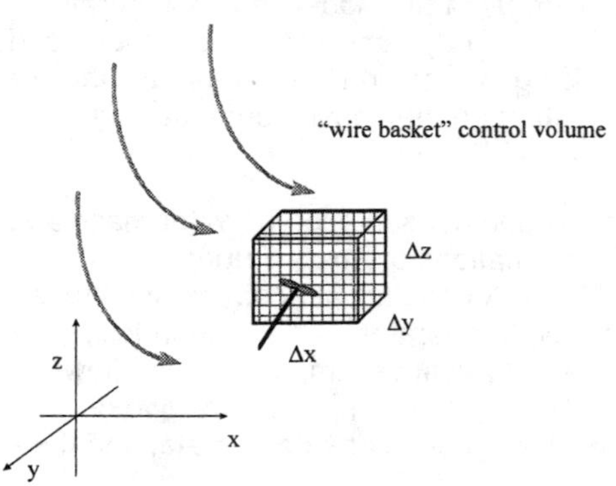

Figure 1. Homogeneous control volume.

The conservation of fluid mass is simply the change in fluid mass within the control volume equaling the sum of inflows to control volume and the sum of outflows from the control volume. The conservation of momentum is based on evaluating the sum of forces acting on a control volume in x, y, or z and equating these to the acceleration of a control volume in x, y, or z as shown in Figure 2. Mathematically, conservation of momentum is described as

$\sum \vec{F} = m\vec{a}$ or in each coordinate direction x, y, and z

$$\sum F_x = ma_x$$

$$\sum F_y = ma_y$$

$$\sum F_z = ma_z$$

where $\vec{F}$: forces acting on control volume

m: mass of control volume

$\vec{a}$: acceleration of fluid within control volume $= \dfrac{d\vec{V}}{dt}$

$\vec{V}$: velocity vector

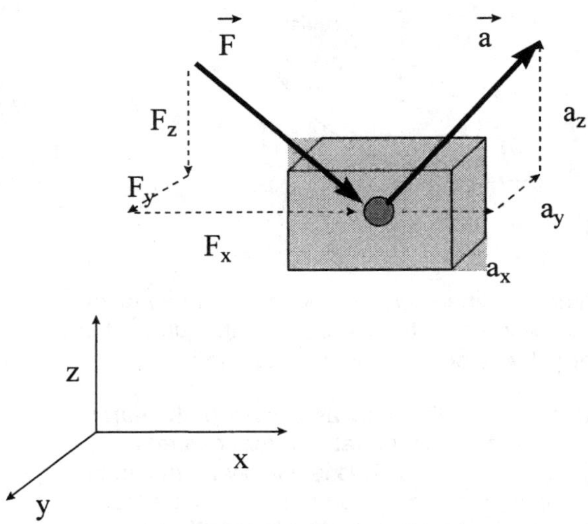

Figure 2 Example of a force acting on a control volume resulting in the acceleration of the fluid within the control volume.

Alternatively, momentum conservation is the net change in momentum of a control volume over time equaling the net efflux of momentum from the control volume and the sum of forces acting on the control volume.

The following section will develop the differential form of these governing mass and momentum equations based on a Cartesian coordinate system.

Coordinate System

The general coordinate system used in the development of the governing equations is shown in **Figure 3**.

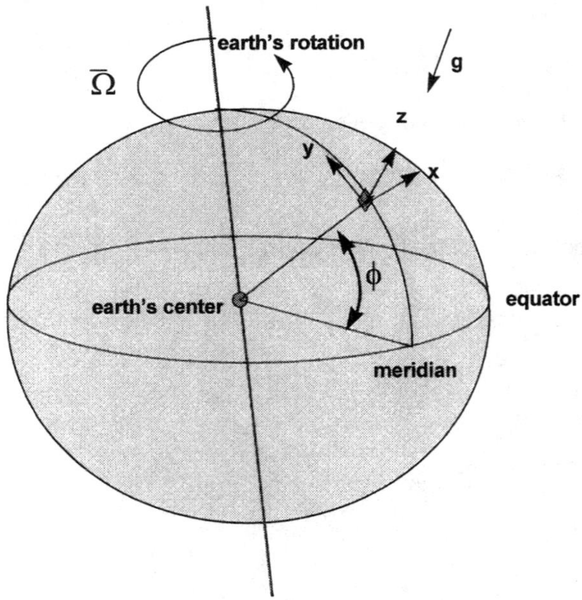

Figure 3. Definition sketch of coordinate system for governing equations where x is oriented east, y is oriented north, and z is oriented upward opposite gravity, Ω is the angular velocity and ϕ is the latitude.

Note that Ω is a vector that represents the angular velocity of the earth spinning on its axis. The rotation of the coordinate system can result in significant horizontal accelerations of fluids. This is usually restricted to large water bodies such as large lakes (such as the Great Lakes in the USA) and oceanic systems. The body force that causes horizontal accelerations because of the spinning coordinate system is termed the Coriolis force.

Turbulent Time-Averaged Equations

The governing equations are obtained by performing a mass and a momentum balance of the fluid phase about a control volume as described in Section 4.1. The resulting equations are the continuity (or conservation of fluid mass) and the conservation of momentum equations for a rotating coordinate system (Batchelor, 1967; Sabersky et al., 1989; Cushman-Roisin, 1994). After using the coordinate system in **Figure 3**, and applying the following assumptions:

- The fluid is incompressible $\frac{\Delta\rho}{\rho} << 1$ where ρ is the fluid density and $\Delta\rho$ is the change in density
- Centripetal acceleration is viewed as a correction to gravitational acceleration
- Apply the Boussinesq approximation (which is related to the incompressibility assumption) to all terms in the momentum equation except those dealing with density gradient induced accelerations

  $\frac{1}{\rho} = \frac{1}{\rho_o + \Delta\rho} \approx \frac{1}{\rho_o}$ where $\rho = \rho_o + \Delta\rho$ where ρ_o is a base value and $\Delta\rho$ has all variations in ρ

- All velocities and pressure are turbulent time averages, i.e., $u = \overline{u} + u'$, where $\overline{u} = \frac{1}{T}\int_t^{t+T} u\,dt$ as shown in Figure 4, and u' is the temporal fluctuation of u about the mean

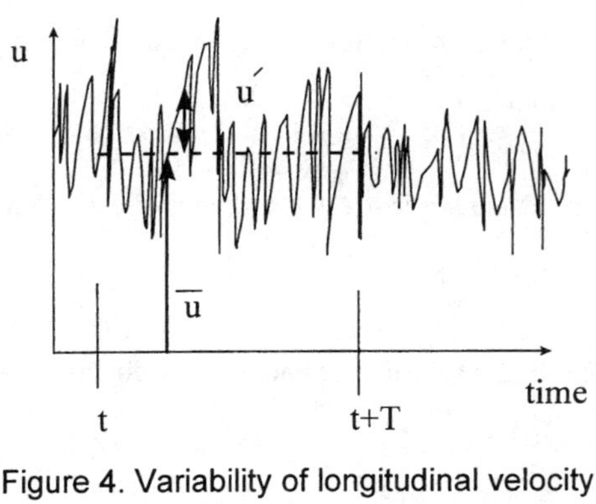

Figure 4. Variability of longitudinal velocity over time.

The governing equations become after simplification:

Continuity

$$\frac{\partial \overline{u}}{\partial x}+\frac{\partial \overline{v}}{\partial y}+\frac{\partial \overline{w}}{\partial z}=0$$

where:

$\overline{u}$:temporal mean velocity in the x-direction

$\overline{v}$: temporal mean velocity in the y-direction

$\overline{w}$:temporal mean velocity in the z-direction

x-Momentum Equation

$$\underbrace{\frac{\partial \overline{u}}{\partial t}}_{\substack{\text{unsteady}\\ \text{acceleration}}}+\underbrace{\overline{u}\frac{\partial \overline{u}}{\partial x}+\overline{v}\frac{\partial \overline{u}}{\partial y}+\overline{w}\frac{\partial \overline{u}}{\partial z}}_{\substack{\text{convective}\\ \text{acceleration}}}\underbrace{-2\Omega_z\overline{v}+2\Omega_y\overline{w}}_{\substack{\text{Coriolis}\\ \text{acceleration}}}=\underbrace{-\frac{1}{\rho}\frac{\partial p}{\partial x}}_{\substack{\text{pressure}\\ \text{gradient}}}+\underbrace{\frac{\mu}{\rho}\left(\frac{\partial^2\overline{u}}{\partial x^2}+\frac{\partial^2\overline{u}}{\partial y^2}+\frac{\partial^2\overline{u}}{\partial z^2}\right)}_{\substack{\text{viscous}\\ \text{stresses}}}$$

$$+\underbrace{\frac{1}{\rho}\left(\frac{\partial \tau_{xx}}{\partial x}+\frac{\partial \tau_{xy}}{\partial y}+\frac{\partial \tau_{xz}}{\partial z}\right)}_{\substack{\text{turbulent}\\ \text{stresses}}}$$

where:

$\tau_{xx} = \rho\overline{u'u'}$ where τ_{xx} is the turbulent shear stress acting in x direction on the x-face of control volume (see Figure 5)

$\tau_{xy} = \rho\overline{u'v'}$ where τ_{xy} is the turbulent shear stress acting in x direction on the y-face of control volume (see Figure 5)

$\tau_{xz} = \rho\overline{u'w'}$ where τ_{xz} is the turbulent shear stress acting in x direction on the z-face of control volume (see Figure 5)

μ = dynamic viscosity

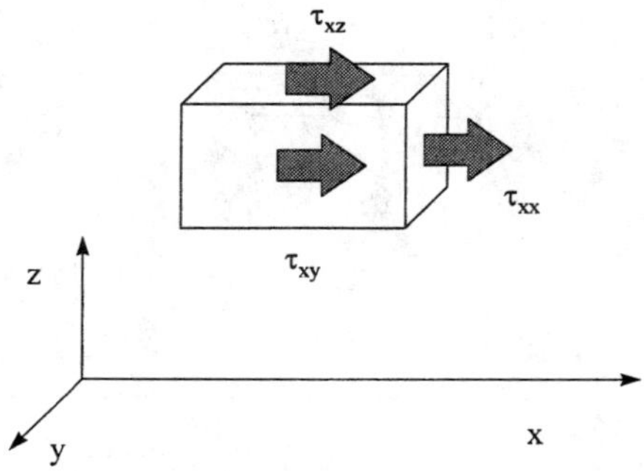

Figure 5. Definition sketch of turbulent shear stresses in x-direction.

Ω= component of Coriolis acceleration where:

$\Omega_z = \Omega_E \sin\phi$

$\Omega_y = \Omega_E \cos\phi$

ϕ = latitude

Ω_E = earth's rotation rate

y-Momentum Equation

$$\frac{\partial \overline{v}}{\partial t} + \overline{u}\frac{\partial \overline{v}}{\partial x} + \overline{v}\frac{\partial \overline{v}}{\partial y} + \overline{w}\frac{\partial \overline{v}}{\partial z} + 2\Omega_z \overline{u} - 2\Omega_x \overline{w} =$$

$$-\frac{1}{\rho}\frac{\partial \overline{p}}{\partial y} + \frac{\mu}{\rho}\left(\frac{\partial^2 \overline{v}}{\partial x^2} + \frac{\partial^2 \overline{v}}{\partial y^2} + \frac{\partial^2 \overline{v}}{\partial z^2}\right) + \frac{1}{\rho}\left(\frac{\partial \tau_{yx}}{\partial x} + \frac{\partial \tau_{yy}}{\partial y} + \frac{\partial \tau_{yz}}{\partial z}\right)$$

where:

$\tau_{yx} = \rho\overline{v'u'}$ where τ_{yx} is the turbulent shear stress acting in y direction on the x-face of control volume (see Figure 6)

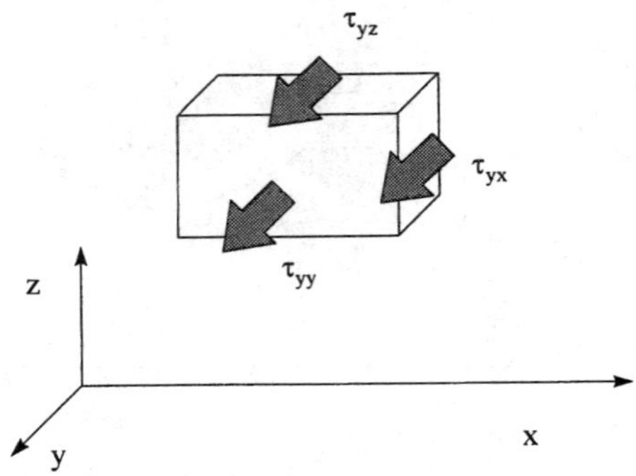

Figure 6. Sketch of turbulent shear stresses in y-direction.

$\tau_{yy} = \rho \overline{v'v'}$ where τ_{yy} is the turbulent shear stress acting in y direction on the y-face of control volume (see Figure 6)

$\tau_{yz} = \rho \overline{v'w'}$ where τ_{yz} is the turbulent shear stress acting in y direction on the z-face of control volume (see Figure 6)

$\Omega_x = 0$

z-Momentum Equation

$$\frac{\partial \overline{w}}{\partial t} + \overline{u}\frac{\partial \overline{w}}{\partial x} + \overline{v}\frac{\partial \overline{w}}{\partial y} + \overline{w}\frac{\partial \overline{w}}{\partial z} - 2\Omega_y \overline{u} + 2\Omega_x \overline{v} = -g$$

$$-\frac{1}{\rho}\frac{\partial \overline{p}}{\partial z} + \frac{\mu}{\rho}\left(\frac{\partial^2 \overline{w}}{\partial x^2} + \frac{\partial^2 \overline{w}}{\partial y^2} + \frac{\partial^2 \overline{w}}{\partial z^2}\right) + \frac{1}{\rho}\left(\frac{\partial \tau_{zx}}{\partial x} + \frac{\partial \tau_{zy}}{\partial y} + \frac{\partial \tau_{zz}}{\partial z}\right)$$

where:

$\tau_{zx} = \rho \overline{w'u'}$ where τ_{zx} is the turbulent shear stress acting in z direction on the x-face of control volume (see Figure 7)

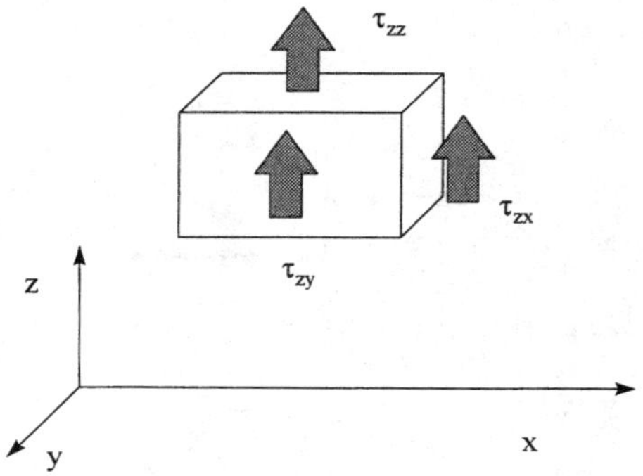

Figure 7. Sketch of turbulent shear stresses in z-direction.

$\tau_{zy} = \rho \overline{w'v'}$ where τ_{zy} is the turbulent shear stress acting in z direction on the y-face of control volume (see Figure 7)

$\tau_{zz} = \rho \overline{w'w'}$ where τ_{zz} is the turbulent shear stress acting in z direction on the z-face of control volume (see Figure 7)

$\Omega_x = 0$

Conservation of Constituent Mass and Heat - The Advective Diffusion Equation

For the control volume shown in Figure 1, the conservation of constituent mass equation is a sum of all the fluxes into and out from the control volume plus sources and sinks (chemistry, biology, physics, withdrawals, inputs) within the control volume. The fluxes include molecular diffusive flux $\vec{J} = -D_x \frac{\partial c}{\partial x}\hat{i} - D_y \frac{\partial c}{\partial y}\hat{j} - D_z \frac{\partial c}{\partial z}\hat{k}$ and advective flux $c\vec{q} = cu\hat{i} + cv\hat{j} + cw\hat{k}$ as shown in **Figure 8** for the x-direction where D_x, D_y, and D_z are diffusion coefficients in x, y, and z, respectively, q is the velocity vector, u, v, and w are the instantaneous velocities in x, y, and z, respectively, and c is concentration of constituent mass.

Total flux in x

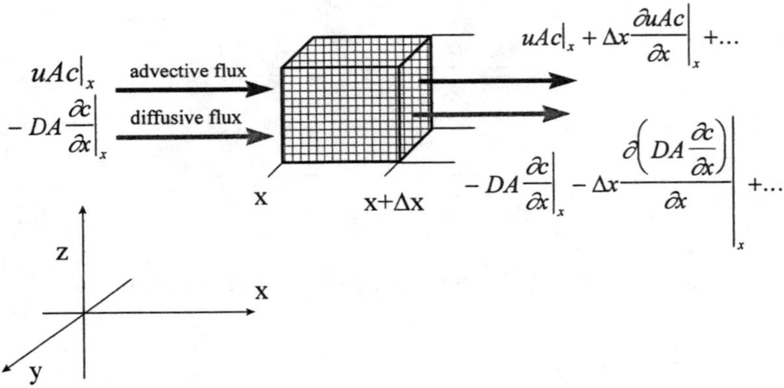

Figure 8. Advective and diffusive fluxes of constituent mass through the control volume in x-direction using Taylor series.

Summing up the fluxes in each direction, assuming that the fluid is incompressible and that the molecular diffusivity, D, is homogeneous and isotropic, the advective diffusion equation becomes

$$\underbrace{\frac{\partial c}{\partial t}}_{\text{unsteady change in concentration}} + \underbrace{u\frac{\partial c}{\partial x} + v\frac{\partial c}{\partial y} + w\frac{\partial c}{\partial z}}_{\text{advective mass transport}} = \underbrace{D\left[\left(\frac{\partial^2 c}{\partial x^2}\right) + \left(\frac{\partial^2 c}{\partial y^2}\right) + \left(\frac{\partial^2 c}{\partial z^2}\right)\right]}_{\text{diffusive mass transport}} + \underbrace{S}_{\text{sources/sinks}}$$

where S is the sources and sinks of reactions occurring in the control volume, or the reaction rate $[ML^{-3}T^{-1}]$.

This equation is a 3-D, unsteady equation that applies to all flow conditions: laminar and turbulent. We use the momentum balance to obtain the velocity field. But rarely can we determine the instantaneous velocity and concentration field. As a result, we usually time average the velocity field, since we can practically predict only the temporal mean velocity.

As in the momentum equation, time-averaged variables for velocity (u, v, and w, see Figure 4) and concentration (see **Figure 9**) are introduced.

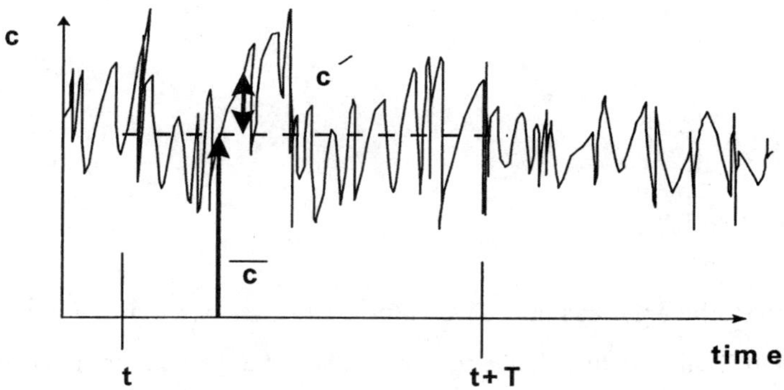

Figure 9. Concentration variability with time.

The instantaneous velocity and concentration are decomposed into a mean and an unsteady component:

$$u(t) = \bar{u} + u'(t) \text{ where } \bar{u} = \frac{1}{T}\int_t^{t+T} u(t)dt$$

Similarly for w, v, and c:

$$v = \bar{v} + v'$$
$$w = \bar{w} + w'$$
$$c = \bar{c} + c'$$

Substituting these into the 3D governing equation and time averaging:

$$\frac{\partial \bar{c}}{\partial t} + \underbrace{\bar{u}\frac{\partial \bar{c}}{\partial x} + \bar{v}\frac{\partial \bar{c}}{\partial y} + \bar{w}\frac{\partial \bar{c}}{\partial z}}_{\text{transport by mean advection}} = \underbrace{D\left[\frac{\partial^2 \bar{c}}{\partial x^2} + \frac{\partial^2 \bar{c}}{\partial y^2} + \frac{\partial^2 \bar{c}}{\partial z^2}\right]}_{\text{molecular diffusive transport}}$$

$$\underbrace{-\frac{\partial}{\partial x}\left(\overline{u'c'}\right) - \frac{\partial}{\partial y}\left(\overline{v'c'}\right) - \frac{\partial}{\partial z}\left(\overline{w'c'}\right) + \bar{S}}_{\text{turbulent mass tranport}}$$

The new terms in the governing equation represent mass transport by turbulent eddies. As the intensity of turbulence increases, turbulent mass transport increases. The turbulent mass fluxes are defined based on the assumption that turbulent mass flux is also a gradient diffusion type process, i.e., the flux is proportional to the mean concentration gradient, as:

$$\left(\overline{u'c'}\right) = -\ E_x \frac{\partial \bar{c}}{\partial x}$$

$$\left(\overline{v'c'}\right) = -\ E_y \frac{\partial \bar{c}}{\partial y}$$

$$\left(\overline{w'c'}\right) = -\ E_z \frac{\partial \bar{c}}{\partial z}$$

where the turbulent mass fluxes would be $\bar{J}_{turbulent-mass} = \left(\overline{u'c'}, \overline{v'c'}, \overline{w'c'}\right)$ and E_x, E_y and E_z are the turbulent eddy diffusivities in x, y and z, respectively.

Substituting into the above equation:

$$\frac{\partial \bar{c}}{\partial t} + \bar{u}\frac{\partial \bar{c}}{\partial x} + \bar{v}\frac{\partial \bar{c}}{\partial y} + \bar{w}\frac{\partial \bar{c}}{\partial z} = \frac{\partial}{\partial x}\left[\left(E_x + D\right)\frac{\partial \bar{c}}{\partial x}\right] + \frac{\partial}{\partial y}\left[\left(E_y + D\right)\frac{\partial \bar{c}}{\partial y}\right] + \frac{\partial}{\partial z}\left[\left(E_z + D\right)\frac{\partial \bar{c}}{\partial z}\right] + \bar{S}$$

In turbulent fluids, E_x, E_y, and $E_z >> D$, and D can be neglected (except at interfaces where turbulence goes to zero). The turbulent diffusion coefficients can be thought of as the product of the velocity scale of turbulence and the length scale of that turbulence. These coefficients are related to the turbulent eddy viscosity. In general, these turbulent diffusion coefficients are non-isotropic and non-homogeneous.

Spatial averaging of these equations leads to the introduction of "dispersion" coefficients which account for the transport of mass from spatial irregularities in the velocity field. Also, these equations are valid for heat transport and temperature modeling by substituting the concentration of heat, $\rho c_p T$, where T is temperature, c_p is the coefficient of specific heat at constant pressure and ρ is the density, such that the governing equation for temperature, T, becomes

$$\frac{\partial \bar{T}}{\partial t} + \bar{u}\frac{\partial \bar{T}}{\partial x} + \bar{v}\frac{\partial \bar{T}}{\partial y} + \bar{w}\frac{\partial \bar{T}}{\partial z} = \frac{\partial}{\partial x}\left[\left(E_x + D\right)\frac{\partial \bar{T}}{\partial x}\right] + \frac{\partial}{\partial y}\left[\left(E_y + D\right)\frac{\partial \bar{T}}{\partial y}\right] + \frac{\partial}{\partial z}\left[\left(E_z + D\right)\frac{\partial \bar{T}}{\partial z}\right] + \frac{\overline{\phi_{net} A}}{V \rho c_p}$$

assuming that the source/sink term can be represented by the net heat flux absorbed into a control volume as ϕ_{net} over a surface area A and a control volume of V.

In this case the transport coefficients for heat are defined as

$$\left(\overline{u'T'}\right) = - E_x \frac{\partial \overline{T}}{\partial x}$$

$$\left(\overline{v'T'}\right) = - E_y \frac{\partial \overline{T}}{\partial y}$$

$$\left(\overline{w'T'}\right) = - E_z \frac{\partial \overline{T}}{\partial z}$$

where the turbulent mass fluxes would be $\overline{J}_{turbulent-heat} = \left(\overline{u'T'}, \overline{v'T'}, \overline{w'T'}\right)$

Generally, heat and mass turbulent eddy diffusivities are of the same order of magnitude and are treated similarly.

Solution of governing equations

The momentum equations differ from the Navier-Stokes equations in the introduction of the terms $-\overline{u'_i u'_j}$ (turbulent shear stresses) and the mass conservation equation differs from its molecular counterpart by the introduction of terms such as $-\overline{u'_i c'}$ or $-\overline{u'_i T'}$ (turbulent mass or heat fluxes). These terms are due to the fluctuating motion and concentrations present in the turbulent flow.

Hence from these 5 equations (continuity or conservation of fluid mass, conservation of momentum in x,y and z, and conservation of constituent mass or heat), we are solving for 5 unknowns: turbulent time average concentration (or temperature) and velocities in x, y, and z, and turbulent time average pressure, i.e., $\overline{c}$ or $\overline{T}, \overline{u}, \overline{v}, \overline{w},$ and $\overline{p}$. The mathematical solution is dependent on specifying the following:

- turbulent shear stresses or Reynolds stresses
- turbulent mass (heat) fluxes
- initial and boundary conditions
- density field
- dynamic viscosity
- Coriolis acceleration terms

Usually the initial and boundary conditions can be specified; dynamic viscosity is a known physical property of the fluid; Coriolis terms are known based on the latitude and the earth's rotation rate; and density field is known through an equation of state defining the density as a function of temperature, salinity and suspended solids. Defining the turbulent shear stresses or the turbulent mass fluxes though has not been so straightforward. This has been termed the problem of "closure".

Since all natural flows in rivers, lakes, reservoirs, and estuaries are all highly turbulent, we can assume that the viscous stress and molecular diffusivity (conductivity) terms are negligible compared to the turbulent terms. This is appropriate everywhere except at interfaces – solid-liquid, air-liquid and sometimes in regions of strong density gradients where turbulence is damped.

For example, the laminar viscous stress acting in the x-direction along the z-face of the control volume is defined as

$$\tau_{\text{laminar}} = \mu \frac{\partial u}{\partial z}$$

The laminar shear stress is the dynamic viscosity, μ, multiplied by the derivative of velocity along the z-axis, which is the differential momentum between adjacent fluid parcels. The dynamic viscosity can be thought of as an empirical constant describing the rate of momentum transfer between adjacent fluid parcels or control volumes. Similarly for turbulent conditions, often an empirical coefficient called the eddy viscosity, $\mu_{\text{turbulent}}$, is used to define the turbulent shear stresses, such that the turbulent velocity fluctuations are determined by gradients in the temporal mean velocity, such as

$$\tau_{turbulent} = \mu_{turbulent} \frac{\partial \overline{u}}{\partial z} = \rho \overline{u'w'}$$

Again, the turbulent eddy viscosity can be thought of as determining the rate of momentum transfer between adjacent fluid parcels.

In a similar way, the molecular diffusive flux in the x-direction has been defined as $J_x = -D \frac{\partial c}{\partial x}$ where D is the molecular diffusion coefficient and c is the concentration. For the turbulent case it is $J_{turbulent\ x} = -E_x \frac{\partial \overline{c}}{\partial x} = -\overline{u'c'}$ where E_x is a turbulent diffusion coefficient in x.

Much effort has been expended in determination of these turbulent shear stresses and the related turbulent mass (heat) fluxes. In many cases though the effort is primarily one of finding the turbulent shear stresses and then using a Schmidt number criterion relating momentum transfer to mass transfer or a Prandtl number criterion relating momentum transfer to heat transfer to determine the related turbulent flux of heat or mass.

The turbulent Schmidt (Sc; for mass transfer) and turbulent Prandtl (Pr; for heat transfer) numbers are defined as

$Sc = \frac{\nu_t}{E}$ and $\Pr = \frac{\nu_t}{K}$, respectively, where E is the turbulent eddy diffusivity, K is the turbulent eddy conductivity, and $\nu_t = \frac{\mu_{turbulent}}{\rho}$ is the turbulent eddy kinematic viscosity.

Most experimental evidence suggests that the Sc and Pr numbers are close to unity for turbulent flows. *For laminar flow* the Prandtl number is about 7 at 20°C, and the Schmidt number is between about 200 to 2000 at 20°C depending on the molecule being diffused (Gebhart et al., 1988). Experimental evidence suggests that the turbulent Sc or Pr numbers vary only little between flows. Even though many models use a constant value of these ratios, buoyancy affects that value (Rodi, 1993).

Turner (1979) shows that the turbulent Pr number increases as the density gradient becomes more stable, i.e., as the Ri (the ratio of gravity to inertial forces) number gets large, and that in the limit of zero density gradient, ν_z and E (and K) are nearly equal, i.e., of order 1. As reported by Turner (1979) of experiments by Ellison and Turner in 1960 of density stratification with salt, the turbulent Sc number was a strong function of Ri. The turbulent Sc number varied from 0.7 to 10 for a variation in Ri from 0.01 to 10. The neutral stability ratio was about 0.7.

Venayagamoorthy et al. (2003) showed used a turbulent Pr number of 1 in neutrally buoyant turbulent flow and a value of between 1 and 1.4 for highly stratified conditions. They used $\Pr_{turbulent} = 0.4\exp(-2.5Fr_k)+1.0$ where Fr_k is the turbulent Froude number defined as $Fr_k = \frac{\varepsilon}{Nk}$, ε is the turbulent dissipation, k is the turbulent kinetic energy, and N is the buoyancy or Brunt-Vaisala frequency= $\sqrt{\frac{-g}{\rho}\frac{\partial \rho}{\partial z}}$.

The determination of the turbulent shear stresses (turbulence closure schemes) are discussed in the following sections.

Overview of Turbulence Closure Modeling Approaches

Turbulence is characterized by a collection of fluid particles exhibiting highly irregular motion. In practice we categorize the intensity of turbulence in a fluid based on the ratio of inertia forces to the stabilizing viscous forces. When this ratio is below some critical value the flow is said to be laminar, the name deriving from the fact that the fluid appears to flow in thin laminae which glide over each other. When the ratio is above the critical value the fluid becomes unstable and the laminae break down into a fluctuating random motion.

The change between the flow regimes of laminar and turbulent flow is not well defined. There is also a large transition region between the two types of flow. Reynolds was the first to quantitatively demonstrate this difference between laminar and turbulent flow and demonstrated that this laminar flow becomes unstable when the non-dimensional ratio $\frac{\rho UL}{\mu}$ (Reynolds number, R_e) is approximately equal to the critical Reynolds number, $R_{e\ crit}$, where U is a velocity scale and L is a length scale. The flow is in a transition regime (in which the turbulence is not fully developed and thus still depends to a lesser extent on the molecular viscosity) between $R_{e\ crit}$ and $R_{e\ FDT}$, the fully developed turbulence Reynolds number.

In the Reynolds number U is a characterizing velocity and L is a characterizing length scale. The magnitude of $R_{e\ crit}$ varies with the flow configuration studied and thus the choice of the characterizing dimensions, however for a river channel, taking L to be the hydraulic radius of the flow channel and U to be the area averaged velocity, $R_{e\ crit}$ is about 500. The flow can usually be considered fully turbulent at about $R_{e\ crit}$ equal to 12500 (French, 1985). When the flow is fully turbulent it is disorderly, unsteady and apparently not amenable to deterministic analysis.

If the flow is turbulent then it will consist of eddies of various sizes ranging from the dimensions of the fluid domain (typically say the flow depth in a reservoir if the system is isothermal) down to the smallest eddy size which may be of the order of hundredths of a millimeter. This is the so-called Kolmogorov length scale ($\left[\frac{\nu^3\delta}{U^3}\right]^{\frac{1}{4}}$) where ν is the kinematic viscosity, which is $\frac{\mu}{\rho}$, and δ is the shear layer width. The range widens with Reynolds number so that at higher Reynolds numbers there is a larger spectrum of eddy sizes to be resolved.

Eddies are constantly moving. Clearly, the computational discretization in space and time required to model such a flow and resolve all the detail would have to be fine enough to resolve the smallest eddy but cover a large enough extent to represent the flow structures of engineering importance. The limitations of computing the effects of turbulence can be demonstrated by considering turbulent flow in a river. For a section of reservoir that is 10 kilometers long, 100 meters wide and 5 meters deep with an average velocity of 0.2 m/s, the smallest eddy containing significant energy may be about 0.1 millimeters. There are therefore, 100,000,000 x 1,000,000 x 50,000 significant eddies to be computed. If we further assume that we could get away with 5 computational points in each co-ordinate direction per eddy (which is probably too coarse a resolution) then there will be $5^3 \times 5\times10^{18}$ computational points in our grid. At each point we would have to store u,v,w and p and compute the one continuity and three momentum equations. Suppose now that we could compute the future values of the variables with no iteration thus we would probably require 100 floating point operations per point per timestep (Emmons, 1970). Thus we would have to do about 6×10^{22} computations per timestep. At the time of this writing, a fast parallel supercomputer may be able to complete tera (10^{12}) floating point operations per second (FLOPS). Thus to complete one time step of our calculation would require 10^{10} s or 32,000 years. Clearly even with highly efficient numerical models this approach is not, at present, an option for practical engineering predictions and is unlikely to become one in the foreseeable future. Despite the daunting example given above, this approach which is termed full simulation or direct numerical simulation has been attempted at least for much lower Reynolds numbers over much smaller domains. For example, Spalart (1988) has simulated the turbulent boundary layer on a flat plate with zero pressure gradient for Reynolds numbers (based on the momentum thickness) up to about 1410 by solving the time dependent Navier-Stokes equations with a spectral method. In this simulation 10 million mesh points were used. When the results at these points were correlated over time and space, Spalart (1988) was able to verify the following results: (a) friction factors accurate to within 5%, (b) velocity profiles that verify the logarithmic wall law and (c) accurate estimates of Reynolds stress and rms fluctuations of both velocity and pressure.

Although the flows considered were barely turbulent, Spalart's results are very important. It is anticipated that this approach will become an important area of research particularly as computers become ever faster and larger. Its usefulness will probably be for calibrating, validating and even designing simpler turbulence models rather than for actual engineering calculations. A less computationally demanding approach is to use a computational grid that resolves only the 'large' eddies and represents the effect of the 'small' sub grid scale eddies with an empirical model. It is argued that as the eddies under consideration get smaller, they tend towards a more isotropic character allowing a simple sub-grid scale model to be applied. Lesieur (1993) gives a state of the art review of LES (Large Eddy Simulation). This approach must be three dimensional and unsteady and so requires very large computational resources. It therefore appears to be limited to simple geometries and boundary conditions. Despite these drawbacks, LES has been able to reproduce some of the large transient turbulent eddies that occur at the floodplain interface in two stage channels (Williams, 1992). Williams (1992) speculates that the ability to compute these motions makes the technique a more effective prediction tool than any other presently used turbulence models. This view is shared by Grass (1992). For practical modeling other techniques are required.

Many different approaches have been used to determine the turbulent shear stresses, such as the following (Nezu and Nakagawa, 1993):

- Eddy viscosity models

Boussinesq in 1877 arbitrarily proposed that the Reynolds stress terms may be assumed instantaneously linearly proportional to the mean strain rate tensor in a way analogous to laminar flows such that $-\overline{u'w'} = \nu_t \frac{\partial \bar{u}}{\partial z}$. The coefficient of proportionality, ν_t, is termed the kinematic eddy viscosity in analogous to the molecular viscosity found in the laminar expression. The problem is now one of specifying the value of the eddy viscosity which is not a constant and is of a higher value than its laminar counterpart.

In stratified water bodies determination of the vertical eddy viscosity and diffusivity has been done by utilizing an empirical model based on the stability of the system is (Munk-Anderson, 1948):

$$\nu_{t-z} = \nu_{t-zo}(1+\beta Ri)^{\alpha}$$

$$E_z = E_{zo}(1+\beta_M Ri)^{\alpha_M}$$

where $Ri = \frac{-g}{\rho}\frac{\frac{\partial \rho}{\partial z}}{\left(\frac{\partial U}{\partial z}\right)^2}$ is the gradient Richardson number (ratio of gravity to inertial forces); E_{zo} and $\nu_{t\text{-}zo}$ are values of E_z and ν_t at neutral stratification where Ri=0, U is the velocity, g is the acceleration due to gravity, ρ is the fluid density, and z is the vertical coordinate. Empirical coefficients which best fit experimental data were α=-0.5; α_M=-1.5; β=10; and β_M=3.33.

In some cases, the vertical viscosity and diffusivity/conductivity are expressed in terms of a functional dependence on the buoyancy or Brunt-Vaisala frequency. These models are of the following form :

$$\nu_{t-z} = a\left(N^{-2}\right)^b$$

where and a and b are empirical parameters, with values of b varying from 0.2 to 2 (Fischer at al., 1979). Holland, et al. (2001) used the following equations for stable and unstable density profiles, respectively:

$$\nu_{t-z} = 0.0004 + (6X10^{-7})(N^{-2})^{0.5} \qquad N^2 > 9.371X10^{-10} s^{-2}$$
$$\nu_{t-z} = 0.02 m^2 s^{-1} \qquad N^2 \leq 9.371X10^{-10} s^{-2}$$

Also, Mamayev (1958) proposed an exponential dependence on the Richardson number such as $\nu_{t-z} = \nu_{t-zo} \exp(-1.5Ri)$ and $E_z = E_{zo} \exp(-3Ri)$ (Lenderstee and Liu, 1975; French, 1985).

Rodi (1993) explains that these relationships tend to fit the majority of experimental data. Many other researchers have used this same functional form and have developed other empirical coefficients which have been fit to experimental data (Martin and McCutcheon, 1999).

In these cases, the vertical viscosity/diffusivity is determined from an empirical model and horizontal diffusivity/viscosity is often assigned a constant value. In these cases, usually the horizontal viscosity is assumed to be the same as the horizontal diffusivity/conductivity (Holland, et al., 2001).

- Mixing length models

Prandtl in the 1920's defined the mixing length as the length a fluid parcel travels before it looses its original identity. Fluid parcels migrated vertically or laterally as they moved downstream. They carried with them their mean velocity and hence their own concentration of momentum.

Prandtl assumed then that the turbulent shear stress can be represented by

$$\overline{u'w'} = \ell^2 \left| \frac{d\overline{u}}{dz} \right|^2$$

such that the turbulent "eddy" viscosity would be defined as

$$\nu_t = \ell^2 \left| \frac{d\overline{u}}{dz} \right|$$

The "mixing length" now is viewed as or interpreted as being proportional to the average size of large eddies. Now the challenge is to predict the distribution of the mixing length. This length is a function of distance from a boundary or wall since the eddy sizes vary as a function of distance from a boundary. This simple model assumes local turbulent equilibrium and no transport of turbulence and is well-suited for calculating simple uni-directional channel flow (Nezu and Nakagawa, 1993). An advantage of this model is its simplicity and economical computational cost. Martin and McCutcheon (1999) show that in estuaries, the transport and history of turbulence can be very important and simple mixing length models may not perform well. Even so, these authors report that in some cases, mixing length models indeed perform adequately (Bloss et al., 1988).

An example of this type of eddy viscosity model in a fully developed infinitely wide open channel is the parabolic distribution, French (1985),

$$\nu_t = \kappa u_* z\left(1-\frac{z}{h}\right)$$

with a maximum,

$$\nu_{t,\max} = \kappa u_* \frac{1}{2}\frac{1}{2} = \frac{\kappa u_*}{4}$$

at mid-depth and average (over the depth),

$$\nu_{t,avg} = \frac{2}{3}\nu_t = \kappa u_* \frac{h}{6} = 0.067 u_* h$$

where z is the distance above the bed. The friction velocity is given by $u_* = \sqrt{\frac{\tau_b}{\rho}}$ where τ_b is the bottom shear stress and may be calculated by some appropriate empirical friction law. Therefore as a first approximation either constant values of the eddy viscosity or depth varying values have been tried even when the flow is not strictly fully developed (Sauvaget and Usseglio-Polatera, 1987).

- One-equation models

Models for the transport of kinetic energy, k, were developed to account for the time-history of turbulent fluctuations and the spatial transport of k. These models though still needed a mixing length approximation to solve the one equation, and for complex flows other than uni-directional channel flows, this has been difficult to prescribe a priori.

In this case, eddy viscosity was estimated using (Martin and McCutcheon, 1999)

$$\nu_t = C_\mu \sqrt{k}\,\ell$$

where ℓ is the length scale of turbulence, k is the turbulent kinetic energy per unit mass and C_μ is an empirical constant. A PDE describing the transport and history of k is solved along with an empirical approximation for the length scale ℓ. The transport model includes horizontal and vertical transport of k, production of k from velocity shear, production or losses of k depending on stable or unstable stratification, and a loss of k from dissipation. The model for k is derived directly from the Navier-Stokes Equations for high Reynolds numbers. This approach was tried by Koutitas and O'Connor (1980) for computing three dimensional wind induced flows in a coastal bay.

Axell and Liungman (2001) used a transport model for k and an algebraic equation for the mixing length, ℓ, that depended on local stratification and distance from a boundary. Their approach matched results of more complicated two-equation models but with less computational effort since only one transport model was solved.

- Two-equation k-ε models

The rate of change of k is governed by the advective transport due to the mean velocity, the diffusive transport due to velocity and pressure fluctuations, the production of k by interaction of the Reynolds stresses and the mean velocity gradients and the dissipation of k by viscous action into heat.

In order to eliminate the specification of a mixing length function, a second equation is often used to specify the transport and history of k and to determine the length scale distribution of turbulence from another transport equation. The length scale of the energy containing eddies is also subject to transport processes in the same way as the kinetic energy. This fact, combined with the desire to avoid having to prescribe the length scale a priori has encouraged researchers to try to formulate a second differential transport equation for the turbulent length scale. In fact the length scale is rarely used itself but rather a variable proportional to it has been adopted. Several different variables have been suggested for the second differential equation with the most popular being ε, the rate of dissipation of the turbulent kinetic energy.

One such model is the k-ε model where ε is the turbulent energy dissipation and is defined approximately as $\varepsilon \approx \dfrac{k^{3/2}}{\ell}$ where ℓ is the length scale for turbulent dissipation (similar to the mixing length of Prandtl). Using this approximation, the equation for the turbulent eddy viscosity is then

$$\nu_t = C_\mu \frac{k^2}{\varepsilon}$$

where C_μ is a new empirical constant. The equations used for the k-ε model are the k equation and an equation for the rate of turbulent dissipation (Rodi, 1993; Nezu and Nakagawa, 1993). The equation for the rate of turbulent dissipation is derived from the Navier Stokes Equation by computing the transport of vorticity. This equation contains terms representing convection, diffusion, generation of vorticity due to vortex stretching connected with the energy cascade and a term representing the viscous destruction of vorticity.

In many of these modeling approaches, only vertical diffusion of TKE and dissipation are modeled even in multi-dimensional models. For example, in a 3-D model of the Baltic Sea, Meier (2001) used the following set of k-ε equations neglecting horizontal and vertical transport and horizontal diffusion:

$$\frac{\partial k}{\partial t} = P - \varepsilon + G + \frac{\partial}{\partial z}\left(\frac{\nu_t}{\sigma_k}\frac{\partial k}{\partial z}\right)$$

$$\frac{\partial \varepsilon}{\partial t} = \frac{\varepsilon}{k}C_{1\varepsilon}P + \frac{\varepsilon}{k}C_{3\varepsilon}G - \frac{\varepsilon^2}{k}C_{2\varepsilon} + \frac{\partial}{\partial z}\left(\frac{\nu_t}{\sigma_\varepsilon}\frac{\partial \varepsilon}{\partial z}\right)$$

where

$P = \nu_t\left[\left(\frac{\partial U}{\partial z}\right)^2 + \left(\frac{\partial V}{\partial z}\right)^2\right]$, the production term,

$G = -\frac{\nu_t}{\sigma_t}N^2$, the buoyancy term,

σ_t =k turbulent Schmidt/Prandtl number =1.0, , σ_e =ε turbulent Schmidt/Prandtl number =1.08, and the others were empirical constants with typical values of C_μ=0.09, $C_{1\varepsilon}$=1.44, $C_{2\varepsilon}$=1.92.

In their calibration of the salinity and temperature stratification, determining the term $C_{3\varepsilon}$ and the turbulent Prandtl number, σ_t, was a source of calibration problems. They concluded that a constant turbulent Prandtl number resulted in too strong an erosion of the halocline and suggested a Richardson number dependence based on Blanke and Delecluse (1993).

Venayagamoorthy et al. (2003) showed that the empirical coefficients of the k-ε models are better defined as functions of the turbulent Froude number rather than as constants when there is weak to moderate stratification.

Typical boundary conditions for k and ε are (Axell and Liungman, 2001):

$k = \left[\frac{u_*^3}{(C_\mu)^3} + \frac{\max(B,0)\kappa d_1}{(C_\mu)^3}\right]$ where u_* is the friction velocity= $\sqrt{\frac{\tau}{\rho}}$, τ is the boundary stress, C_μ=0.5562,d_1 is the distance from the boundary, κ=0.41, and B is the buoyancy flux= $\frac{-g}{\rho}\left(\frac{\partial \rho}{\partial T}\frac{Q_H}{\rho c_p} + \frac{\partial \rho}{\partial S}Q_S\right)$, Q_H is the heat flux, and Q_S

is the salt flux which is related to precipitation and evaporation and
$$\varepsilon = (C_\mu)^3 \frac{k^{3/2}}{\kappa d_1}.$$

Craig and Banner (1994) showed the importance of breaking gravity surface waves controlling the flux of turbulent kinetic energy into the interior. A flux boundary condition, rather than the fixed or Dirichlet boundary condition shown above, was proposed (Mellor and Blumberg, 2004).

- Reynolds stress equation models (second-order closure models)

These equations do not assume an eddy viscosity assumption to determine the turbulent shear stresses. Another set of equations is developed to determine the stresses directly. The procedure for determining this equation (Hinze, 1975) is to time average the Navier-Stokes momentum equation and subtract it from the time-dependent Navier-Stokes Equation for the x_i and x_j momentum equations. Then these equations are cross-multiplied by u_i and u_j, summed, and time averaged. But these higher-order equations also need to be "closed" by making appropriate assumptions. To solve the above equation, simplifying assumptions have to be made in the diffusion, pressure-strain and dissipation terms (Rodi, 1993) where additional empirical coefficients are required to close the mathematical solution. In the solution of the higher order stresses there are 6 components $\overline{u_i u_j}$, 3 components of the scalar flux $\overline{u_i \phi}$, and one equation for the scalar fluctuation $\overline{\phi^2}$. This requires the solution of (at a minimum) 10-coupled PDEs to determine the turbulence field (Rodi, 1993). Because of this high computational cost, efforts were made to introduce algebraic simplifications of these terms. This is the motivation for development of the algebraic stress models presented in the next section.

- Algebraic stress models

By making assumptions about the unsteady, convection and diffusion terms in the shear stress equations, these terms are converted to algebraic equations to simplify the computational cost of solving the PDEs. The basic idea of this approach is to eliminate the gradients in the unsteady, convective, and diffusive terms so that the resulting PDE is an algebraic equation. Some researchers have merely neglected the unsteady and convective transport terms, which appears to be a reasonable assumption in many cases (Rodi, 1993). The algebraic model then consists of solving a k-ε model together with complex algebraic equations for the individual stresses $\overline{u_i' u_j'}$, the scalar flux $\overline{u_i' \phi'}$, and the scalar fluctuation $\overline{\phi'^2}$. Unfortunately, these models are still computationally expensive.

Modeling Approaches

1-D Lake/Reservoir Model Development

There exist several 1-D lake or reservoir models, such as CE-QUAL-R1 (Corps of Engineers, 1986a), DYRESM (Imberger and Patterson, 1981, Antenucci and Imerito, 2000), MIT (Octavio, et al., 1977), WQRRS (Smith, 1978), and HEC-5 (Corps of Engineers, 1986b), that model the lake or reservoir as a series of horizontally homogeneous layers as shown in Figure 10.

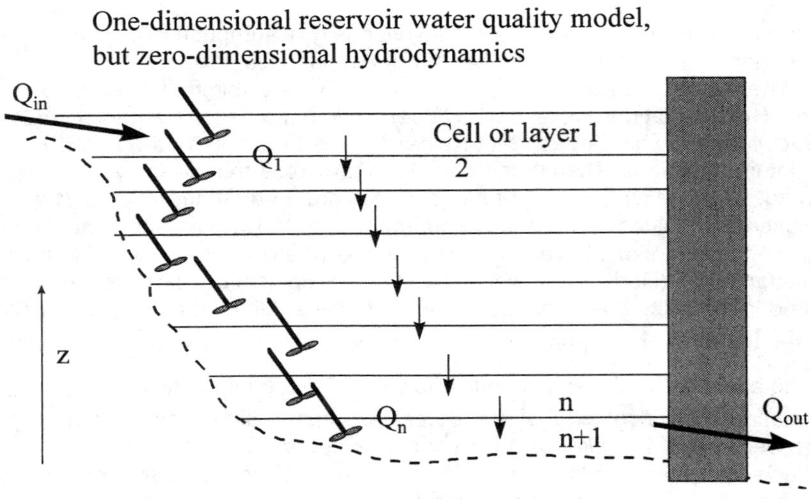

Figure 10. Typical layout of a one-dimensional homogeneous slab reservoir model.

These models are not 'hydrodynamic' in the sense that they do not solve the momentum equations to predict flows. They use a water balance approach.

A global water balance is done over the reservoir in order to compute the new water surface elevation, i.e.,

$$\frac{dV_{reservoir}}{dt} = \sum Q_{in} - \sum Q_{out}$$

After this, a local flow balance is applied to each control volume starting at the top and moving down through the cells (or vice-versa) in order to compute the vertical flow (and thus the vertical velocity knowing the horizontal area between cells). For example, for cell 1 this becomes simply:

$$Q_1 = \frac{dV_{reservoir}}{dt} - Q_{in} = Q_{out}$$

For cells 2 through n this becomes simply

$$Q_n = Q_{n-1} = Q_{out}$$

The vertical velocity between a layer n with horizontal area A is then

$$V_n = \frac{Q_n}{A_{horizontal}}.$$

This concept can be applied for any combination of inflow locations or outflow locations. Most of these models do though have special algorithms to improve the predictive capability of the flow modeling, such as selective withdrawal, vertical inflow placement based on density of the inflow, and wind-mixing algorithms.

The constituent mass equation (or temperature equation) is derived by formally averaging the 3-D equation over the horizontal slab such that the governing equation is typically:

$$\frac{\partial T}{\partial t} + \frac{1}{A}\frac{\partial}{\partial z}\left(Q_v T\right) = \frac{1}{A}\frac{\partial}{\partial z}\left(AE_z \frac{\partial T}{\partial z}\right) + \frac{BU_i T_i}{A} - \frac{BU_0 T}{A} - \frac{1}{\rho c_p A}\frac{\partial\left(A\Phi_z\right)}{\partial z}$$

where: Q_v is the vertical flow rate, A is the horizontal area and ϕ is the heat flux,

$$\Phi_z = \Phi_{sn}\left(1-\beta\right)e^{-\eta(z_s - z)}$$
$$Q_v = A(z)W(z)$$
$$A(z) = B(z)L(z)$$

ϕ_{sn} is the surface short wave solar radiation, L is the length of a layer, B is the width, W is the vertical velocity, β is the fraction of long-wave radiation absorbed at the surface, and η is the light extinction coefficient.

Boundary conditions were specified for the surface and bottom:

- at the surface $z = z_s$ $E_z \frac{\partial T}{\partial z} = \beta\Phi_{sn} + \Phi_{an} - \Phi_{br} - \Phi_e - \Phi_c$ where
- at the lake/reservoir bottom at $z = 0$ a no flux condition was used: $\frac{\partial T}{\partial z} = 0.$

Horizontal velocities were computed from inflow and outflow rates assuming Gaussian input and outlet distributions. Vertical velocities (or flow rates) were computed from continuity for each element using

$$Q_v = B\int_0^z U_i dz - B\int_0^z U_0 dz$$

Vertical mixing and erosion of the thermocline due to wind is usually accomplished using a mixed-layer approach where the turbulent kinetic energy from the wind is used to entrain fluid from lower layers. This approach considers an energy balance in the mixed layer, termed an integral energy approach (Spigel, et al., 1986).

1-D Riverine Model Development

One-dimensional models are used to study the hydraulics and water quality of impacted river systems. One-dimensional (cross-sectional, area-averaged) models typically represent the geometry of a river as a series of prismatic channels at cross-sections, separated by some distance. Area-averaged models may be differentiated based on the manner in which they provide for representation of the cross-sectional characteristics. The representations are typically based on a simplified cross-sectional shape (rectangular, trapezoidal, etc.), equations relating cross-sectional characteristics to water depth, or look-up tables specifying water surface elevations as a function of distance across the channel. These representations work well for most applications. However, representation of complex cross-sectional geometry is a difficulty in one-dimensional (area-averaged) models. For example, representing a main river channel as well as a flood plain presents difficulties. Rapid variations in channel characteristics (such as abrupt changes in width) can often result in numerical instabilities in area-averaged models. Some available models allow for specification of a main conveyance area as well as left and right overbank areas to partly accommodate for differences in flows in these sections. Some available models also provide for off-channel storage to partially allow for simulation of the impacts of flooded areas. Some models can not handle large differences in flow regimes, such as hydraulic jumps.

One-dimensional models can also be differentiated in the way they handle multiple channel configurations. Some models only allow simulation of a single section of a river while others allow for more complex channel configurations.

A well-known example of a one-dimensional, steady-state flow in an open channel (Figure 11) model would be the use of Manning's Equation (in SI units),

$$Q = \frac{1}{n} R_H^{2/3} S^{1/2} A.$$

where n: Manning's friction factor

A: cross-sectional area

R_H: hydraulic radius (cross-sectional area over wetted perimeter)

S: slope or change in head over distance

The steady-state continuity equation is flow in = flow out, or

$$\sum Q_{in} = \sum Q_{out} .$$

This is the approach used in the EPA QUAL2E model (Brown and Barnwell, 1987). In QUAL2E, the continuity or volume balance determines the flow in each river section, but the Manning's equation (or a different empirical relationship) is used to compute the depth in the river. Typical empirical relationships for flow (Q) and depth (d) are

$$U = aQ^b$$
$$d = \alpha Q^\beta$$

where a, b, α, β are empirical coefficients. The HEC-2 and the HEC-RAS model are also 1-D steady-state hydraulic models.

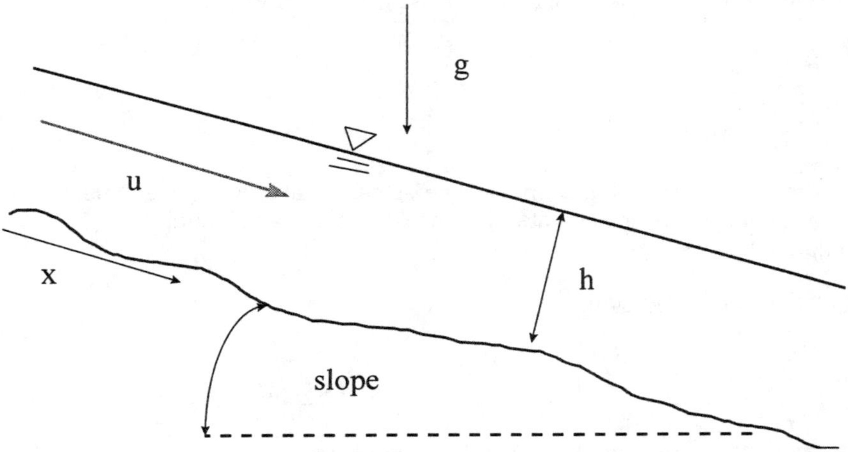

Figure 11. Steady-state flow down a slope where there is a balance between resistive shear stresses and gravity.

For one-dimensional dynamic models, the full 1-D continuity and 1-D conservation of momentum are required. The type of application could be a river channel, vertically and laterally well-mixed estuary, or flow in a pipe where boundary conditions are changing over time. Examples of 1-D dynamic hydraulic models include

- EPA DYHYD (Ambrose, et al. 1988) model for rivers and estuaries
- MIKE11 Model (a Danish Hydraulics Institute proprietary model)
- EPA SWMM model for pipe or open channel flow hydraulics (Roesner et al. 1988)
- Corps WQRRS (Smith, 1978) or HEC-5 models only for the river sections
- Corps UNET: 1-D river model, (Corps of Engineers, 2001, Barkau, 1997)
- EPA HSPF (Donnigen et al., 1984) model (for open channel or pipe flow sections)
- HEC RAS steady-state model for river channels and a more recent unsteady model based on the UNET model (Brunner, 2002)
- Corps CE-QUAL-RIV1 (Environmental Laboratory, 1995)

The governing equations for 1-D dynamic models are termed the St. Venant equations and are often written as follows (Abbott, 1992):

Continuity

$$\frac{\partial A}{\partial t}+\frac{\partial Q}{\partial x}=q$$

This equation states that changes in cross-sectional area, A, over time, t, are a result of changes in flow-rate, Q, over distance, x, and lateral inflows/outflows, q.

Longitudinal Momentum

$$\underbrace{\frac{\partial Q}{\partial t}}_{\text{unsteady\_acceleration}}+\underbrace{\frac{\partial\left(\alpha Q^2/A\right)}{\partial x}}_{\text{convective\_acceleration}}=\underbrace{-gA\frac{\partial h}{\partial x}}_{\text{longitudinal\_pressure\_gradient}}\underbrace{-\frac{gn^2Q|Q|}{AR^{4/3}}}_{\text{friction}}+\underbrace{u\cos\varphi q}_{\text{momentum\_lateral\_inflows}}$$

where q is the lateral inflow under an angle φ with a mean velocity of u, α is a momentum correction factor, Q is the flow rate, h is the channel depth, A is the cross-sectional area, u is the longitudinal velocity, n is Manning's friction factor, R is the hydraulic radius, t is time, and x is the longitudinal coordinate. If the equation is written in terms of the Chezy friction factor, C, the friction term becomes $-\frac{gQ|Q|}{C^2AR}$.

Another term that is often ignored in 1-D models is the acceleration due to changes in longitudinal density, a longitudinal pressure gradient, which is important for situations where there are strong longitudinal temperature or salinity gradients.

Each of the terms describes processes that affect changes in flow rate over time and space: (1) represents changes in flow as a result of changes in cross-sectional area (convective acceleration term); (2) represents changes in flow from pressure gradients as a result of changes in water surface elevation differences; (3) represents changes in velocity as a result of frictional losses from bed shear; (4) represents momentum from side tributaries as they intersect the main branch and (5) represents the acceleration of the flow from pressure gradients as a result of longitudinal density differences.

The cross-sectionally averaged constituent mass transport equation (or temperature) is

$$\frac{\partial c}{\partial t}+\frac{1}{A}\frac{\partial}{\partial x}(UAC)=\frac{1}{A}\frac{\partial}{\partial x}(AE_L\frac{\partial c}{\partial x})+r$$

In some cases, researchers (Yen, 1973; Bencala and Walters, 1983) have included side channel transient storage, such that the mass conservation equations become

$$\frac{\partial c}{\partial t}+\frac{1}{A}\frac{\partial}{\partial x}(UAC)=\frac{1}{A}\frac{\partial}{\partial x}(AE_L\frac{\partial c}{\partial x})+r+k_1(s-c)$$ Scalar - main channel

$$\frac{\partial s}{\partial t}=-k_2(s-c)$$ Scalar – transient storage zone

where k_1, is an exchange rate with an off-channel storage area, and s, the concentration of the constituent in the off channel storage. Presently, relatively few models include the off channel, or transient, storage terms. For water quality models for the main channel and side channel are solved for each water quality constituent, which can range from a single to more than 30 constituents depending on the model formulation. The transport information and hydraulic characteristics (e.g. A, Q, and E_L) apply equally to all constituents. What distinguishes the individual constituent transport equations are the terms that enter into r, the source and sink terms. The formulation of these terms is the focus of Chapter 5.

Note that these equations are based on averaging over a cross-section (area averaged), implying that variations in velocity are ignored except in the direction of flow, so that u represents a cross-sectionally averaged velocity. In these models, the friction factors (Mannings, n, or Chezy, C) and α are empirical parameters which must be specified. The momentum correction factor, α, accounts for the non-uniform velocity distribution over the cross section. These variables represent the effects of turbulence in a lumped parameter sense. Boundary conditions usually consist of specified flow hydrographs at upstream boundaries and specified water stages or rating curves at downstream boundaries although alternative boundary conditions are possible. These equations, when furnished with suitable initial conditions may be solved by a numerical method to give flows and stages throughout the river, such as in Cunge, Holly and Verwey (1989).

The one dimensional river models are well proven and appear to work well for what they were designed, i.e., calculating the stage and flow throughout river and reservoir systems. They do not, however, give much detailed information at local points on the river. At certain points of the river, e.g., bridges or local expansions or contractions, the equations as given above are not representative and local empirical relationships must be applied. These relationships are typically included in models designed for flood routing or analyses, but are often not included in models designed for other purposes. Area-averaged models have difficulty representing the effects of complex geometry, such as complex valley flows, and in handling transitions between subcritical and supercritical flows.

The acceleration due to local inflows (conservation of longitudinal momentum at a tributary intersection) is typically not included in available models. Few models include the term for longitudinal accelerations as a result of longitudinal density differences. For example, to include density effects, a third equation would typically be included for the conservation of heat energy (to predict temperature) or salinity (for estuarine applications). The density would then be related to the predicted temperatures or salinities through an equation of state. These relationships are typically not included in one-dimensional (area averaged) models, so the applicability of these models is limited in areas where density induced flows are significant.

2-D Longitudinal-Vertical Model development

A widely-used 2-D hydrodynamic and water quality model is the Corps of Engineers model CE-QUAL-W2 (Cole and Wells, 2003). Other similar model types were reviewed in ASCE (1989) and Johnson (1981). Other reservoir 2-D models include the TVA BETTER model, which is 2-D but uses a flow balance for solving the velocity field rather than the momentum equations. The EPA model WASP can be 2-D or 3-D, but it does not determine the hydrodynamic regime. WASP usually uses the output from the 1-D DYNHYD model for its 1-D flow field.

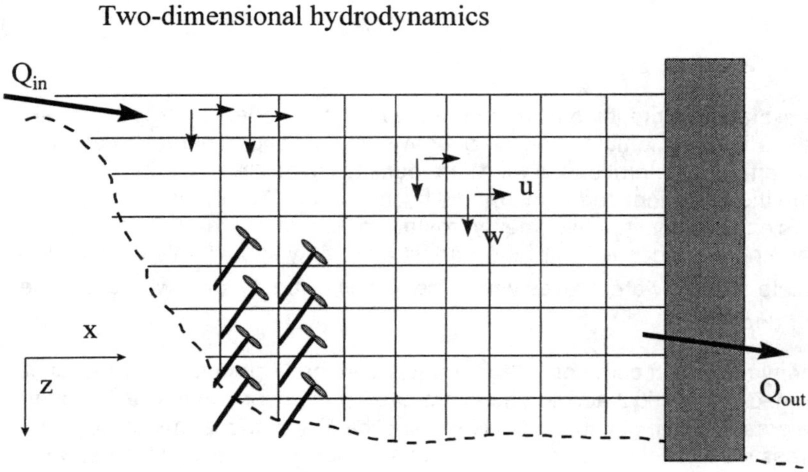

Figure 12. Typical reservoir grid for 2-D model.

A typical schematic of a 2-D reservoir grid is shown in Figure 12.

Laterally averaged models are a simplification of the quasi-three dimensional models with the flow integrated laterally across the channel width and varying vertically with depth. The initial development of the laterally averaged model for reservoirs was accomplished by Edinger and Buchak (1978a). The quasi-three dimensional equations were averaged over the channel width as shown in Edinger and Buchak (1978b). These formulations became the basis for the Version 2 of CE-QUAL-W2 (Cole and Buchak, 1995). CE-QUAL-W2 Version 3 was expanded to allow river basin modeling connecting river, reservoir and estuary segments (Wells, 2002a, 2002b, Cole and Wells, 2003).

If one integrates over the y-direction (therefore assuming the net velocity in y is zero) and assumes that the horizontal length scale is much greater than vertical length scale, it can be shown by using scaling arguments that the Coriolis acceleration forces are negligible (Cushman-Roisin, 1994). Hence, prior to lateral averaging, the Coriolis acceleration terms are neglected. Also, the coordinate system is oriented along an arbitrary channel slope such that if the system slope is zero, z is in the same direction as gravity.

If the longitudinal length scale is much greater than the vertical length scale, then this makes all vertical velocities << horizontal velocities. A result of this assumption is that vertical velocities are very small such that the z-momentum equation becomes the hydrostatic equation:

$$\frac{1}{\rho}\frac{\partial \overline{p}}{\partial z} = g\cos\alpha \tag{1}$$

This assumption prevents using the hydrodynamic model to predict vertical accelerations of the fluid because of convective cooling at night and other such vertical accelerations. If there are density instabilities, mechanical mixing though is induced in the model by increasing the vertical mixing coefficient. Models that assume hydrostatic pressure are termed shallow water models since, as it will be demonstrated, they should only be used to simulate shallow water waves whose celerity is given by $\sqrt{gh}$ where h is the water depth.

In many models in common usage today, even three-dimensional models, it is assumed that vertical accelerations are small in comparison to gravitational acceleration. This, of course, limits the applicability of these models to address the impact of, for example, buoyant jets, submersed diffusers, and the near-field flow regime in the vicinity of outlet structures that impart a significant vertical momentum.

The governing equations were formally laterally averaged after decomposing all velocities and pressure into a lateral average and a deviation from the lateral average. The lateral, longitudinal, and vertical velocities and pressure are defined as follows:

$$\overline{v} = \overline{\overline{v}} + v''; \quad \overline{u} = \overline{\overline{u}} + u''; \quad \overline{w} = \overline{\overline{w}} + w''; \quad \overline{p} = \overline{\overline{p}} + p''$$

where:

$$\overline{\overline{u}} = \frac{1}{B}\int_{y_1}^{y_2} \overline{u}\, dy$$

B = control volume width, *m*

y_1 = left bank coordinate

y_2 = right bank coordinate

The double overbars represent the spatial average of the temporal average quantity. The double prime represents the deviation from the lateral average and is a function of y. This is shown in Figure 12 for the longitudinal velocity.

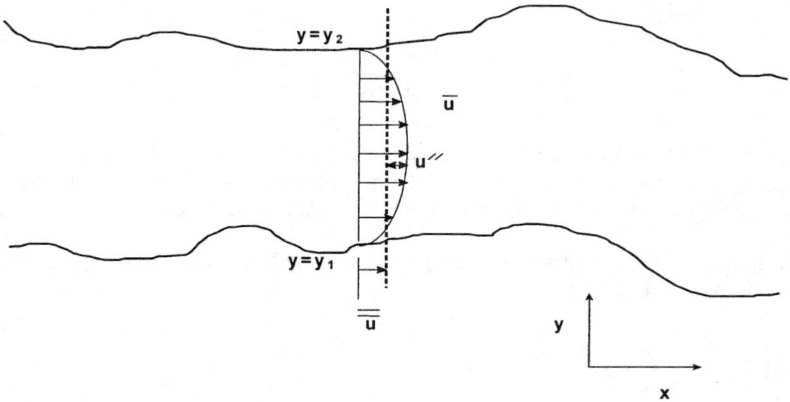

Figure 13. Lateral average and deviation from lateral average components of longitudinal velocity.

These definitions are substituted into the turbulent time-average governing equations and then laterally averaged.

Also in the development of the x-momentum equation, a term for longitudinal momentum conservation at branch intersections was also added.

The shear stress terms will be assumed lateral averages and the double overbars will be dropped for convenience. Making these simplifications, the governing equations become:

Continuity Equation

$$\frac{\partial UB}{\partial x}+\frac{\partial WB}{\partial z}=qB$$

x-Momentum Equation

$$\frac{\partial \mathrm{UB}}{\partial \mathrm{t}}+\frac{\partial \mathrm{UUB}}{\partial \mathrm{x}}+\frac{\partial \mathrm{WUB}}{\partial \mathrm{z}}=gB\sin\alpha-\frac{B}{\rho}\frac{\partial P}{\partial x}$$
$$+\frac{1}{\rho}\frac{\partial B\tau_{xx}}{\partial \mathrm{x}}+\frac{1}{\rho}\frac{\partial \mathrm{B}\tau_{xz}}{\partial \mathrm{z}}+qBU_x$$

z-Momentum Equation

$$\frac{1}{\rho}\frac{\partial P}{\partial z} = g \cos \alpha$$

where dropping the overbars for convenience, *U*, *W*, and *P* are the lateral average longitudinal and vertical velocity and pressure, respectively, q is the lateral inflow per unit volume of cell, and B is the width of a cell.

Substituting the z-momentum equation into the longitudinal pressure gradient in the momentum equation,

$$-\frac{1}{\rho}\frac{\partial P}{\partial x} = -\frac{1}{\rho}\frac{\partial P_a}{\partial x} + g \cos \alpha \frac{\partial \eta}{\partial x} - \frac{g \cos \alpha}{\rho}\int_{\eta}^{z} \frac{\partial \rho}{\partial x} dz$$

where η is the water surface deviation from a fixed datum and P_a is the atmospheric pressure. The first term on the RHS is the atmospheric pressure term (accelerations due to atmospheric pressure changes over the water surface), the second is the barotropic pressure term (accelerations due to water surface variations), and the third is the baroclinic pressure term (accelerations due to density driven currents).

In CE-QUAL-W2, the atmospheric pressure variability over x is neglected. The pressure term then becomes:

$$-\frac{1}{\rho}\frac{\partial P}{\partial x} = g \cos \alpha \frac{\partial \eta}{\partial x} - \frac{g \cos \alpha}{\rho}\int_{\eta}^{z} \frac{\partial \rho}{\partial x} dz$$

The revised form of the x-momentum equation is then:

$$\frac{\partial \mathrm{UB}}{\partial \mathrm{t}} + \frac{\partial \mathrm{UUB}}{\partial \mathrm{x}} + \frac{\partial \mathrm{WUB}}{\partial \mathrm{z}} = gB\sin\alpha + g\cos\alpha B\frac{\partial \eta}{\partial x} - \frac{g\cos\alpha B}{\rho}\int_{\eta}^{z}\frac{\partial \rho}{\partial x}dz +$$

$$\frac{1}{\rho}\frac{\partial B\tau_{xx}}{\partial \mathrm{x}} + \frac{1}{\rho}\frac{\partial \mathrm{B}\tau_{\mathrm{xz}}}{\partial \mathrm{z}} + qBU_x$$

Effectively, pressure has been removed from the unknowns by combining the z-momentum and x-momentum equations, but η has been added as an unknown. Hence another equation is necessary for η.

Free Water Surface Equation

This equation is a simplification of the continuity equation. The continuity equation integrated over the depth from the water surface to the bottom is called the free water surface equation. Figure 14 is a definition sketch of the computational grid with an arbitrary channel slope.

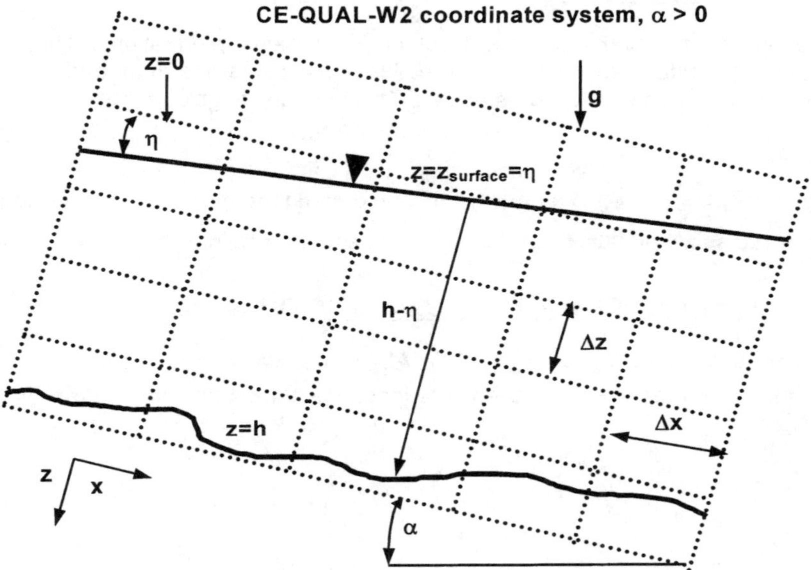

Figure 14. Coordinate system with channel slope.

The continuity equation is integrated over the depth and simplified such that

$$B_\eta \frac{\partial \eta}{\partial t} = \frac{\partial}{\partial x}\int_\eta^h UBdz - \int_\eta^h qBdz$$

where B_η is the width at the surface.

In order for the laterally averaged hydrodynamics, or even the quasi-three dimensional hydrodynamics to be of any use, semi-implicit numerical solution techniques had to be found that could overcome the Courant gravity wave limitation on explicit solutions. The methods for developing numerical forms that satisfied these conditions were given in Edinger and Buchak (1978b), where the horizontal momentum balance was combined with the vertically integrated continuity relationship to give a surface elevation relationship that could be solved implicitly using the Thomas algorithm. This method allowed the use of computationally long time steps in very deep water bodies.

However, as shown by Wells (2002a) the surface wave and internal wave for stratified cases is highly dampened unless the time step is exactly the Courant condition.

Equation of State

The density must be known for solution of the momentum equations. The equation of state is an equation that relates density to temperature and concentration of dissolved substances. This equation is given by:

$$\rho = f(T_w, \Phi_{TDS}, \Phi_{ISS})$$

where $f(T_w, \Phi_{TDS}, \Phi_{ISS})$ is a density function dependent upon temperature, total dissolved solids or salinity, and inorganic suspended solids.

Turbulence Closure Model Used in CE-QUAL-W2

In order to solve for τ_{xx} *and* τ_{xz}, CE-QUAL-W2 uses an eddy viscosity formulation for the shear stresses. The longitudinal turbulent shear stress is defined as

$$\frac{\tau_{xx}}{\rho} = \nu_{t-x} \frac{\partial U}{\partial x}$$

where ν_{t-x} is the longitudinal turbulent viscosity or the longitudinal eddy viscosity, which is a user-defined constant in the model.

The algorithm for the vertical shear stress is

$$\frac{\tau_{xz}}{\rho} = \nu_{t-z} \frac{\partial U}{\partial z}$$

The CE-QUAL-W2 model allows the user to specify one of several mixing length models for ν_{t-z} such that $\nu_t = \ell^2 \left| \frac{dU}{dz} \right|$ where ℓ is the mixing length. Also, the mixing length is decreased as a result of stratification effects by using a Richardson number criterion. How does one know which turbulent closure scheme or model to use for τ_{xz}? According to Hamblin and Salmon (1975) "the vertical diffusion of momentum is probably the most important internal parameter" for predicting internal circulation patterns. Because of the "disarray in the literature" over which formulation is best, Shanahan and Harleman (1982) suggested that the modeler "use theory and literature as a guide to develop alternative viscosity functions and then test those functions in calibration runs against field data." In the absence of expensive-to-obtain current velocity data, the use of temperature profiles is often used to test the adequacy of the hydrodynamic regime against different formulations.

2-D Water Quality Transport

After introducing spatial averages of the velocity and concentration field and formally laterally averaging the 3-D conservation of constituent mass equation, the governing equation is

$$\frac{\partial B\overline{\overline{c}}}{\partial t}+\frac{\partial B\overline{\overline{u}}\,\overline{\overline{c}}}{\partial x}+\frac{\partial B\overline{\overline{w}}\,\overline{\overline{c}}}{\partial z}=qB+\frac{\partial}{\partial x}\left(D_x B\frac{\partial\overline{\overline{c}}}{\partial x}\right)+\frac{\partial}{\partial z}\left(D_z B\frac{\partial\overline{\overline{c}}}{\partial z}\right)+\overline{\overline{S}}B$$

where double overbars represent time and lateral averaged variables. The dispersion terms, D_x and D_z, are a result of lateral averaging of the velocity field. In general, except at an interface, $D_x >> E_x >> D$ and similarly for $D_z >> E_z >> D$.

If the overbars are dropped and replaced with capitals, then the following equation is obtained:

$$\frac{\partial BC}{\partial t}+\frac{\partial UBC}{\partial x}+\frac{\partial WBC}{\partial z}-\frac{\partial\left(BD_x\frac{\partial C}{\partial x}\right)}{\partial x}-\frac{\partial\left(BD_z\frac{\partial C}{\partial z}\right)}{\partial z}=q\ B+S\ B$$

where C is the laterally averaged constituent concentration, q is the lateral inflow or outflow mass flow rate of constituent per unit volume, and S is the laterally averaged source/sink term.

Note that this can be concentration or temperature since the concentration of heat can be determined to be $\rho c_p T$ where ρ is the fluid density, c_p is the specific heat of water, and T is the temperature.

In order to solve for C or T, the following is required:

1. Water surface solution

2. U,W, laterally-averaged velocity field from momentum equations

2. Boundary and initial conditions

3. D_x is a user-defined constant and using the Reynolds' analogy $D_z = \alpha \nu_{t-z}$

4. Laterally-averaged source/sink terms, S

2-D (x-y) Horizontal Models

These models are applicable for systems that are wide and are vertically well-mixed (see Figure 15) which can have significant effect of lateral fluid motions. A typical example of this type of model is VEST (Celik and Rodi, 1984).

Other examples of these models include (see also ASCE, 1988):

- RMA-2 model (King and Norton, 1978)
- Mike21 model (a Danish Hydraulics Institute proprietary model)

The first step in developing the governing equations is to assume vertical homogeneity and to decompose the velocity and scalar variable as a vertical average (double overbar) and a deviation from the vertical average (double primed variable), such as:

$$\overline{u} = \overline{\overline{u}} + u'' \qquad \overline{v} = \overline{\overline{v}} + v'' \qquad \overline{T} = \overline{\overline{T}} + T'' \qquad \overline{c} = \overline{\overline{c}} + c''$$

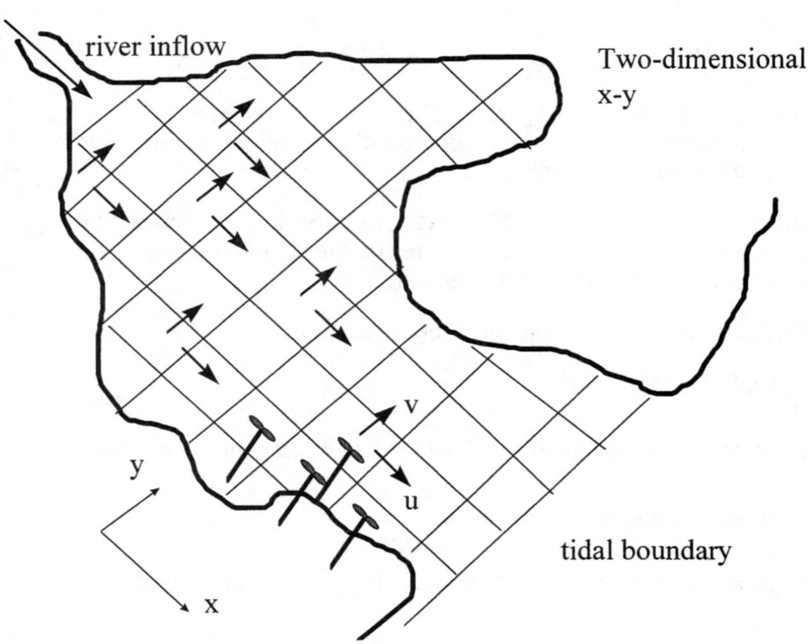

Figure 15. Typical 2-D horizontal grid assuming vertical homogeneity.

where $\overline{\overline{u}} = \frac{1}{h}\int_{z_1}^{z_2} \overline{u}dz \qquad \overline{\overline{T}} = \frac{1}{h}\int_{z_1}^{z_2} \overline{T}dz \qquad \overline{\overline{\rho}} = \frac{1}{h}\int_{z_1}^{z_2} \overline{\rho}dz$

$\overline{\overline{v}} = \frac{1}{h}\int_{z_1}^{z_2} \overline{v}dz \qquad \overline{\overline{c}} = \frac{1}{h}\int_{z_1}^{z_2} \overline{c}dz$

and the temporal averaged is designated as a single overbar, u is x-velocity, v is y-velocity, c is concentration, θ is the temperature, and ρ is the density.

Substituting these into the governing 3-D equations and formally vertically averaging them, we obtain the vertically averaged continuity, momentum, and scalar equations:

Mass conservation

$$\frac{\partial h}{\partial t} + \frac{\partial \overline{\overline{u}} h}{\partial x} + \frac{\partial \overline{\overline{v}} h}{\partial y} = 0$$

Note that we have assumed that h (the depth of water) >>η (the water surface deviation from the free surface).

Momentum conservation

x-momentum

$$\frac{\partial \overline{\overline{u}}}{\partial t} + \overline{\overline{u}}\frac{\partial \overline{\overline{u}}}{\partial x} + \overline{\overline{v}}\frac{\partial \overline{\overline{u}}}{\partial y} - 2\Omega\overline{\overline{v}} = -\frac{1}{\rho_o}\frac{\partial p_a}{\partial x} - g\frac{\partial \eta}{\partial x} - \frac{1}{2}\frac{gh}{\rho_o}\frac{\partial \overline{\overline{\rho}}}{\partial x} - \frac{1}{h}\overbrace{\left[\frac{\partial h\overline{\overline{u}}''^2}{\partial x} + \frac{\partial h\overline{\overline{u''v''}}}{\partial y}\right]}^{\text{dispersion term}} + \frac{\tau_{2x}}{\rho_o h} - \frac{\tau_{1x}}{\rho_o h}$$

y-momentum

$$\frac{\partial \overline{\overline{v}}}{\partial t} + \overline{\overline{u}}\frac{\partial \overline{\overline{v}}}{\partial x} + \overline{\overline{v}}\frac{\partial \overline{\overline{v}}}{\partial y} + 2\Omega\overline{\overline{u}} = -\frac{1}{\rho_o}\frac{\partial \overline{p}_a}{\partial y} - g\frac{\partial \eta}{\partial y} - \frac{1}{2}\frac{gh}{\rho_o}\frac{\partial \overline{\overline{\rho}}}{\partial y} - \frac{1}{h}\underbrace{\left[\frac{\partial h\overline{\overline{u''v''}}}{\partial x} + \frac{\partial h\overline{\overline{v''^2}}}{\partial y}\right]}_{\text{dispersion term}} + \frac{\tau_{2y}}{\rho_o h} - \frac{\tau_{1y}}{\rho_o h}$$

where the subscripts 1 and 2 refer to the bottom and top of the layer, respectively, and τ is the shear stress either at the surface (wind stress) or at the bottom (bottom shear stress from bottom friction).

Note that we have assumed that $\frac{\partial \bar{p}}{\partial x} \cong \frac{\partial \bar{\bar{p}}}{\partial x}$ and $\frac{\partial \bar{p}}{\partial y} = \frac{\partial \bar{\bar{p}}}{\partial y}$. Also we need now a turbulence closure assumption for the new horizontal transport terms, $\overline{\overline{u''v''}}$ and $\overline{\overline{u''^2}}$. Generally, these terms are parameterized as follows,

$\overline{\overline{u''v''}} = -\nu_{xy}\left(\frac{\partial \bar{\bar{u}}}{\partial y} + \frac{\partial \bar{\bar{v}}}{\partial x}\right)$ and $\overline{\overline{u''^2}} = -\nu_{xx} 2\frac{\partial \bar{\bar{u}}}{\partial x}$ where ν_{xy} and ν_{xx} are the horizontal eddy viscosities. Often the turbulence closure scheme is unimportant for 2-D horizontal flows because the numerical solution of the governing equations can introduce enough numerical diffusion to dominate the horizontal eddy transport terms. But with higher order numerical techniques, these terms are often included for numerical smoothing since they act like a diffusion coefficient for the velocity field.

Heat Conservation and Dissolved Mass Conservation

The heat conservation equation becomes

$$\frac{\partial \bar{\bar{T}}}{\partial t} + \bar{\bar{u}}\frac{\partial \bar{\bar{T}}}{\partial x} + \bar{\bar{v}}\frac{\partial \bar{\bar{T}}}{\partial y} = \underbrace{-\frac{\varphi_{net}}{\rho c_p h}}_{\text{radiation transfer}} \underbrace{-\frac{1}{h}\left[\frac{\partial h\overline{\overline{u''T''}}}{\partial x} + \frac{\partial h\overline{\overline{v''T''}}}{\partial y}\right]}_{\text{horizontal turbulent diffusion}}$$

where the turbulent horizontal flux terms become $\overline{\overline{v''T''}} = -\mathrm{E}_y \frac{\partial \bar{\bar{T}}}{\partial y}$ and $\overline{\overline{u''T''}} = -\mathrm{E}_x \frac{\partial \bar{\bar{T}}}{\partial x}$, φ_{net} is the net radiation input at the water surface, E_x and E_y are turbulent dispersion coefficients for mass and heat in x and y directions, respectively. These equations also assume that there is no net mass transfer of water at the surface.

$$\frac{\partial \bar{\bar{c}}}{\partial t} + \bar{\bar{u}}\frac{\partial \bar{\bar{c}}}{\partial x} + \bar{\bar{v}}\frac{\partial \bar{\bar{c}}}{\partial y} = -\overbrace{c'w'\big|_2 - c'w'\big|_1}^{\substack{\text{net mass transfer at water} \\ \text{surface} \equiv \text{evaporation - precipitation}}} \underbrace{-\frac{1}{h}\left[\frac{\partial h\overline{\overline{u''c''}}}{\partial x} + \frac{\partial h\overline{\overline{v''c''}}}{\partial y}\right]}_{\text{horizontal turbulent diffusion}} + r$$

where the turbulent horizontal flux terms become $\overline{\overline{v''c''}} = -\mathrm{E}_y \frac{\partial \bar{\bar{c}}}{\partial y}$ and $\overline{\overline{u''c''}} = -\mathrm{E}_x \frac{\partial \bar{\bar{c}}}{\partial x}$

In order to complete the set of equations required to determine $\overline{\overline{u}},\overline{\overline{v}},\overline{\overline{\theta}},\overline{\overline{c}},\overline{\overline{\rho}},\eta$, an equation of state is also required.

Three-Dimensional Models

Three-dimensional models evaluate the continuity and momentum equations in all coordinate directions (Figure 16). They are applied to systems that are wide and deep and where flow patterns are complex (ASCE, 1989, Johnson, 1981).

The equations used are the continuity equation, the x and y momentum equations and usually the hydrostatic relation. Most applications of these models so far have been for water bodies with large horizontal dimensions compared to the depth, such as lakes and estuaries. Three dimensional modeling is being tested for reservoir applications where it is now recognized that detailed hydrodynamics and transport are required to explain the water quality interactions between main channel regions and extensive over bank regions and complex wind-driven horizontal circulation.

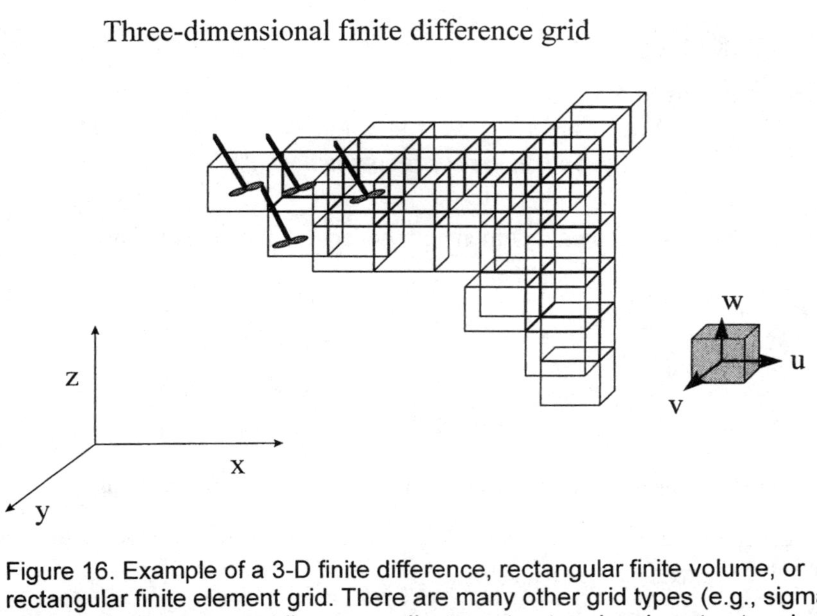

Figure 16. Example of a 3-D finite difference, rectangular finite volume, or rectangular finite element grid. There are many other grid types (e.g., sigma-stretch coordinates, curvi-linear coordinates, structured and unstructured finite volume grids).

Liggett (1969) has applied this type of model to computing lake circulation. Falconer, et al. (1991) also describe an application of this type of model to a shallow, homogeneous lake. Leendertse and Liu (1975) have applied this type of model to computing flow and water quality in seas and estuaries. Van der Kuur et al. (1989) present a version of Leendertse's basic scheme in a boundary fitted coordinate system and have used it for simulating flows in tidal bays. Koutitas and O'Connor (1980) have used a quasi-three dimensional model to investigate wind driven circulations in Thessalonki Bay. Blumberg and Mellor (1983) have proposed a model of this type for computing flow and water quality in open seas, and Galperin and Mellor (1990) have applied the same model to predicting river-estuary system circulations. Casulli and Stelling (1998) and Cheng and Casulli (2001) have developed a semi-implicit 3-D finite difference scheme for an unstructured grid solving the shallow water equations.

Blumberg et al. (1992) have further applied three dimensional hydrodynamics to computing velocity profiles in river channels. Shimizu et al. (1990) have applied a hydrostatic pressure three dimensional model to predicting flow and sediment transport for in-bank meandering channels. Their results highlight the importance of computing the pressure driven secondary motions for sediment transport.

Equations

The equations used are,

$$\frac{\partial u}{\partial x}+\frac{\partial v}{\partial y}+\frac{\partial w}{\partial z}=0 \qquad \text{Conservation of water mass (local)}$$

Conservation of water momentum in both horizontal directions (x and y):

$$\frac{\partial u}{\partial t}+\frac{\partial u^2}{\partial x}+\frac{\partial uv}{\partial y}+\frac{\partial uw}{\partial y}+\frac{1}{\rho_0}\frac{\partial p}{\partial x}=\frac{\partial}{\partial x}\left[\nu_t\frac{\partial u}{\partial x}\right]+\frac{\partial}{\partial y}\left[\nu_t\frac{\partial u}{\partial y}\right]+\frac{\partial}{\partial z}\left[\nu_t\frac{\partial u}{\partial z}\right]$$

$$\frac{\partial v}{\partial t}+\frac{\partial vu}{\partial x}+\frac{\partial v^2}{\partial y}+\frac{\partial vw}{\partial y}+\frac{1}{\rho_0}\frac{\partial p}{\partial y}=\frac{\partial}{\partial x}\left[\nu_t\frac{\partial v}{\partial x}\right]+\frac{\partial}{\partial y}\left[\nu_t\frac{\partial v}{\partial y}\right]+\frac{\partial}{\partial z}\left[\nu_t\frac{\partial v}{\partial z}\right]$$

assuming isotropic turbulence, where ν_t is the turbulent eddy viscosity.

Conservation of z-momentum assuming hydrostatic pressure becomes as previously derived:

$$p=\rho_0 g(\eta-z)+\rho_0 g\int_z^{\eta}\frac{\Delta\rho}{\rho_0}dz$$

Hence, we can re-write the pressure gradient terms as:

$$\frac{1}{\rho_0}\frac{\partial p}{\partial x} = g\frac{\partial \eta}{\partial x} + g\frac{\partial}{\partial x}\left[\int_z^{\eta}\frac{\Delta\rho}{\rho_0}dz\right]$$

$$\frac{1}{\rho_0}\frac{\partial p}{\partial y} = g\frac{\partial \eta}{\partial y} + g\frac{\partial}{\partial y}\left[\int_z^{\eta}\frac{\Delta\rho}{\rho_0}dz\right]$$

such that the horizontal momentum equations become:

$$\frac{\partial u}{\partial t} + \frac{\partial u^2}{\partial x} + \frac{\partial uv}{\partial y} + \frac{\partial uw}{\partial y} + g\frac{\partial \eta}{\partial x} + g\frac{\partial}{\partial x}\left[\int_z^{\eta}\frac{\Delta\rho}{\rho_0}dz\right] = \frac{\partial}{\partial x}\left[\nu_t\frac{\partial u}{\partial x}\right] + \frac{\partial}{\partial y}\left[\nu_t\frac{\partial u}{\partial y}\right] + \frac{\partial}{\partial z}\left[\nu_t\frac{\partial u}{\partial z}\right]$$

$$\frac{\partial v}{\partial t} + \frac{\partial vu}{\partial x} + \frac{\partial v^2}{\partial y} + \frac{\partial vw}{\partial y} + g\frac{\partial \eta}{\partial y} + g\frac{\partial}{\partial y}\left[\int_z^{\eta}\frac{\Delta\rho}{\rho_0}dz\right] = \frac{\partial}{\partial x}\left[\nu_t\frac{\partial v}{\partial x}\right] + \frac{\partial}{\partial y}\left[\nu_t\frac{\partial v}{\partial y}\right] + \frac{\partial}{\partial z}\left[\nu_t\frac{\partial v}{\partial z}\right]$$

Conservation of water mass (global):

$$\frac{\partial \eta}{\partial t} + \frac{\partial}{\partial x}\left[\int_{-h}^{\eta}udz\right] + \frac{\partial}{\partial y}\left[\int_{-h}^{\eta}vdz\right] = 0$$

Note that positive z was assumed to be aligned opposite the direction of gravity. Conservation of scalar quantity (e.g., sediment, dissolved mass, or heat):

$$\frac{\partial c}{\partial t} + \frac{\partial uc}{\partial x} + \frac{\partial vc}{\partial y} + \frac{\partial wc}{\partial y} = \frac{\partial}{\partial x}\left[\frac{\nu_t}{\sigma_c}\frac{\partial c}{\partial x}\right] + \frac{\partial}{\partial y}\left[\frac{\nu_t}{\sigma_c}\frac{\partial c}{\partial y}\right] + \frac{\partial}{\partial z}\left[\frac{\nu_t}{\sigma_c}\frac{\partial c}{\partial z}\right] \pm Source/Sink$$

where σ_c is the turbulent Schmidt number$=\frac{\nu_t}{D_t}$ and D_t is the turbulent eddy diffusivity.

There are a variety of numerical solutions schemes (finite difference, finite volume, finite element; explicit versus implicit formulations; eulerian versus semi-lagrangian approaches) that have been used in multi-dimensional models. Most of the available models useful for water quality modeling have a finite-difference basis. Some available models allow options for the use of different numerical solution techniques. In some cases, different solutions schemes are used in the horizontal direction than in the vertical. For example, some models (e.g., CE-QUAL-W2, Cole and Wells, 2003) use an explicit time differencing scheme in the horizontal and an implicit scheme in the vertical.

A number of the quasi three-dimensional (hydrostatic) models use a mode-splitting scheme to solve the governing equations. The governing equations contain terms for fast moving external gravity waves and slow moving internal gravity waves. It is often considered desirable in terms of computer economy to separate the vertically integrated equations (external mode) from the vertical structure equations (internal mode). Essentially, the two-dimensional depth-averaged equations are solved first to determine the impact of the gravity waves (barotropic conditions), followed by a separate calculation for the density-induced (baroclinic) variations. Some models allow for separate barotropic and baroclinic time-steps (such as POM, Blumberg and Mellor, 1987), while some (such as ECOM, a modified version of POM, Blumberg, et. al., 1993) incorporate schemes for solving the gravity wave so that the need for separate time steps is eliminated.

Computational time is still an issue in running 3-D hydrodynamic models, even though eventually all modeling efforts will be in 3-D as computational ability increases.

Non-Hydrostatic 3-D Models

Due to high computing overhead, many studies have not retained all the terms in the Reynolds Averaged Navier-Stokes equations where they could reasonably be argued to be negligible. Therefore the model equations were not fully elliptic in all directions. Parabolic and partially parabolic equations have been employed more often than the fully elliptic ones. Another device which allows these type of models to be studied with less computational effort is to study the case of a flow in an infinitely wide channel when the problem reduces to a two dimensional problem in the vertical plane. It, however, retains many of the characteristics of the fully three-dimensional problem, notably the non-hydrostatic pressure distribution and the interaction of all the turbulent stresses. In non-hydrostatic models the vertical momentum equation is retained in its entirety. This introduces two problems: (1) how do you solve for pressure? and (2) how do you treat the free surface ?

Pressure Solution

Various algorithms for the pressure solution have been proposed in the past. Patankar and Spalding (1972) and Patankar (1980) describe what has come to be known as the SIMPLE algorithm. The algorithm has several offshoots (SIMPLER, SIMPLEC, PISO) but they work on approximately the same principle. Fletcher (1988) describes the method. Using a guessed pressure field the momentum equations are solved to estimate the velocity field. The amount by which this velocity field deviates from satisfying the continuity equation is then used as a source for calculating a correction to the guessed pressure field which in turn is used to make a better attempt at solving the momentum equations. This procedure is repeated until the solution converges.

4.2. Mathematical solution techniques, computational efficiency and hydrodynamic and water quality model linkages

Solution Techniques

After the theoretical development presented in this chapter, the next challenge is to develop solutions to the partial differential equations (PDEs). In most cases, the researcher can either solve the equations analytically or numerically. Unfortunately analytical solutions exist for only a small subset of useful problems that are linear.

Eliason and Bourgeois (1997) showed analytical solutions to the shallow water momentum and continuity equations (where the average basin depth, h, is much greater than deviations in water surface elevation,η). With the following assumptions: frictionless bottom and side walls, no surface shear stress, hydrostatic and Boussinesq approximations, negligible non-linear terms in the momentum equation, no longitudinal density gradients, and no turbulent or viscous fluid stresses, the continuity and x momentum equations become:

$$\frac{\partial u}{\partial x} = -\frac{1}{h}\frac{\partial \eta}{\partial t}$$

$$\frac{\partial u}{\partial t} = -g\frac{\partial \eta}{\partial x}$$

where u is the cross-sectionally average velocity in the x-direction, g is the acceleration due to gravity. After cross-differentiation of each equation and simplifying, the "familiar" wave equation is obtained,

$$\frac{\partial^2 \eta}{\partial t^2} - (gh)\frac{\partial^2 \eta}{\partial x^2} = 0$$

The solution of the governing equations (continuity and x-momentum) resulted in a wave equation that has the following analytical solution for the case of an initial water level disturbance and no wind,

$$\eta = \eta_o \cos\left(\frac{\pi x}{L}\right)\cos\left(\frac{\pi c_o t}{L}\right)$$

where L is the closed basin length, η is the water surface elevation, η_o is the amplitude of the surface elevation disturbance, c_o is the gravity wave speed or $\sqrt{gh}$, and h is the basin depth. This represents a seiche that continues ad infinitum since there is no frictional resistance.

The value of analytical solutions has more to do with numerical model testing and validation than with actual use in the practice of water quality and hydrodynamic modeling. For example, a comparison of the analytical model with the CE-QUAL-W2 numerical model with similar assumptions as the analytical model is shown in Figure 17 (Wells, 2002).

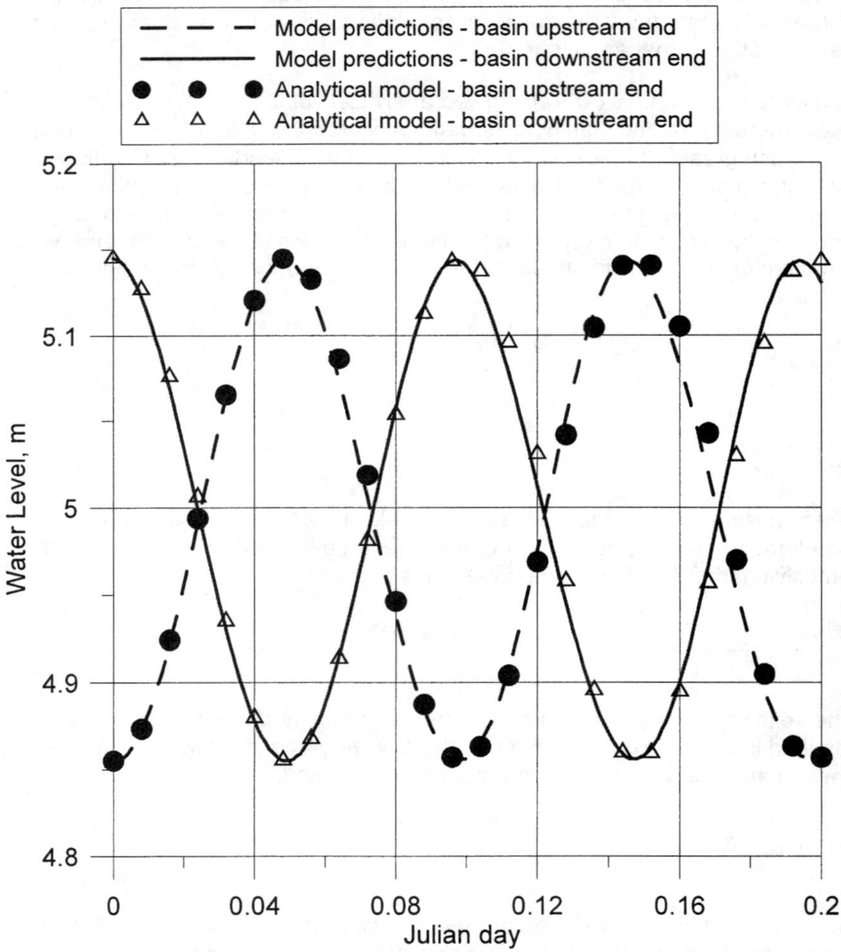

Figure 17. Comparison of analytical solution of an infinite seiche in a narrow long basin to the prediction of a numerical model CE-QUAL-W2 (Wells, 2002).

Numerous analytical models exist for solving the advective diffusion equation (Fischer et al., 1979, Chapra, 1997). For example, solution to the 1-D advective diffusion model with first order decay,

$$\frac{\partial c}{\partial t}+u\frac{\partial c}{\partial x}=D\frac{\partial^2 c}{\partial x^2}-kc$$

for an instantaneous injection of mass, M, at x=0 and t=0, is

$$c=\frac{M}{A\sqrt{4\pi Dt}}\ exp - \left[\frac{[x - ut]^2}{4Dt}+kt\right]$$

where A is the cross-sectional area and k is the first order decay coefficient. This analytical model though is restricted to the following: constant cross-sectional area, no flux conditions at channel side-walls, bottom and surface, constant advective velocity u. As in the case of the analytical solution using the fluid continuity and x-momentum equation for a seiche, these analytical models for pollutant concentration are useful in screening analyses and in comparison of numerical solutions to exact analytical solutions since the initial and boundary conditions limit their usefulness.

Numerical models allow the user to be flexible in application of boundary and initial conditions and in formulation of the model geometry or the grid. Numerical models which solve the partial differential equations can be generally broken into three main categories: finite difference, finite element and finite volume methods.

Our ability to use numerical methods to solve complex problems is based on the large computer advances in the last few decades. With the advent of computer technology (high speed digital computers started becoming available in the late 1960's), the traditional problem solving strategies (experimental and theoretical approaches) started to evolve toward a numerical approach. Computer models were able to solve problems numerically that were not able to be solved analytically. Examples of the first numerical models were upgraded Streeter-Phelps models for river systems that lead to the QUAL II model in 1973 (Brown and Barnwell, 1987). This model was a steady-state one-dimensional river model with cross-sectionally varying area. Since then, attention focused on reservoir, lake, estuary, and coastal models of water quality that are multi-dimensional and dynamic. As our computational power continues to increase (and the relative cost of computing), we begin applying more and more complex models to real systems. Over the last several decades, the cost of computing has decreased by 10 X every 8 years! But the cost of gathering field data for model set-up, calibration/verification has not followed the trend of computation. Typical of the advances in computational power is Figure 18 which shows the floating point performance of desktop PC chips since the introduction of the Intel 386-25 MHz chip in 1989.

This is not meant to be exhaustive review of numerical methods but to provide basic background material for understanding how the governing equations are solved numerically.

Governing Equations, Initial and Boundary Conditions

The governing equations were represented earlier and can include:

1. Fluid continuity

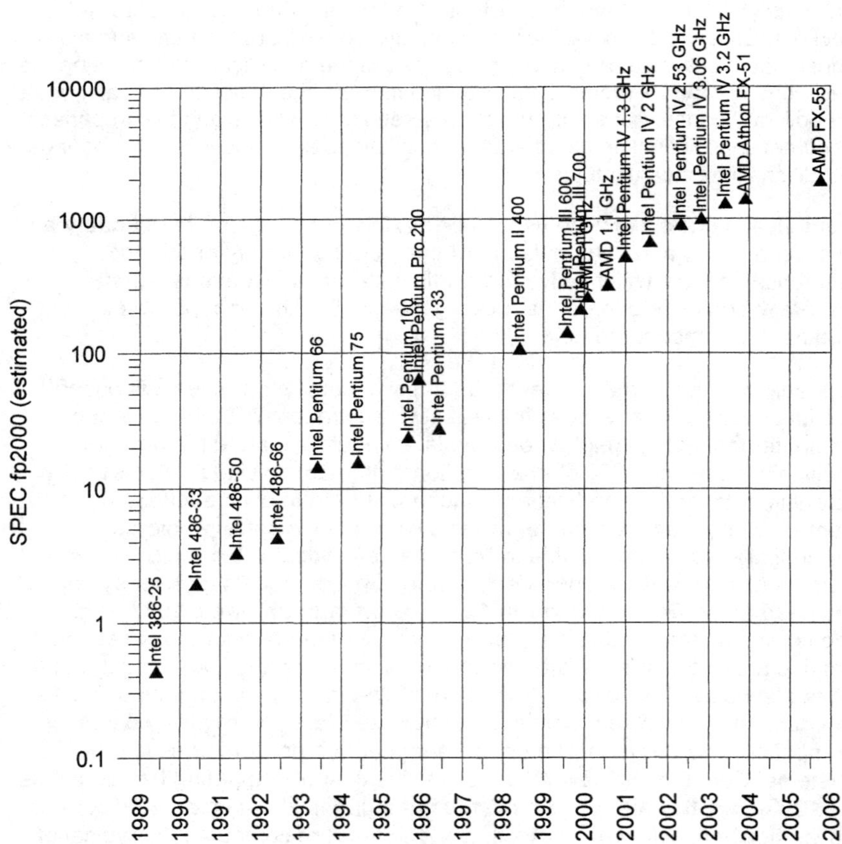

Figure 18. History of the evolution of floating point performance with desktop PC chips since 1989.

2. Momentum equations in x, y, and z

3. Conservation of mass of pollutant and/or temperature

In order to specify properly the mathematical solution of these equations, initial and boundary conditions are required.

Initial conditions are specification of the variable of interest (state variable) within the mathematical domain of the problem, such as the initial concentration field and/or velocity field. Boundary conditions are used to specify values of the state variables or derivatives of the state variables for all boundaries. The boundary conditions for a state variable ϕ can be classified as

a. Dirichlet BC or BC of 1st kind where $\varphi = f(t)$ is specified on the boundary where t is time, or

b. Neumann BC or BC of the 2nd kind where $\frac{\partial \varphi}{\partial n} = f(t)$ is specified on the boundary where n is the unit normal vector to the surface, or

c. Mixed BC or BC of the 3rd kind where $a\varphi + b\frac{\partial \varphi}{\partial n} = f(t)$ is specified on the boundary where a and b are constants.

With the mathematical problem properly specified, the next step is to adopt a solution technique.

Finite Difference Methods

Taylor Series Expansions

Consider an approximation to the 2-D transport equation of constituent φ with constant diffusivity, D,

$$\frac{\partial \varphi}{\partial t} + u\frac{\partial \varphi}{\partial x} + v\frac{\partial \varphi}{\partial y} = D(\frac{\partial^2 \varphi}{\partial x^2} + \frac{\partial^2 \varphi}{\partial y^2})$$

with u and v known and boundary conditions (either ϕ or $\partial\phi/\partial n$) specified on a 2-D domain. From the domain, we construct a grid of Δx, Δy, and Δt such that the grid points can be defined as (x_i, y_i) or ($i\Delta x$, $j\Delta y$) at a time $n\Delta t$, where i, j, and n define the x, y, and t levels, respectively. For this example, consider equal grid point spacing as shown in Figure 19.

Then the notation for the state variable defining the spatial location and time level is

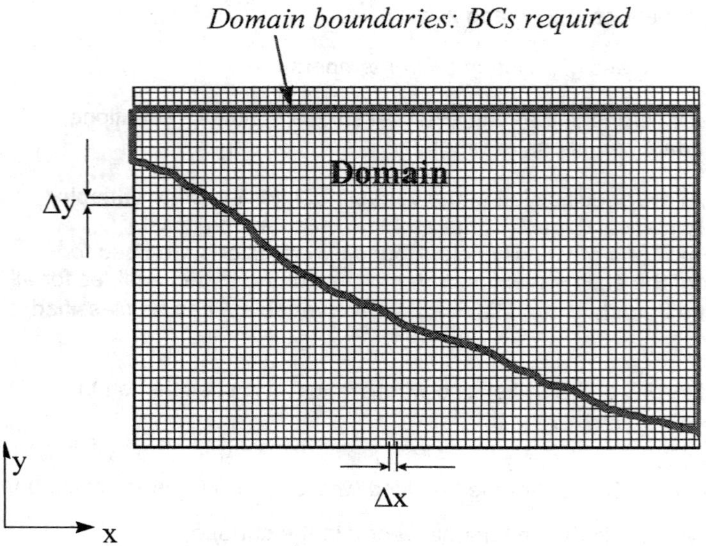

Figure 19. Grid example for 2-D transport equation.

$$\varphi_{ij}^{n} \equiv \varphi(x_i, y_j, t_n) \equiv \varphi(i\Delta x, j\Delta y, n\Delta t).$$

The derivatives of φ in the PDE are approximated by discrete or finite differences by using a Taylor Series Expansion (TSE). For example, Figure 20 shows the variation of ϕ at several grid points in x.

TSEs for the value of φ at grid point i-1 and i+1 in terms of φ at i for equal grid point spacing Δx are then:

$$\varphi_{i-1,j}^{n} = \varphi_{i,j}^{n} - \Delta x \left.\frac{\partial \varphi}{\partial x}\right|_{i,j}^{n} + \frac{\Delta x^2}{2!}\left.\frac{\partial^2 \varphi}{\partial x^2}\right|_{i,j}^{n} - \frac{\Delta x^3}{3!}\left.\frac{\partial^3 \varphi}{\partial x^3}\right|_{i,j}^{n} + \ldots + \frac{\Delta x^k}{k!}\left.\frac{\partial^k \varphi}{\partial x^k}\right|_{i,j}^{n} + \ldots$$

$$\varphi_{i+1,j}^{n} = \varphi_{i,j}^{n} + \Delta x \left.\frac{\partial \varphi}{\partial x}\right|_{i,j}^{n} + \frac{\Delta x^2}{2!}\left.\frac{\partial^2 \varphi}{\partial x^2}\right|_{i,j}^{n} + \frac{\Delta x^3}{3!}\left.\frac{\partial^3 \varphi}{\partial x^3}\right|_{i,j}^{n} + \ldots + \frac{\Delta x^k}{k!}\left.\frac{\partial^k \varphi}{\partial x^k}\right|_{i,j}^{n} + \ldots$$

Using Equation (1), we can solve for ∂φ/∂x for i,j,n:

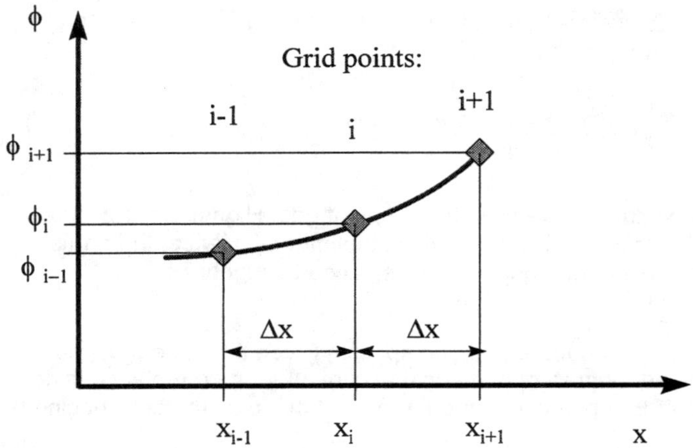

Figure 20. Variation of state variable in space with equal grid spacing.

$$\left.\frac{\partial\varphi}{\partial x}\right|_{i,j}^{n} = \frac{\varphi_{i,j}^{n} - \varphi_{i-1,j}^{n}}{\Delta x} + \frac{\Delta x}{2!}\left.\frac{\partial^2\varphi}{\partial x^2}\right|_{i,j}^{n} - \frac{\Delta x^2}{3!}\left.\frac{\partial^3\varphi}{\partial x^3}\right|_{i,j}^{n} + H.O.T.s$$

where HOT's means "higher order terms" of the TSE which are of the form:

$$\frac{\Delta x^k}{k!}\frac{\partial^k\varphi}{\partial x^k}$$

This approximation for the derivative of φ with respect to x is termed a backward difference.

Or using Equation (2) we can construct a forward difference approximation to the first derivative as

$$\left.\frac{\partial\varphi}{\partial x}\right|_{i,j}^{n} = \frac{\varphi_{i+1,j}^{n} - \varphi_{i,j}^{n}}{\Delta x} + \frac{\Delta x}{2!}\left.\frac{\partial^2\varphi}{\partial x^2}\right|_{i,j}^{n} - \frac{\Delta x^2}{3!}\left.\frac{\partial^3\varphi}{\partial x^3}\right|_{i,j}^{n} + H.O.T.s$$

If we subtract Eq. 1 from 2 and solve for $\partial\varphi/\partial x$ for i,j,n we obtain a central difference approximation:

$$\left.\frac{\partial\varphi}{\partial x}\right|_{i,j}^{n} = \frac{\varphi_{i+1,j}^{n} - \varphi_{i-1,j}^{n}}{2\Delta x} - \frac{\Delta x^2}{3!}\left.\frac{\partial^3\varphi}{\partial x^3}\right|_{i,j}^{n} + H.O.T.s$$

A 2nd derivative can also be derived similarly by adding Eq. 1 and 2 and solving for $\partial^2\varphi/\partial x^2$ for i,j,n:

$$\left.\frac{\partial^2 \varphi}{\partial x^2}\right|_{i,j}^{n} = \frac{\varphi_{i+1,j}^{n} - 2\varphi_{i,j}^{n} + \varphi_{i-1,j}^{n}}{\Delta x^2} - \frac{2\Delta x^2}{4!}\left.\frac{\partial^4 \varphi}{\partial x^4}\right|_{i,j}^{n} + H.O.T.s$$

The above approximations are valid for a 3 point grid. Higher order accuracy can be derived by using more grid points in the interval. Hence, the partial derivatives in the governing equation are replaced by algebraic approximations by truncating the TSE.

This method does not require equal grid spacing (even though equal grid spacing has a higher degree of numerical accuracy). For example, consider Figure 21 and a TSE at points i+1 and i-1, except that now the cell spacing is variable:

$$\varphi_{i-1,j}^{n} = \varphi_{i,j}^{n} - \Delta x_1 \left.\frac{\partial \varphi}{\partial x}\right|_{i,j}^{n} + \frac{\Delta x_1^{\,2}}{2!}\left.\frac{\partial^2 \varphi}{\partial x^2}\right|_{i,j}^{n} - \frac{\Delta x_1^{\,3}}{3!}\left.\frac{\partial^3 \varphi}{\partial x^3}\right|_{i,j}^{n} + \ldots + \frac{\Delta x_1^{\,k}}{k!}\left.\frac{\partial^k \varphi}{\partial x^k}\right|_{i,j}^{n} + \ldots$$

$$\varphi_{i+1,j}^{n} = \varphi_{i,j}^{n} + \Delta x_2 \left.\frac{\partial \varphi}{\partial x}\right|_{i,j}^{n} + \frac{\Delta x_2^{\,2}}{2!}\left.\frac{\partial^2 \varphi}{\partial x^2}\right|_{i,j}^{n} + \frac{\Delta x_2^{\,3}}{3!}\left.\frac{\partial^3 \varphi}{\partial x^3}\right|_{i,j}^{n} + \ldots + \frac{\Delta x_2^{\,k}}{k!}\left.\frac{\partial^k \varphi}{\partial x^k}\right|_{i,j}^{n} + \ldots$$

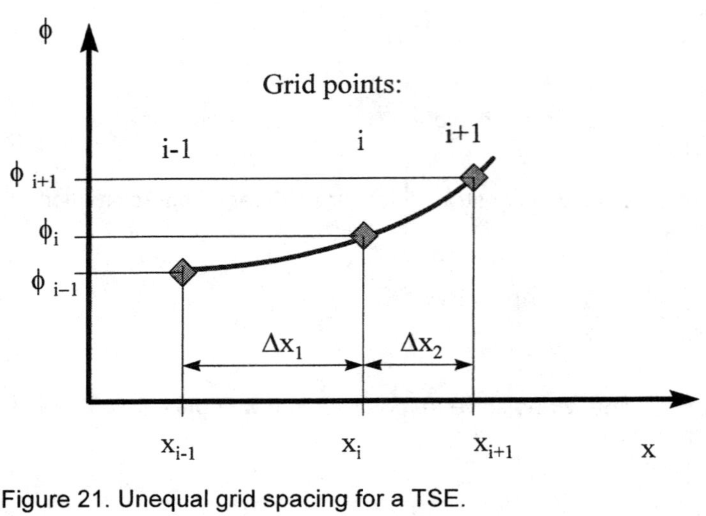

Figure 21. Unequal grid spacing for a TSE.

The technique for determining the forward and backward difference are the same as for equal grid spacing. Subtracting Equation 3 from 4 results in a central difference for unequal grid spacing:

$\left.\frac{\partial \varphi}{\partial x}\right|_{i,j}^{n} = \frac{\varphi_{i+1,j}^{n} - \varphi_{i-1,j}^{n}}{\Delta x_1 + \Delta x_2} + \frac{(\Delta x_1^2 + \Delta x_2^2)}{2(\Delta x_1 + \Delta x_2)} \left.\frac{\partial^2 \varphi}{\partial x^2}\right|_{i,j}^{n} + H.O.T.s\ldots$, which is of lower numerical accuracy than equal grid spacing.

Truncation Error

The definition of truncation error (TE) is the difference between the original PDE and the finite difference equation used to approximate the PDE.

We can truncate the TSE at any point using the Mean Value Theorem of calculus, which says that

$$\varphi_{i+1,j}^{n} = \varphi_{i,j}^{n} + \Delta x \left.\frac{\partial \varphi}{\partial x}\right|_{i,j}^{n} + \frac{\Delta x^2}{2!} \left.\frac{\partial^2 \varphi}{\partial x^2}\right|_{i,j}^{n} + \frac{\Delta x^3}{3!} \left.\frac{\partial^3 \varphi}{\partial x^3}\right|_{i,j}^{n} + \ldots + \frac{\Delta x^k}{k!} \left.\frac{\partial^k \varphi}{\partial x^k}\right|_{\xi,j}^{n}$$

where $i \le \xi < i+1$ and ξ lies within the interval of i and i+1

Consider again the forward difference formulation, it has 1st order truncation error, or the error is of order Δx, i.e., O(Δx).

$$\left.\frac{\partial \varphi}{\partial x}\right|_{i} = \frac{\varphi_{i+1} - \varphi_i}{\Delta x} \underbrace{- \frac{\Delta x}{2} \left.\frac{\partial^2 \varphi}{\partial x^2}\right|_{\xi}}_{\text{first order TE } O(\Delta x)}$$

This leading truncation error is termed the leading error of the approximation.

Consider an approximation to our original 2-D PDE using a 3-point approximation in t, x, and y, as shown in Figure 22, which is termed a 7 point molecule:

There are many choices for possible schemes for each partial derivative. For example, for the $\frac{\partial}{\partial t}$, we can choose forward, central or backward difference in time. Similarly, for

$\frac{\partial}{\partial x}$ and $\frac{\partial}{\partial y}$, we can choose either forward, central or backward difference in space. For

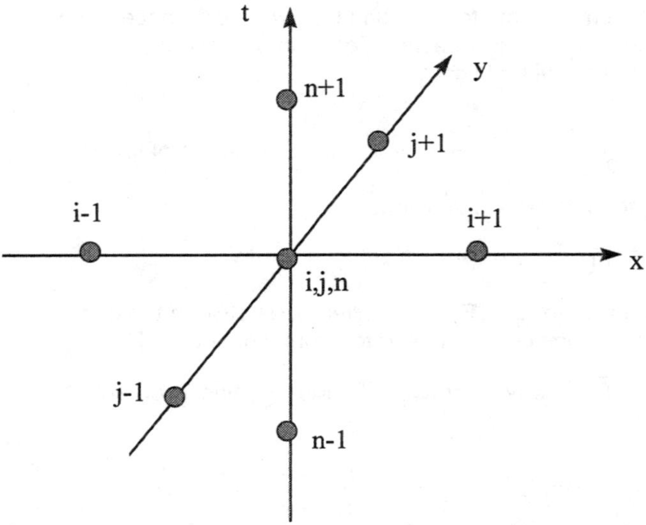

Figure 22. 7-point finite difference molecule for solving the 2-D advective diffusion equation.

$\frac{\partial^2}{\partial x^2}$ *and* $\frac{\partial^2}{\partial y^2}$ for a 3 point grid in space we can only choose a central difference approximation.

One possible finite difference approximation (FDA) to this PDE is a Forward Time Centered Space, FTCS, approximation where forward time and centered space derivatives are chosen. Substituting these and their leading truncation errors into the original PDE, we obtain the truncation error as

$$TE = -\frac{\Delta t}{2}\frac{\partial^2 \varphi}{\partial t^2}\bigg| - u\frac{\Delta x^2}{6}\frac{\partial^3 \varphi}{\partial x^3}\bigg| + D\frac{\Delta x^2}{12}\frac{\partial^4 \varphi}{\partial x^4}\bigg| - v\frac{\Delta y^2}{6}\frac{\partial^3 \varphi}{\partial y^3}\bigg| + D\frac{\Delta y^2}{12}\frac{\partial^4 \varphi}{\partial y^4}\bigg|$$

These terms are the truncated TSE evaluated at some point in the neighborhood of i, j, and n. The TE affects numerical accuracy and stability. A useful method for examining the effect of the TE on the mathematical solution is to examine what is called the modified equation.

The modified equation is the equation satisfied by the numerical approximation when the leading truncation error is included. The modified equation of the 2-D advective-diffusion equation with forward time, central space formulation is:

$$\frac{\partial\varphi}{\partial t}+u\frac{\partial\varphi}{\partial x}+v\frac{\partial\varphi}{\partial y}-D(\frac{\partial^2\varphi}{\partial x^2}+\frac{\partial^2\varphi}{\partial y^2})-\underbrace{\left\{\begin{array}{l}-\frac{\Delta t}{2}\frac{\partial^2\varphi}{\partial t^2}\Big| -u\frac{\Delta x^2}{6}\frac{\partial^3\varphi}{\partial x^3}\Big| +D\frac{\Delta x^2}{12}\frac{\partial^4\varphi}{\partial x^4}\Big| - \\ v\frac{\Delta y^2}{6}\frac{\partial^3\varphi}{\partial y^3}\Big| +D\frac{\Delta y^2}{12}\frac{\partial^4\varphi}{\partial y^4}\Big| \end{array}\right\}}_{\text{truncation error}}=0$$

This equation represents the PDE we are actually solving when we use the FDA, and hence φ represents the solution to the modified equation, not the exact solution of the original PDE. By solving the equation with these additional terms, the TE, we introduce numerical errors to the solution.

The types of numerical errors that can be introduced include what are termed dissipative errors and dispersive errors. Dissipative errors are introduced from even spatial derivatives in the TE and can lead to numerical diffusion where mass or momentum is spread faster than the actual governing equation specifies. If this dissipative error is negative, it can lead to anti-diffusion where mass is concentrated leading to an unstable solution.

Dispersive error is a result of odd spatial derivative terms in the TE. These errors are lagging or leading depending on the sign of the term and causes characteristic "wiggles." Numerical diffusion is a term often used to indicate the combined effect of dissipative and dispersive error.

Stability

Stability is a description of whether or not local errors introduced by TE or round-off errors will grow with time or decay with time. There are several formal techniques to evaluate numerical stability for linear PDEs. These techniques are (1) von Nuemann (Fourier) method, (2) matrix method, and (3) the energy method. See Anderson et al. (1984), Smith (1978), and Roache (1976) for further information about these formal approaches for stability.

Note that these methods are only applicable for linear PDEs. For non-linear PDEs, one usually uses stability criteria for the linearized PDE. Fortunately stability is exponential in nature and shows up quite quickly.

Explicit Schemes

An explicit scheme is a FDA that can be solved explicitly for the unknown variable at the next time step and grid point location. All information is known to solve this problem explicitly.

Consider the 1-D advective-diffusion equation with no diffusion and no reaction:

$$\frac{\partial c}{\partial t}+u\frac{\partial c}{\partial x}=0$$

The FDA to this equation using a central difference in space and a forward difference in time is:

$$c_i^{n+1}=c_i^n-\frac{\Delta t}{2\Delta x}u\left(c_{i+1}^n-c_{i-1}^n\right)$$

The computational stencil of this scheme is shown in Figure 23.

The modified equation for this FDA becomes:

$$\frac{\partial c}{\partial t}+u\frac{\partial c}{\partial x}=-\frac{\Delta t}{2}u^2\frac{\partial^2 c}{\partial x^2}+u\frac{\Delta x^2}{3!}\frac{\partial^3 c}{\partial x^3}+\ldots$$

The truncation error of this FDA is $O(\Delta x^2,\Delta t)$. The FDA though is always unstable since the sign of the dissipative term (even derivative term) is negative and acts like anti-dissipation building up mass rather than diffusing it.

If the dissipative term was positive, the FDA would exhibit signs of numerical diffusion which would be of order of magnitude of the coefficient of the even derivative. For example, an upwind FDA (a backward difference in space and forward in time) for the same governing equation would be:

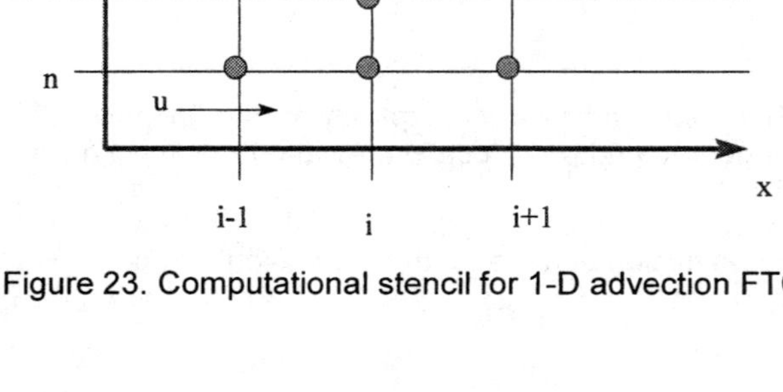

Figure 23. Computational stencil for 1-D advection FTCS scheme.

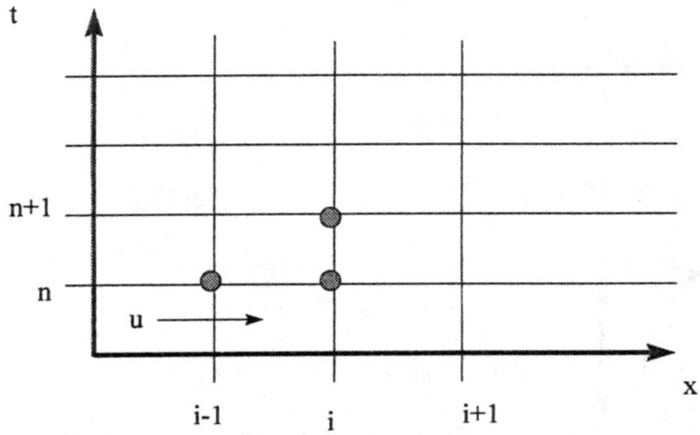

Figure 24. Computational stencil for 1-D advection upwind scheme.

$$c_i^{n+1} = c_i^n + \frac{\Delta t}{\Delta x} u\left(c_i^n - c_{i-1}^n\right)$$

The computational stencil of this scheme is shown in Figure 24.

The modified equation for this FDA is:

$$\frac{\partial c}{\partial t} + u\frac{\partial c}{\partial x} = u\frac{\Delta x}{2}\left(1 - \frac{u\Delta t}{\Delta x}\right)\frac{\partial^2 c}{\partial x^2} + \ldots$$

Hence, the scheme is stable but with predominant dissipative error. Even though the original PDE did not have any explicit diffusion in the governing equation, the numerical "diffusion" coefficient is of magnitude

$$D_{numerical} = \frac{u\Delta x}{2}$$

The magnitude of the numerical error is proportional to the grid spacing Δx and velocity u. Note also that for numerical stability, $\left(1 - \frac{u\Delta t}{\Delta x}\right) > 0$ or $\Delta t < \frac{\Delta x}{u}$. Stability and the amount of numerical diffusion is a function of the non-dimensional group called the Courant number, defined as $\frac{u\Delta t}{\Delta x}$ as shown in Figure 25.

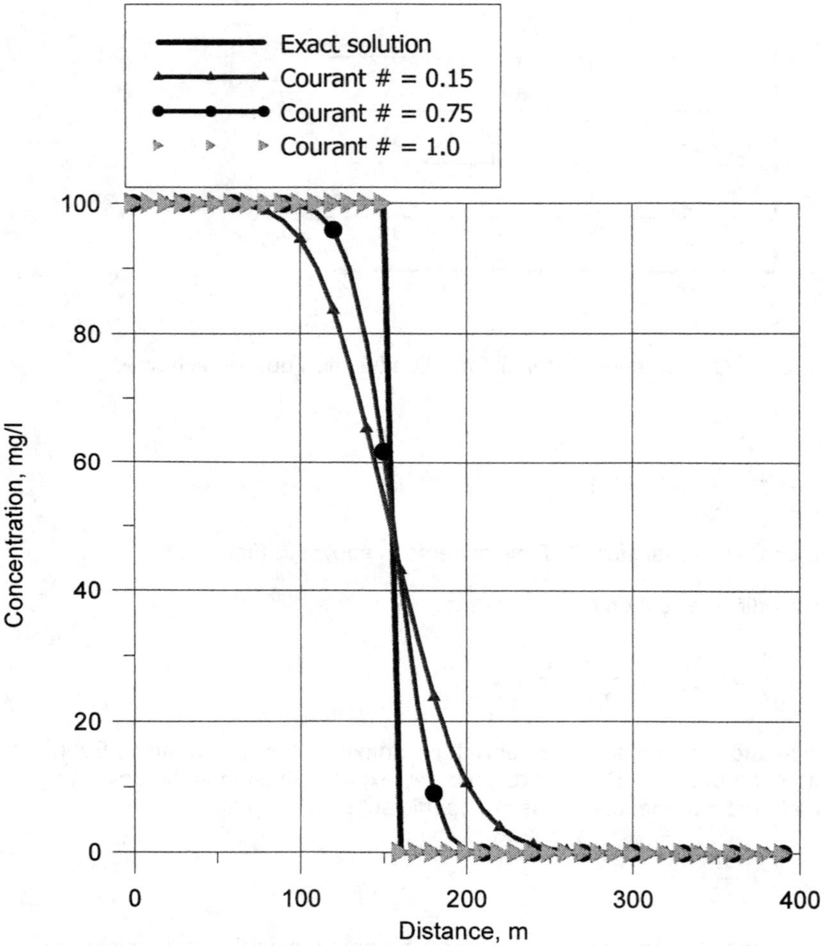

Figure 25. Numerical solution of unsteady advection equation with forward time and backward space differencing after 1000 s with u=0.15 m/s.

Implicit Schemes

Implicit schemes are written in such a way that the unknown state variable cannot be solved for explicitly. The resulting FDA must be solved implicitly. This usually results in solving simultaneous equations for the unknowns.

The time step necessary for stability is usually greater with implicit schemes than with explicit schemes, but this does not necessarily imply more accuracy with implicit schemes.

Using as an example the 1-D advective transport equation for a conservative tracer with no diffusion, the governing equation is

$$\frac{\partial c}{\partial t} + u\frac{\partial c}{\partial x} = 0$$

A general differencing scheme using a forward time approximation for the time derivative and a time weighted centered space difference for the advective term would be:

$$c_i^{n+1} = c_i^n - (1-\Theta)\frac{\Delta t}{2\Delta x}u\left(c_{i+1}^n - c_{i-1}^n\right) - (\Theta)\frac{\Delta t}{2\Delta x}u\left(c_{i+1}^{n+1} - c_{i-1}^{n+1}\right)$$

where Θ is the degree of explicitness or implicitness. For $\Theta=0$, we have the earlier FTCS explicit method. For $\Theta=1$, we have a fully implicit FTCS method. The computational molecule for this formulation is shown in Figure 26.

The equation above is implicit in that the unknown concentration at the n+1 time level cannot be solved for directly unless $\Theta=0$. The implicit FDA is applied at all points in the domain such that n simultaneous equations are generated for the n unknowns at the n+1 time level. Often the Thomas algorithm is used to solve the resulting tri-diagonal matrix solution (see Anderson et al., 1984).

For the explicit scheme, the FTCS FDA was unstable. The modified equation for the general time-weighting scheme though is:

$$\frac{\partial c}{\partial t} + u\frac{\partial c}{\partial x} = \left(\Theta - \frac{1}{2}\right)u^2\Delta t\left(\frac{\partial^2 c}{\partial x^2}\right) + O(\Delta x^2, \Delta t^2)$$

For this scheme, only when $\Theta<0.5$ is the sign on the leading truncation error negative - implying unconditionally unstable. But for $\Theta\geq0.5$, the sign of the leading TE is positive and hence stable (unless the other higher order even derivatives are negative). For $\Theta=0.5$ (termed the Crank-Nicholson scheme), this numerical dissipation is minimized (it may still exist somewhat in higher order even derivatives of the TE though).

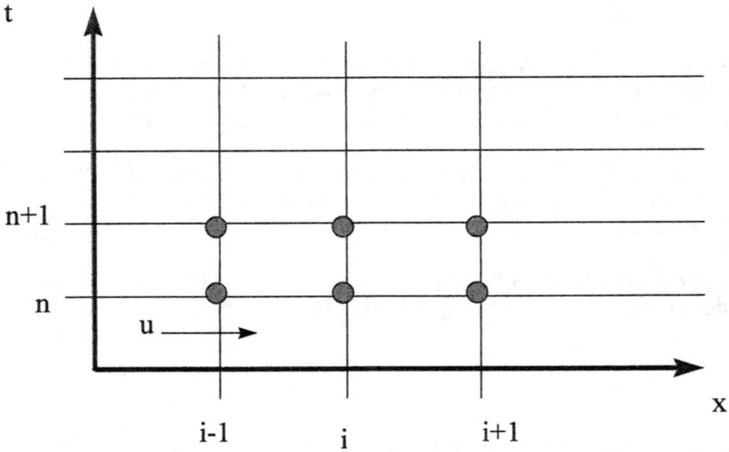

Figure 26. Computational stencil for general explicit-implicit solution to 1-D advection.

Summary of Finite Difference Method

Finite difference equations are used in many numerical codes that are used in river and reservoir/lake/estuary models, such as CE-QUAL-W2 (Cole and Wells, 2003), WASP (Ambrose, et al., 1988), Princeton Ocean Model (Blumberg and Mellor, 1987), CE-QUAL-RIV1 (Environmental Laboratory, 1995), and EFDC (Hamrick, 1996).

Often a modeler sees what are termed "wiggles" in the mathematical solution of a FDA. These wiggles are not necessarily instabilities leading to an anti-dissipative scheme since often they do not grow with time. Wiggles are associated with phase errors or dispersive numerical errors and are common features of sharp front solutions for FTCS, Crank-Nicholson, and even QUICKEST numerical schemes (Vreugdenhil, 1989). In these cases, wiggles are related to numerical accuracy rather than to instability. The use of upwind differencing of course smooths out the wiggles, but the resulting solution is very inaccurate. If one is using upwind differencing one must exercise caution in its use.

Press et al. (1992) show that numerical errors can be either amplitude error (leading to instability by increasing mass, termed earlier "anti-dissipative," or smoothing out mass differences, termed earlier "dissipative"); phase errors (termed earlier "dispersive" errors), or non-linear stability (occurring when a non-linear PDE is being solved and the non-linear terms interact to produce a "numerical shock"). According to Press et al. (1992) to model transport properties "better", they recommend using the upwind scheme even though it annoys mathematicians. They contend that “the goal of numerical simulations is not always 'accuracy' in a strictly mathematical sense, but sometimes 'fidelity' to the underlying physics in a sense that is looser and more pragmatic. In such contexts, some kinds of error are much more tolerable than others."

Many of the advances in numerical methods are to find more accurate (minimize numerical diffusion) and less computationally expensive methods for solving the FDAs. Some of these techniques have focused on resolving how to determine the best numerical approximation to advective terms in the advective-diffusion and momentum equations. In many cases grid refinement is too computationally "expensive" and higher order approximations are used. An example of a higher order method is the QUICKEST algorithm by Leonard (1979). This scheme was derived using an upstream weighted quadratic interpolation function defined over three uniformly spaced grid points.

One of the problems though with the QUICKEST scheme is that it can give rise to spurious oscillations at the leading and trailing edge of a sharp front or gradient. This can occur where there are fresh water – salt water interfaces or cases of strong temperature stratification. Even though the UPWIND scheme eliminated spurious oscillations about a sharp front, it also introduced numerical diffusion, which unrealistically smoothed these gradients.

An improvement was introduced by Leonard (1991) called the ULTIMATE scheme which eliminated spurious oscillations but preserved the higher order solution scheme of QUICKEST. This technique is a universal (in the sense that it can be applied to numerical schemes other than QUICKEST) limiter for maintaining monotonic profiles in the vicinity of a gradient.

According to tests preformed by Lin and Falconer (1997), the QUICKEST-ULTIMATE scheme, with splitting of the diffusion and source/sink terms, provided complete elimination of numerical oscillations and was able to fully conserve mass. Leonard (1991) also recommended that the QUICKEST scheme coupled with the ULTIMATE scheme was numerically accurate and cost-effective in terms of computational cost.

To illustrate the differences between ULTIMATE-QUICKEST, QUICKEST, and an UPWIND scheme, Figure 27 shows the results of a square pulse of 100 mg/l moving downstream with no diffusion. The UPWIND scheme has significant, numerical diffusion, whereas the QUICKEST scheme has non-physical oscillations about the leading and trailing edge of the solution. ULTIMATE-QUICKEST eliminated the spurious oscillations.

For any case of unequal grid spacing the order of accuracy diminishes. Leonard (1991) recommends that if the ULTIMATE scheme is used with non-uniform grid spacing, that the formal accuracy of the method is only preserved practically if the grid does not expand or contract more than about 125%.

Even though the ULTIMATE scheme can be used with any suitable numerical technique, Leonard (1991) recommends that when coupled with a 3rd order QUICKEST scheme and a second order central difference diffusion operator, that the results are virtually indistinguishable from still higher-order advective solvers.

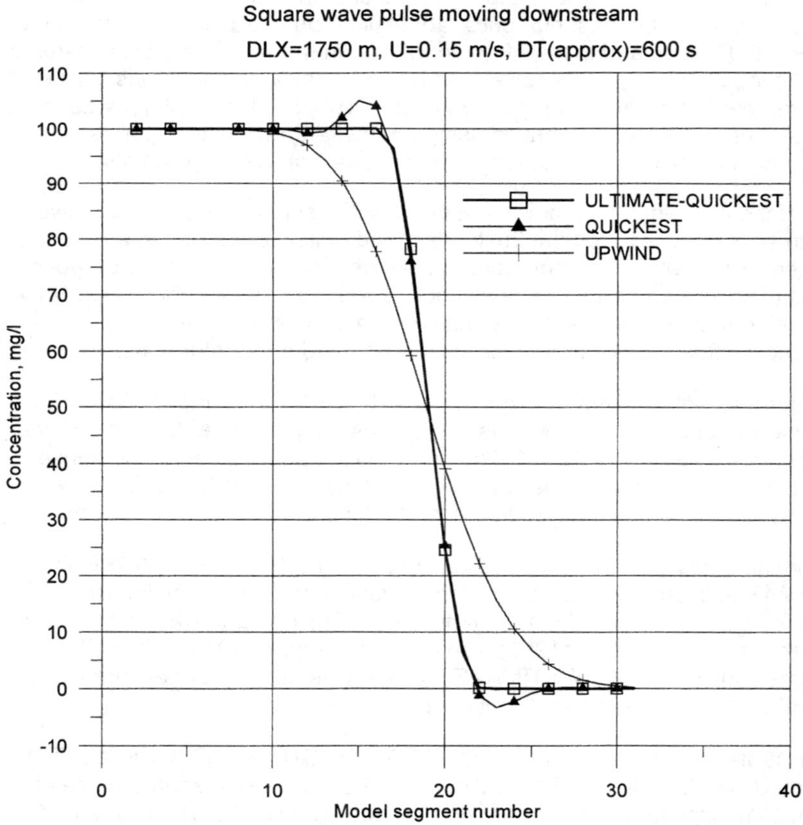

Figure 27. Comparison of ULTIMATE-QUICKEST, QUICKEST, and UPWIND scheme for CE-QUAL-W2 model for conservative tracer transport with no diffusion with 1 active vertical layer.

The Finite Element Method

The concept of the finite difference method was to approximate derivatives at a point. Finite element method (FEM) solutions are found by integrating the PDEs over the control volume of interest rather than approximating the derivatives at a point. The finite volume method is the simplest form of a general set of methods called the FEM. The finite volume method evaluates fluxes into and out of a control volume regardless of its shape. The FEM is a mathematical procedure to satisfy the PDE in an average sense over the finite element or control volume.

The FEM technique usually follows this procedure (Lapidus and Pinder, 1982):

1. Choose elements (discretize the problem domain into geometric elements or control volumes)

2. Choose interpolation shape functions for the elements (describe the spatial variation of the state variable across the element or control volume)

3. Formulate the element equations (integrate the governing equation over each element applying internal boundary conditions)

4. Form global equations by assembly of individual elements (assemble the integral statements of each element into a set of simultaneous equations)

5. Solve global equations (solve the matrix of simultaneous equations for the state variable)

Discretization

In this case, we take the domain, termed R, and subdivide R into subvolume or finite elements. The possible element shapes are linear in 1-D, 3-node triangular, 6-node triangular, or rectangular elements in 2-D, and rectangular, triangular, or arbitrary shaped volumes in 3-D. The advantage of the finite element method is that one can choose element shapes that more accurately reflect the boundary conditions of the problem. Even though the finite difference approximation can approximate similar curved or irregular boundaries, there is more work involved with using unequal spaced finite differences and making the orthogonal finite difference grid match such a boundary.

The complete specification of an element requires knowing the element shape (for example, linear, triangular, rectangular), the number and location of the nodes (these correspond to the location of the grid points for solution of the state variable), and the type of interpolation function (these specify how the state variable is interpolated between nodes).

Interpolation Functions

The interpolation functions are a postulate of the spatial distribution of the state variable. These functions depend on the particular element used for the discretization and the number of nodes in the element.

As an example, let's take the 1-D advective-diffusion equation:

$$\frac{\partial c}{\partial t}+u\frac{\partial c}{\partial x}=D\frac{\partial^2 c}{\partial x^2}$$

In this case we assume that the actual solution to the PDE, c, can be approximated by an assumed form of the solution, $\hat{c}$, such that

$$c(x,t)\cong\hat{c}(x,t)=\sum_{i=1}^{N}P_i(x)c_i(t)$$

where c_i are the nodal values of concentration, P_i are the polynomial interpolation functions for the element, i is the nodal location, and N is the number of nodes.

Generally, polynomials are used to describe how the nodal concentrations are distributed in space. For 3-node triangular elements, the shape functions are usually linear, i.e., the linear functions P_i vary from unity at the node *i* to 0 at its nearest neighbors.

Integral Representations and Galerkin's Method

For the 1-D advective-diffusion equation, we can write this as

$$\frac{\partial c}{\partial t}+u\frac{\partial c}{\partial x}-D\frac{\partial^2 c}{\partial x^2}=0$$

Substituting the assumed form of solution, $\hat{c}$, into the governing equation, then

$$\frac{\partial \hat{c}}{\partial t}+u\frac{\partial \hat{c}}{\partial x}-D\frac{\partial^2 \hat{c}}{\partial x^2}=\in$$

where $\in$ is the residual or error between the original PDE and the assumed form of the solution.

Integrating this equation over the each element, R', of the domain, R, and setting this result equal to zero, we obtain

$$\int_{R'}\left(\frac{\partial \hat{c}}{\partial t}+u\frac{\partial \hat{c}}{\partial x}-D\frac{\partial^2 \hat{c}}{\partial x^2}\right)dR=\int_{R'}\in dR=0$$

Applying this requirement to each element is equivalent to requiring that the average error in each element be zero.

The method of weighted residuals is a general procedure for obtaining approximate solutions PDEs where a weighting function, W, is used in the integral of the residual, such that

$$\int_{R'} W_i \in dR = 0 \text{ where i=1, 2, ... N}$$

If the weighting function choice is the interpolation function, $P_i(x)$, then this method is known as Galerkin's Method. Galerkin's Method for our example becomes

$$\int_{R'} P_i \left(\frac{\partial \hat{c}}{\partial t} + u \frac{\partial \hat{c}}{\partial x} - D \frac{\partial^2 \hat{c}}{\partial x^2} \right) dR = 0 \quad \text{for} \quad i = 1,2 \ldots N$$

Assembly and Solution

The element equations are then assembled by using Galerkin's method for each element. For the region of the element, R^e, the integral is determined for each element such that

$$\int_{R^e} P_i^e \left[\frac{\partial \hat{c}^e}{\partial t} + u \frac{\partial \hat{c}^e}{\partial x} - D \frac{\partial^2 \hat{c}^e}{\partial x^2} \right] dR^e = 0 \quad \text{for} \quad i = 1,2 \ldots N_e$$

where N_e is the number of nodes in the element.

The calculation proceeds by substituting the interpolation function for the assumed solution,

$$\hat{c} = \sum_{i=1}^{N} P_i \left(\vec{X} \right) c_i \left(t \right),$$

into the above element equation and evaluating this integral for every element. Integration by parts is used to simplify the integrals before they are added together to form a global equation. The global equation is then a matrix solution of a set of simultaneous linear equations.

Summary of the FEM

There are several good reference works on FEMs such as Mitchell and Wait (1977) and Lapidus and Pinder (1982). Some numerical hydrodynamic and transport codes that use the FEM include the RMA-2 model (King and Norton, 1978), RMA-11 (King, 1998), and the finite volume CH3D model (Sheng, 1987). In some cases, the FDA and the FEM solutions can be identical. The main advantage of the FEM compared to the FDA is that elements of various shapes can be used to match irregular boundaries. The finite difference method can handle unusual boundary shapes by using unequal grid spacing, but still must have a fine grid to map an irregular boundary with high resolution. Also, the FEM can have local areas of refined grid, whereas the finite difference method must have much larger areas of grid refinement even if grid refinement was required in one location.

Computational Efficiency

Increased computing capabilities have resulted in dramatic changes in the techniques available in hydropower reservoir engineering and environmental management and will continue to do so. However, computational efficiency remains an issue in hydrodynamic and water quality modeling. As with most things, the more we have the more we use. As computational efficiency increases, so do the complexity of the models that are developed and applied.

The importance of the issue of computational efficiency varies with the problems that the hydrodynamic models are designed to solve. For example, typical applications to problems that many hydrodynamic models are designed for, such as analysis of flooding and scour, require simulations ranging from seconds to weeks. However, the application of models to address environmental issues can require many simulations for periods ranging from months to years for conventional pollutants (nutrients and dissolved oxygen) and years to decades for hydrophobic contaminants (PCBs, dioxins, etc.). The coupling of chemical equilibrium models and those for solute transport is an additional area that necessitates highly efficient numerical algorithms if solutions are to be returned within realistic timescales. In addition, a general trend is to get away from deterministic models and go toward quoting uncertainties and model confidence limits. This usually entails a Monte Carlo or other multi-simulation based procedures. Presently, the computational resources required to do the many runs and build databases often deter this practice in all but the simplest models (usually those not coupled with a transport model). However, this is expected to change in the future as our computational resources permit.

In addition to the complexity of the model, hydrodynamic models may vary in computational efficiency due to the efficiency of the numerical algorithms, the efficiency of the computer coding of those algorithms, and the ability of the models to take advantage of modern computing capabilities. For example, many existing models were initially developed one or more decades ago and not designed to take advantage of multiple processors, and or may limited to running on specific computer platforms.

Model Inputs

Model provisions to allow the development of model input and the interpretation of model output could also be considered an issue in the efficiency of a model. There has been a tendency in recent years to concentrate as much on supporting software as on the structure of the models themselves. The supporting software developed for some models has greatly simplified the model application. For example, the development of software to support development of a numeric grid has, for some models, reduced the time required for grid generation from formerly weeks to months to presently hours to days. The development of supporting software has made many models easier to use, and reduced input errors, so that model users can now concentrate more on the model and its application, rather than the format and error checking of some highly structured input data files.

Use and Linkages for Water Quality Modeling

Often available hydrodynamic models, and model applications, have been designed for purposes other than transporting water quality constituents or the analysis of water quality conditions. Since the time and space scales of processes impacting water quality vary from those addressed by traditional hydrodynamic applications, the applicability of those models and/or applications may be limited. For example, hydrodynamic models have traditionally been designed and used in the design of engineering structures, and control and storage of water. Therefore, the time and space scales to which hydrodynamic models have typically been applied often vary from those of importance in the analysis of water quality. Traditional applications may also be to conditions, such as high flows, which are not those critical for the analysis of water quality.

One other consideration is that water quality modelers have historically not been trained in hydrodynamics, and vice versa. Conflicts in model and study design often occurred where the needs of both fields were not considered. Fortunately, there is an increasing trend for development and design of hydrodynamic models with consideration to environmental modeling, as exemplified by the emerging field of environmental hydraulics.

Many of the hydrodynamic models, particularly those designed for analysis of flows in rivers, have no direct water quality capabilities. That is, they do not include equations for the transport of constituents. However, for multi-dimensional models in estuaries, lakes, and reservoirs, it is usually necessary to include the impact of density differences, due to variations in salinity and/or temperature, on the hydrodynamics. The impact is predicted by including transport equations for salinity and/or temperature, with the predicted values related to density through an equation of state. So, at a minimum, these models can be used to predict water temperatures and salinities. A number of these models also solve the transport equation for other water quality constituents. The approach can be called the direct linkage approach since the hydrodynamic and water quality equations are solved within the same computer code. An alternative approach is to write the hydrodynamic predictions to a file, which is then read by the computed code designed to predict water quality variations (an indirect linkage). Both of these approaches are commonly used and each have advantages and disadvantages.

Models such as the EPA EFDC (a 3-D hydrodynamic model, Hamrick, 1996) are linked to the 3-D EPA model WASP (Ambrose et al., 1988). This linkage is one-way in the sense that the hydrodynamics are solved independently of the water quality. This is also the case with the Corps hydrodynamic model CH3D (Sheng, 1987) and the associated linked 3-D water quality model ICM (Cerco and Cole, 1994). Other models, like CE-QUAL-W2 (Cole and Wells, 2003) have both the water quality and hydrodynamics linked internally in one code. This carries a computational cost, but allows for impacts of water quality on hydrodynamics through the equation of state (e.g., a turbidity inflow or self-shading from algae affecting light extinction and hence the temperature regime and ultimately affecting the hydrodynamic field).

One consideration in the linkage of hydrodynamic and water quality models is that a number of models designed purely for analysis of hydrodynamics do not exactly conserve mass. For example, in riverine models that solve both the momentum and continuity equations at a point (cross-section), mass is not exactly conserved. While the errors do not appreciably impact hydrodynamic predictions, they may substantially impact water quality predictions, which are based upon the conservation of mass. Hydrodynamic models designed for water quality purposes such as CE-QUAL-W2 (Cole and Wells, 2003) use solutions schemes designed for mass conservation. The ability to guarantee that solute mass will not be artificially created or destroyed by the numerical solution is a crucial test for all water quality models, whose sole purpose is to predict the ultimate fate of species mass (Manson and Wallis 2000).

4.3. REFERENCES

Abbott, M. B. (1992) Computational Hydraulics Elements of the Theory of Free Surface Flows, Ashgate, Brookfield, VT.

Ambrose, R. B.; Wool, T.; Connolly, J. P.; and Schanz, R. W. (1988) "WASP4, A Hydrodynamic and Water Quality Model: Model Theory, User's Manual, and Programmer's Guide," Envir. Research Lab., Envir. Protection Agency, EPA 600/3-87/039, Athens, GA.

Anderson, D.; Tannehill, J. , and Pletcher, R. (1984) Computational Fluid Mechanics and Heat Transfer, McGraw-Hill, NY.

Antenucci, J. and Imerito, A. (2000) The CWR Dynamic Reservoir Simulation Model DYRESM User Manual, Report WP 1571 JA, Centre for Water Research, The University of Western Australia.

ASCE (1988) "Turbulence Modeling of Surface Water Flow and Transport," Journal of Hydraulic Engineering, ASCE Committee on Turbulence Models in Hydraulic Computations, 114(9), 970-1073.

Axell, L. B. and Liungman, O. (2001) A One-Equation Turbulence Model for Geophysical Applications: Comparison with Data and the k-ε Model, Environmental Fluid Mechanics, 1: 71-106.

Barkau, R. (1997) "UNET One-Dimensional Unsteady Flow Through a Full Network of Open Channels, User's Manual, US Army Corps of Engineers, Hydrol. Engr. Center, Davis, CA.

Batchelor, G. K. (1967) An Introduction to Fluid Dynamics, Cambridge University Press, NY.

Bencala, K.E. and R.A. Walters (1983) Simulation of solute transport in a mountain pool-and-riffle stream: a transient storage model, Water Resources Research, 19(3):718-724.

Blanke, B. and Delecluse, P. (1993) Variability of the Tropical Atlantic Ocean simulated by a general circulation model with two different mixed layer physics, J. Phys. Oceanogr., 23, 1363-1388.

Bloss, S., Lehfeldt, and Patterson, J. C. (1988) Modeling Turbulent transport in a stratified estuary, ASCE, J. Hydr. Engr., 114(9), 1115-1133.

Blumberg, A. F. and Mellor, G.L. (1987) A description of a three-dimensional coastal circulation model, in Three-Dimensional Coastal Ocean Models , Vol. 4, edited by N. Heaps, American Geophysical Union, Washington, D.C.

Blumberg, A.F., Galperin, B., O'Connor, D.J. (1992) Modeling Vertical Structure of Open-Channel Flows, *Journal of Hydraulic Engineering*, ASCE, 118 (8).

Blumberg, A. F., Signell, R. P. and Jenter, H. L. (1993) “Modeling transport processes in coastal ocean,” J. Mar. Envir. Eng., 1:31-52.

Blumberg, A. F. and Mellor, G. L. (1983) Diagnostic and Prognostic Numerical Circulation Studies of the South Atlantic Bight, Journal of Geophysical Research, Vol88, No C8, pp4579-4592, 1983.

Brown, L. C. and Barnwell, T. O. (1987)"The Enhanced Stream Water Quality Models QUAL2E and QUAL2E-UNCAS: Documentation and Users Manual," Envir. Res. Lab., USEPA, EPA/600/3-87/007, Athens, Ga.

Brunner, G. W. (2002) HEC-RAS, River Analysis System User's Manual, Report No. CPD-68, Hydrologic Engineering Center, Davis, CA.

Casulli, V. and Stelling, G. S. (1998) Numerical simulation of 3D quasi-hydrostatic free-surface flows, ASCE, J. Hydr. Engr., Vol 124, 678-686.

Celik, I. and Rodi, W. (1984) “Simulation of free-surface effects in turbulent channel flow,” Physiochemical Hydrodynamics, 5(3/4), 217-227.

Cerco, C. and Cole, T. (1994) Three-Dimensional Eutrophication Model of Chesapeake Bay, Technical Report EL-94-4, Waterways Experiment Station, Vicksburg, MS.

Chapra, S. (1997) Surface Water Quality Modeling, McGraw-Hill, NY.

Cheng, R. T. and Casulli, V. (2001) Evaluation of the UnTRIM Model for 3-D Tidal Circulation, Proceedings 7th International Conference on Estuarine and Coastal Modeling, St Petersberg, Florida, 628-642.

Cole, T. and Buchak, E. (1995) CE-QUAL-W2: A Two-Dimensional, Laterally Averaged, Hydrodynamic and Water Quality Model, Version 2.0, Technical Report El-95-May 1995, Waterways Experiments Station, Vicksburg, MS.

Cole, T. M. and Wells, S. A. (2003) "CE-QUAL-W2: A two-dimensional, laterally averaged, Hydrodynamic and Water Quality Model, Version 3.1," Instruction Report EL-03-1, US Army Engineering and Research Development Center, Vicksburg, MS.

Corps of Engineers (1986a) “CE-QUAL-R1: A Numerical One-Dimensional Model of Reservoir Water Quality,” Waterways Experiment Station, Instruction Report e-82-1, Vicksburg, MS.

Corps of Engineers (1986b) "HEC-5 Simulation of Flood Control and Conservation Systems," CPD-5Q, Hydrologic Engineering Center, Davis, CA.

Corps of Engineers (2001) UNET: One-Dimensional Unsteady Flow Through a Full Network of Open Channels, User's Manual, HEC, Davis, CA.

Craig, P.D. and M.L. Banner (1994) Modeling wave-enhanced turbulence in the ocean surface layer. J. Phys. Oceanogr., 24, 2546-2559.

Cunge, J.A., Holly, F.M., Verwey, A. (1989) Practical Aspects of Computational River Hydraulics, Pitman Advanced Publishing Program.

Cushman-Roisin, B. (1994) Introduction to Geophysical Fluid Dynamics, Prentice-Hall, Englewood Cliffs, NJ.

Donigian, A.S., Jr., J.C. Imhoff, B.R. Bicknell and J.L. Kittle, Jr. (1984) "Application Guide for Hydrological Simulation Program Fortran (HSPF)," EPA-600/3-84-065, U.S. Envir. Prot. Agency, Athens, GA.

Edinger, J.E., and Buchak, E.M. (1978a). Reservoir Longitudinal and Vertical Implicit Hydrodynamics, Environmental Effects of Hydraulic Engineering Works, Proceedings of an International Symposium, Knoxville, TN.

Edinger, J. E. and Buchak, E. M. (1978b). "Numerical hydrodynamics of estuaries."Estuarine and Wetland Processes with Special Emphasis on Modeling, edited by P. Hamilton and K. B. MacDonald, Plenum Press, NY, 115-146.

Eliason, D. and Bourgeois, A. (1997) "Validation of Numerical Shallow Water Models for Stratified Seiches", International Journal for Numerical Methods in Fluids, 24, 771-786.

Environmental Laboratory (1995) "CE-QUAL-RIV1: A Dynamic, One-Dimensional (Longitudinal) Water Quality Model for Streams: User's Manual," Instruction Report EL-95-2, USACE Waterways Experiments Station, Vicksburg, Ms.

R.A.Falconer D.G.George P.Hall, Three Dimensional Numerical Modelling of Wind-Driven Circulation in a Shallow Homogeneous Lake, Journal of Hydrology, Vol124 pp59-79, 1991.

Fischer, H. B., List, J., Koh, R., Imberger, J., and Brooks, N. (1979) Mixing in Inland and Coastal Waters, Academic Press, New York.

Fletcher, C. A. J. (1988) Computational Techniques for Fluid Dynamics (2 Volumes), Springer-Verlag.

French, R. H. (1985) Open-Channel Hydraulics, McGraw-Hill, New York.

Galperin, B. and Mellor, G. L. (1990) A Time-dependent, Three-dimensional Model of the Delaware Bay and River System. Part 1: Description of the Model and Tidal Analysis, Estuarine, Coastal and Shelf Science, Vol 31(3) Sept, pp 231-253.

Grass, A.J. (1992) "Turbulence - What are we trying to model? Is Turbulence Modelling of Any Use?," IAHR Seminar April 9th.

Hamblin, P. F. and Salmon, J.R. (1975) "On the Vertical Transfer of Momentum in a Lake," Memoires Societe Royale des Sciences de Liege, 6a serie, tome VII, page 211.

Hamrick, J.M. (1996) User's Manual for the Environmental Fluid Dynamics Computer Code, Special Report No. 331 in Applied Marine Science and Ocean Engineering, School of Marine Science, Virginia Institute of Marine Science, The College of William and Mary, Goucester Point, VA

Hinze, J.O. (1975) *Turbulence*, 2nd ed., McGraw-Hill, NY.

Holland, P. R., Kay, A., and Botte, V. (2001) A numerical study of the dynamics of the riverine thermal bar in a deep lake, Envir Fluid Mech., 1: 311-332.

Imberger, J. and Patterson, J. C. 1981. A dynamic reservoir simulation model - DYRESM:5. In "Transport Models for Inland and Coastal Waters". H.B. Fischer (ed). Academic. pp310-361.

Johnson, B. H. (1981) "A Review of Numerical Reservoir Hydrodynamic Modeling," Technical Report E-81-2, Corps of Engineers, Waterways Experiment Station, Vicksburg, MS.

King, I.P. and W.R. Norton (1978) "Recent Application of RMA's Finite Element Models for Two Dimensional Hydrodynamics and Water Quality," Finite Elements in Water Resources, Pentech Press, London.

King, I. P. (1998) RMA-11 -- A Three Dimensional Finite Element Model For Water Quality In Estuaries And Streams, Dept. of Civil and Environmental Engineering University of California, Davis, Ca.

Koutitas, C. and O'Connor, B. (1980) Modeling Three-Dimensional Wind-Induced Flows, Journal of the Hydraulics Division, ASCE, Vol.106 No.HY11 Nov.

Lapidus, L. and Pinder, G. (1982) Numerical Solution of Partial Differential Equations in Science and Engineering, John Wiley, NY.

Leendertse, J. J., and Liu, S. K. (1975) A three dimensional model for estuaries and coastal seas. II: Aspects of computation. Rand Report R-1764-OWRT, The Rand Corporation, Santa Monica, CA.

Leonard, B. P. (1979) "A Stable and Accurate Convective Modelling Procedure based on Quadratic Upstream Interpolation," Computer Methods in Applied Mechanics and Engineering, 19, 59-98.

Leonard, B. P. (1991) "The ULTIMATE conservative difference scheme applied to unsteady one-dimensional advection," Computer Methods in Applied Mechanics and Engr, 88 (1991), 17-74.

M.Lesieur, Advance and State of the Art on Large Eddy Simulations Reynolds-Stress Turbulence Closures, Refined Flow Modelling and Turbulence Measurements, Proc of the 5th International Symposium, Paris, France, 1993.

Liggett, J. A. (1969) Cell Method for Computing Lake Circulation, ASCE Journal of Hydraulics Division, March HY3 Vol 96.

Lin, B., and Falconer, R. A. (1997) "Tidal Flow and Transport Modeling Using ULTIMATE QUICKEST Scheme", Journal of Hydraulic Engineering, ASCE, 123(4), 303-314.

Mamayev, O. I. (1958) "The influence of stratification on vertical turbulent mixing in the sea, Izv. Geophys. Ser., 870-875.

Manson, J. R. and Wallis, S. G. (2000) A conservative, semi-Lagrangian fate and transport model for fluvial systems—I. Theoretical development, Water Res 34(15), 3769-3777.

Martin, J. L., McCutcheon, S.C. (1999) Hydrodynamics and Transport for Water Quality Modeling, Lewis Publishers, NY.

Meier, H. E. M. (2001) On the parameterization of mixing in 3D Baltic Sea models. J. Geophys. Res. – Oceans, 106(C12), 30,997-31.

Mellor, G. and Blumberg, A. (2004) "Wave Breaking and Ocean Surface Layer Thermal Response, Journal of Physcial Oceanography, 34:693-698.

Mitchell, R. and Wait, R. A. (1977) The Finite-Element Method in Partial Differential Equations, Wiley, NY.

Munk, W. H. and Anderson, E. R. (1948) "Notes on the Theory of the Thermocline," Journal of Marine Research, Vol 1, p.276.

Nezu, I. And Nakagawa, H. (1993) Turbulence in Open-Channel Flows, A.A. Balkema, Rotterdam.

Octavio, S., Jirka, G., and Harleman, D. (1977) "Vertical Heat Transport Mechanisms," MIT, Department of Civil Engineering, Technical Report 227, R. M. Parsons Laboratory, Cambridge, Ma.

Patankar, S.V. (1980) Numerical Heat Transfer and Fluid Flow, Hemisphere, NY.

Patankar, S. V. and Spalding, D. B. (1972) A Calculation Procedure for Heat, Mass and Momentum Transfer in Three-Dimensional Parabolic Flows, Int.J.Heat Mass Transfer, Vol15 pp1787-1806.

Press, W.; Teukolsky, S.; Vetterling, W.; and Flannery, B. (1992) Numerical Recipes: The Art of Scientific Computing, Cambridge University Press, NY.

Roache, P. (1982) Computational Fluid Dynamics, Hermosa, Albuquerque, NM.

Roesner, L. A., J. A. Aldrich, and R. E. Dickinson (1988). "Storm water management model user's manual version 4: EXTRAN addendum 1 EXTRAN." Cooperative Agreement CR-811607, U.S. Environmental Protection Agency, Cincinnati, Ohio.

Rodi, W. (1993) Turbulence Models and Their Application in Hydraulics, 3rd edition, IAHR, A.A. Balkema, Rotterdam.

Sabersky, R.; Acosta, A., and Haupmann, E. (1989) Fluid Flow A First Course in Fluid Mechanics, Macmillan Publishing Co., NY.

Sauvaget, P. and Usseglio-Polatera, J.M. (1987) Numerical Simulation of Stratified Flows in Estuaries and Reservoirs, Proceedings of the Third International Symposium on Stratified Flows, Pasadena, California, Eds: E.J.List G.H.Jirka.

Shanahan, P. and Harleman, D. R. F. (1982) "Linked Hydrodynamic and Biogeochemical Models of Water Quality in Shallow Lakes," MIT Technical Report 268, Cambridge, Ma.

Sheng, Y.P. (1987) "On Modeling Three-Dimensional Estuarine and Marine Hydrodynamics," Three-Dimensional Models of Marine and Estuarine Dynamics (J.C.J. Nihoul and B.M. Jamart, Eds.), Elsevier Oceanography Series, Elsevier, pp. 35-54.

Shimuzi, Y., Yamaguchi, H., and Itakura, T. (1990) Three Dimensional Computation of Flow and Bed Deformation, Journal of Hydraulic Engineering, ASCE, Vol.116 No.9 Sept.

Smith, D.J. (1978) "WQRRS, Generalized computer program for River-Reservoir systems," User's Manual 401-100, 100A, USACE Hydrol. Engr. Center, Davis, CA., 210 p.

Smith, G. D. (1985) Numerical Solution of Partial Differential Equations: Finite Difference Methods, Oxford University Press, London.

Spalart, P.R. (1988) Direct Simulation of a Turbulent Boundary Layer Up to Re=1410, Journal of Fluid Mechanics, Vol.187.

Spigel, R. H., Imberger, J., and Rayner, K. N. (1986) "Modeling the Diurnal Mixed layer," Limnology and Oceanography, 31(3), 533-556.

Venayagamoorthy, S. K., Koseff, J. R., Ferziger, J. H., and Shih, L. H. (2003) Testing of RANS turbulence models for stratified flows based on DNS data, Center for Turbulence Research, Annual research report briefs 2003, 127-138.

Van der Kuur, P., Roelfzema, A., Verboom, G.K. (1989) The Three Dimensional Programme TRISULA with Curvilinear Orthogonal Co-ordinates, Advances in Water Modelling and Measurement, pp135 Editor-M.H.Palmer.

Vreugdenhil, C. (1989) Computational Hydraulics, Springer-Verlag, NY.

Wells, S. (2002a) "Validation of the CE-QUAL-W2 Version 3 River Basin Hydrodynamic and Water Quality Model," Proceedings, HydroInformatics 2002, IAHR, Cardiff, England.

Wells, S. (2002b) "Basis of the CE-QUAL-W2 Version 3 River Basin Hydrodynamic and Water Quality Model," Proceedings, 2nd Federal InterAgency Hydrologic Modeling Conference, Las Vegas, July 28-Aug 1, 2002.

Williams, J.J.R. (1992) Large Eddy Simulation, Is Turbulence Modelling of Any Use?, IAHR Seminar April 9th.

Yen, B.C. (1973) Open Channel Flow Equations Revisited, Journal of the Engineering Mechanics Division, ASCE, Vol.99 No.EM5 Oct.

5. WATER QUALITY MODELING THEORY

John E. Edinger

James L. Martin

Gregory J. Pelletier

5.1. Introduction

As in all engineering practice, the management of our water resources requires the ability to assess the changes that may result, such as in water quality conditions, from the implementation of some management alternative. For example, to address many of the regulatory issues discussed in Chapter 2, it may be necessary to estimate the maximum loading a waterbody can assimilate and still meet water quality standards or to estimate the impact of removal of some fraction of a contaminant.

The assessment of the impact of management alternatives typically requires establishment of a quantitative link between the cause of change and potential impact or effect. For water quality that link is typically established using mechanistic water quality models that integrate available data, information and knowledge of the physical/chemical/biological mechanisms impacting the aquatic system.

Water quality models are mechanistic in that they attempt to include each of the underlying mechanisms causing change in the aquatic system. The underlying principals for water quality models are the laws of conservation. For example, models for the prediction of temperature variations, discussed later in this Chapter, are based upon the conservation of energy. Models for the prediction of chemical concentrations are based upon the law conservation of mass, and are often referred to as mass balance models. The approach requires that the quantities of that conservative property (energy or mass) entering the system, less quantities stored, transformed or degraded in the system, must be equal to the quantities leaving the system. Thus the implementation of the approach based upon laws of conservation is simply a budget, or a "book keeping" of all the processes (mechanisms) affecting that conservative property in the system. Once the budget has been established, the approach can be used to provide quantitative estimates of the effects of changes in that budget.

The implementation of the budget approach is illustrated in the following section (See Equation 5-1) by the basic transport equation used to predict the rate of change in a water quality constituent. In the constituent transport equation, the rate of change in a water quality constituent in a waterbody (such as dissolved oxygen, a toxic chemical concentration, etc.) is equated to the transport of that constituent (via advection or mixing processes) through the waterbody and the sources and sinks of that constituent (e.g. due to processes such as degradation) within the waterbody. The transport equation may be one, two or three dimensional as discussed in Chapter 4. Typically, the transport terms of the transport equation are derived from the solution of the hydrodynamic equations, also discussed in Chapter 4.

The basic form of the transport equation (see Equation 5-1), and values of the transport terms (advection and mixing) used in that equation, is typically the same regardless of the water quality process being modeled. What will distinguish between the transport equations for specific water quality constituents (e.g. ammonia, dissolved oxygen, a toxic chemical, etc.) are the transformation and degradation processes included in the source/sink term. In addition, a separate transport equation must be solved for each constituent, so models will also vary in the number of equations solved (number of state variables). The number of equations solved and degradation/transformation reactions included will depend upon the particular water quality process the model is designed to depict, as well as the assumptions on which that model is based. There are three general classes of water quality processes, and water quality models, discussed in this Chapter, eutrophication (Section 5.2), toxic chemicals (Section 5.4) and heat or temperature (Section 5.5).

As discussed in Section 5.2, excessive loads of nutrients and carbon (expressed either in carbon or dissolved oxygen equivalents) to reservoirs and from their releases may result in excessive plant growth, taste and odor problems, low dissolved oxygen concentrations, or other undesirable impacts. For many waterbodies, such as reservoirs, there is a high degree of interaction between state variables and internal production (autochthonous) and cycling must be considered. Models developed to predict these impacts (models of eutrophication, Section 5.2) must typically include a relatively large number of state variables and kinetic reactions in order to depict the processes impacting the cycling of nutrients, dissolved oxygen, and carbon (living and non-living). In addition, the accumulation and decay (diagenesis) of materials in bottom sediments must also be considered as part of the internal cycling. The modeling of sediment diagenesis is the subject of Section 5.3 of this Chapter.

Toxic chemical modeling, or contaminant modeling, is discussed in Section 5.4 of this Chapter. These chemicals may include organic chemicals (such as PCB's or Dioxin), metals (such as copper, zinc), or organo-metallics (such as mercury). The chemicals may produce acute toxicity (such as unionized ammonia or some metals) or chronic effects (such as carcinogenic impacts) in humans or wildlife. For many of these contaminants, the form of the chemical (free metal concentration, unionized ammonia concentration, etc.) affects the resulting toxicity and must be considered in modeling. The degradation and transformation processes impacting the contaminant, and degree of interaction between contaminants varies with the type of chemical, impacting the structure of contaminant models. Some contaminants (e.g. hydrophobic organic chemicals) will strongly sorb to organic solids that may settle, resulting in the contaminant becoming sequestered in sediments so that the stability of the sediments becomes a modeling issue. For some contaminants (such as PCBs, mercury, etc.), food chain accumulation may also pose a risk to humans and wildlife. Due to food chain accumulation, the consumption of fish in many areas has been prohibited or restricted. As a result, models have been developed to predict change in contaminant concentrations in biota as a function of water column and sediment concentrations, as discussed in Section 5.4.

Modeling of heat, or water temperature, as discussed in Section 5.5 of this Chapter, is important for multiple reasons. Water temperatures impact density, and as a result temperature predictions must be included in the solution of the multi-dimensional hydrodynamic equations in order to predict such impacts as density stratification in reservoirs. Water temperatures affect the rates of degradation and transformation of conventional (as in eutrophication) and toxic pollutants, and must be considered in these models as well. Temperature is also considered a pollutant (see Chapter 2) and excessive temperatures may result in adverse impacts on aquatic biota. Models of heat must typically include sources of heat to the water column, such as thermal loading and processes impacting heat exchange at the water's surface, internal impacts on the sorption and transmission of heat, and heat exchange between the water column and sediment, as discussed in Section 5.5.

5.2. Eutrophication Modeling

Excess nutrients in a waterbody can have many detrimental effects on designated or existing uses, including drinking water supply, recreational use, aquatic life use, and fishery use (see Chapter 2, EPA 1999). For example, a eutrophic system typically contains an undesirable abundance of plant growth, particularly phytoplankton, periphyton, and macrophytes which can clog filters at water treatment plants, hinder navigation, decrease water clarity, produce undesirable taste and odors and interfere with the recreational use and aesthetics of waterbodies (Chapra 1997, EPA 1999). Excessive plant growth can also affect a systems chemistry. Of particular note is the impact on dissolved oxygen, which is directly impacted by plant production and respiration, and the decay of organic materials produced. Low dissolved oxygen levels may impair aquatic life and fisheries. Breakdown of dead organic matter in water also can produce un-ionized ammonia, which can adversely affect aquatic life. Changes in the chemistry along with changes in the aquatic plant life may alter the biological composition of the waterbody.

Modeling studies often focus on the impacts of excessive nutrient loadings on the eutrophication of a waterbody, and the resulting impacts on concentrations of dissolved oxygen and other water quality parameters. These studies may be required, for example, for the effective management of the waterbody or to insure the waterbody achieves compliance with water quality standards, as discussed in Chapter 2.

The eutrophic state of a waterbody is evaluated by the concentration of nutrients, dissolved oxygen and phytoplankton within it at different locations throughout the water column. Eutrophication water quality models formulate the inter-relationships of nutrients, phytoplankton, zooplankton and dissolved oxygen as they are affected by inflows and outflows of a waterbody, sediment exchange and surface exchange.

Eutrophication Processes

There are many different formulations of eutrophication models. Most agree about the necessary state variables and the general inter-relationship among them. Chosen to illustrate the general properties of eutrophication models is one that includes the major nutrient constituents, carbon, nitrogen and phosphorous, with their organic components in both dissolved and particulate form. The process diagram for the eutrophication model being examined here is presented in Figure 5.1. Examining the breakdown of the nutrients, it shows that organic carbon (expressed here as CBOD), decays or oxidizes to some simpler form (expressed here as CO_2), and in the process consumes dissolved oxygen. The organic nitrogen (ON) first mineralizes to ammonia (NH_3) that in turn nitrifies to nitrate (NO_3) and consumes dissolved oxygen (DO) in the process. The organic phosphorous mineralizes to phosphate (PO4).

The role of phytoplankton (PHYT) in eutrophication models is illustrated in Figure 5.1. It shows that phytoplankton recycle nutrients to their dissolved and particulate organic forms by excretion and death, and to ammonia and phosphate by respiration. Phytoplankton produce dissolved oxygen by photosynthesis when sufficient light is available, and consumes dissolved oxygen by respiration otherwise. At the same time, the levels of ammonia, nitrate and phosphate, and the light level at different locations within the water column determine the rate of growth of phytoplankton. A major process controlling phytoplankton densities and recycling nutrients is zooplankton grazing. It recycles nutrients obtained from grazing on phytoplankton to particulate organic carbon, organic nitrogen and organic phosphorous.

Additional processes are settling of the particulate forms of organic carbon, nitrogen and phosphorous and of dead phytoplankton and exchanges with the sediment. The latter are sediment oxygen demand, and the release of ammonia and phosphate from the sediment.

Eutrophication models differ in degrees of complexity. For some models, the organic components may be further divided into constituents with fast and slow decay and mineralization rates, and multiple phytoplankton species and zooplankton species may be included.

<u>Model Formulation and Computational Properties</u>

The computational structure for the eutrophication model shown in Figure 5.1 is given in Figure 5.2 in a matrix form showing sources and sinks entering the constituent transport relationships given in Chapter 4 as:

Equation 5-1

$$\frac{\partial C_n}{\partial t} + \frac{\partial UC_n}{\partial x} + \frac{\partial VC_n}{\partial y} + \frac{\partial WC_n}{\partial z} \quad \frac{-\partial(A_x C_n/\partial x)}{\partial x}$$

$$-\frac{\partial A_y(C_n/\partial y)}{\partial y} - \frac{\partial(A_z(C_n/\partial z)}{z} - \frac{\partial(C_n/\partial z)}{\partial z} = S_n$$

where C_n is the concentration for constituent "n" and S_n is its source/sink term. As in Chapter 4, U, V and W are the velocity components in each of the coordinate directions, and A_x, A_y and A_z are the constituent dispersion coefficients in each of the coordinate directions.

The source/sink term varies with time as affected by constituent concentrations, sediment exchange processes and grazing. The rate parameters in the matrix relating the constituents to the source/sink term are not constant over time. The matrix determines how a constituent source/sink changes over time at a point within a water body due interaction with the remaining constituents. In the matrix for example, the source/sink relationship for dissolved oxygen (DO) is:

Equation 5-2

$$S_{DO} = a_{6,1}NH_3 + a_{6,3}PHY + a_{6,4}BODD + a_{6,5}BODP + a_{6,6}DO - SOD + RearSat$$

S_{DO} depends on the most constituents, five, of any of those comprising the water quality model. Going down a column, NH_3 enters four source/sink terms, and NO_3, BODD and PO_4 enter three source/sink terms. The OND, ONP, OPD and OPP enter two source/sink terms.

The source/sink matrix in Figure 5.2 shows that every source/sink is affected directly by the phytoplankton density either as a source recycling the nutrient or as a sink taking up the nutrient for phytoplankton growth. On the other hand, the fate of phytoplankton within the matrix depends on the single parameter, $a_{3,3}$, which is basically the difference in phytoplankton growth and death rates and can lead to a continuous growth or decay of phytoplankton unless it is balanced by grazing.

The rate parameters in Figure 5.2 depend on the formulation of the individual processes shown in Figure 5.1. The formulation of the individual processes for each rate parameter related to each constituent is listed in Table 5.1. The definitions of the kinetic coefficients for each formulation are given in Table 5.2 along with the ranges over which it varies. Most of the kinetic coefficients are dependent on temperature as indicated by the temperature coefficients in Table 5.2. Some of the matrix rate parameters shown in Table 5.1 depend on the concentration of dissolved oxygen and phytoplankton density making the solution to the constituent matrix shown in Figure 5.2 not only highly interactive among constituents, but also non-linear through the rate parameters.

Evaluation of Kinetic Coefficients

The most complete source of information about the rate coefficients in Table 5.2 is EPA(1985) “Rates and Constants”. A number of the rate parameters given in Table 5.1 depend on coefficients in Table 5.2 that are relatively straightforward in terms of the processes they describe. These include the rate of oxidation of NH_3 to NO_3, k_{12}; the rate of oxidation of CBODD and CBODP (dissolved and particulate CBOD), k_d; the rate of mineralization of OND and ONP (dissolved and particulate ON) to NH3, k_{71}; and the rate of OPD and OPP (dissolved and particulate OP) to PO4, k_{83}. These coefficients are modified by a Michalis-Menton type of rate limitation for the oxidation rates of BODP and the mineralization of ONP and OPP.

More complex rate processes are the phytoplankton growth rates, the sediment source-sink terms including settling, and zooplankton grazing.

Phytoplankton Growth and Mortality Rates

The term controlling phytoplankton growth and mortality rates, $a_{3,3}$ in Figure 5.2, incorporates a number of complex processes including the temperature, light intensity and nutrient concentration dependent growth rate throughout the water column, the temperature dependent respiration rate and the death rate. It can be written as:

Equation 5-3

$$a_{3,3} = [(1-f_e)Gp - D_R - D_D]$$

where:

Gp = Phytoplankton growth rate

D_R = Phytoplankton respiration rate

D_D = Phytoplankton death rate

f_e =Excretion fraction of phytoplankton

The phytoplankton growth rate can be written as:

Equation 5-4

$$G_p = k_{1c}\Theta^{(T-20)}\left[(\frac{I}{Is})\exp\left(1-\frac{I}{Is}\right)\right]f(NH_3, NO_3, PO_4)$$

Where k_{1c} is the phytoplankton maximum growth rate and Θ is the temperature coefficient as defined in Table 5.2, I is the incident light at a given depth and time, and I_s is the saturation light value and $f(NH_3, NO_3, PO_4)$ is the nutrient limitation function. The k_{1c} and I_s have specific values for specific species of phytoplankton.

The nutrient limitation function is defined as:

Equation 5-5

$$f(NH3, N03, P04) = \min\left[\frac{(NH3+NO3)}{(K_{mN}+NH3+NO3)}, \frac{fD3PO4}{(K_{mP}+fD3PO4)]}\right]$$

with the coefficients defined in Table 5.2.

The light extinction function can be expressed as:

Equation 5-6

$$I = I_a(t)\exp[-(0.38 + 0.005S(PHY(Z))*Z)]$$

Where Ia(t) is the light intensity at the water surface, PHY(Z) is expressed in units of Chlorophyll-a, and Z is the depth below the water surface.

The phytoplankton respiration is expressed as:

Equation 5-7

$$D_R = k_{1R}\Theta^{(T-20)}$$

Where k_{1R} is defined in Table 5.2. The death rate, D_D, is often taken as a constant defined as k_{1D} in Table 5.2.

If $a_{3,3}$ is positive, phytoplankton will increase in density, and if negative it will decrease in density. The result of all the processes is that $a_{3,3}$ will vary seasonally and diurnally with light intensity and water temperature as indicated by the above relationships. It is the most complex rate process in the interactive constituent relationships shown in Figure 5.2. As indicated in Figure 5.1 and Figure 5.2, phytoplankton death feeds dissolved and particulate CBOD, ON and OP, and respiration feeds NH_3 and PO_4. At the same time, respiration algal can remove dissolved oxygen. The relationships as expressed here are somewhat over-simplistic in that they assume that phytoplankton growth and respiration can take place at the same time.

Zooplankton Grazing

Different formulations of zooplankton grazing are examined in Edinger, et al. (2003) for their influence on phytoplankton density variations over time. Analytical solutions to the phytoplankton density relationship were developed to illustrate how each grazing relationship would maintain phytoplankton densities over time.

There are a number of different zooplankton grazing rate relationships that have been used in different water quality eutrophication models. The most elementary relationship would be a constant grazing rate. Its formulation of the form:

Equation 5-8

$$G_{zr} = C_{gr}$$

where G_{zr} is the grazing rate in g $m^{-3}day^{-1}$ and C_{gr} is the constant grazing rate with the same units.

A linear grazing rate formulation is presented in EPA (1985) and is an option in the eutrophication submodel of WASP (Ambrose, et al., 1993). It is of the form:

Equation 5-9

$$G_{zr} = k_{lg}\ C_{phy}$$

where k_{lg} is the constant growth rate of zooplankton, day^{-1}; and, C_{phy} is the phytoplankton density. This relationship assumes that the growth of the zooplankton population
is a constant that is proportional to k1g, and the rate at which
phytoplankton are consumed is Gzr

Density dependent grazing rates have recently been reviewed by Gentleman, et al. (2000) for food web systems, but have received little examination for their use in traditional eutrophication types of water quality models. A simple density dependent grazing relation can be formulated as:

Equation 5-10

$$G_{zr} = (k_{dg}\ C_{phy}\ )\ C_{phy}$$

where k_{dg} is the density dependent grazing coefficient, $m^3g^{-1}day^{-1}$. The rational basis of this relation is that the growth of the zooplankton population is proportional to the phytoplankton density or available food, $k_{dg}\ C_{phy}$, and the rate at which phytoplankton are consumed is G_{zr}. Ambrose (2005) indicates that this density dependent grazing rate which is proportional to the square of the phytoplankton density is a heuristic function that may include double counting of the consumption of phytoplankton since the latter are being consumed to increase the zooplankton population and the increased population may not be consuming phytoplankton in proportion to the zooplankton density. However, Ambrose (2005) further suggests that zooplankton grazing my be some power of the phytoplankton density between 1 and 2.

The phytoplankton densities over time based on the choice of grazing relationship can be examined using a simplified form of the phytoplankton component of the water quality model. The change in phytoplankton density with time can be written as:

Equation 5-11

$$\frac{dC_{phy}}{dt} = \left[(1\text{-}fe)G_p(N,P,I) - D_D\ - D_r\right]C_{phy}\ -\ G_{zr}$$

where Gp(N,P,I) is the phytoplankton growth rate that is limited by the concentrations of nitrogen and phosphorous constituents and by light as presented in the previous section. The phytoplankton death rate is D_D, and D_r is the phytoplankton respiration rate.

Letting k=[Gp(N,P,I) – D_D –D_r] for convenience (a3,3 in Figure 5.2), the solution to Equation 5-11 when there are no time varying inputs to the system and C = Co at t=0 for the constant grazing rate given in Equation 5-8 is:

Equation 5-12

$$C(t) = \left(C_o + \frac{C_{gr}}{k}\right) e^{-kt} + \frac{C_{gr}}{k}$$

The solution to Equation 5-11 for the linear grazing rate given in Equation 5-9 is:

Equation 5-13

$$C(t) = C_o \; e^{[-(k-k_{lg})t]}$$

Equation 5-13shows that the phytoplankton density will continually decrease with time if (k-k_{lg}) is positive, and will continually increase with time if (k-klg) is negative. Ambrose (2005) points out that because of real time cycling of nutrients through phytoplankton these extremes are not reached because, as indicated earlier, the growth rate Gp(N,P,I) varies with the nutrients and light, and when C_{phy} builds up, it depletes the N and P, and shades the light (I). This leads to a smaller Gp at a higher C_{phy} reducing their density, which in turn in followed by an increase in the N and P. It would be worthwhile to design a simple nutrient-phytoplankton "tank test" to study the cycling process further.

The solution to Equation 5-11 for the density dependent grazing rate given in Equation 5-10 is:

Equation 5-14

$$C(t) = \left(C_o + \frac{k}{k_{dg}}\right) - \frac{k}{k_{dg}} \frac{e^{-kt}}{\left(1 + e^{-kt}\right)}$$

Equation 5-14 shows that over time the phytoplankton density, the phytoplankton density approaches the value of C_o if k is positive, and approaches the value of (C(t) =C_o - |k|/k_{dg} if k is negative. The density dependent rate relationship has the property that it maintains a stable phytoplankton density through time in comparison to the constant or linear grazing rate formulations.

The density dependent grazing formulation overcomes the singularity that develops in the phytoplankton equation when k=0 during a period when the phytoplankton growth rate, G_p, is equal to the phytoplankton respiration rate and death rate (D_r +D_D), and the tendency for the phytoplankton density to increase or decease over time depending on the difference between G_p and (D_r + D_D). Applications of the density dependent growth formulations in water body examples are illustrated in Edinger (2001).

Plankton Vertical Migration

One of the processes known to produce chlorophyll and DO maxima below the water surface is vertical movement of phytoplankton. Some phytoplankton locate at a position in the water column where the penetrating light intensity is nearly equal to their optimal light saturation value. Some phytoplankton adjust their density to achieve and maintain this location. Vertical migration of dinoflagellate was found to be an important process in maintaining the vertical dissolved oxygen profile in the studies of the Budd Inlet (Edinger, et al 2003) and the details of the formulations used in that study are presented here.

The adaptation in response to light of vertical migration as formulated for example for dinoflagellate and is known as diel vertical migration (DVM). One formulation of DVM has been developed by Kamykowski et al. (1988) given below. This model is based on experiments of the affects of temperature and light on the dinoflagellate swimming speed. The model uses three equations that represent temperature and light dependence, and sinking due to buoyancy using Stokes' law. These equations vectorially add up to the total swimming speed, which is applied in the vertical direction. The temperature (Equation 5-15) and light dependent equations (Equation 5-16) are given below.

Equation 5-15

$$V_T = V_{TMAX}\,(1 - \exp[-b\,\{T - T_L\}])\,\,(1 - \exp[-c\,\{T_H - T\}]) - V_{off}$$

where:

V_T = temperature dependent velocity (m/s)

V_{TMAX} = maximum temperature dependent velocity (m/s)

b, c = empirical constants (b=0.632, c=2.0)

T = temperature (°C)

T_L = lower temperature constant (°C) (T_L=11.68)

T_H = higher temperature constant (°C) (T_H= 33.0)

V_{off} = swimming speed enhancement due to light during experiments (m/s) (V_{off}=0.35e-4)

Equation 5-16

$$V_L = V_{LMAX} \tanh\left[\alpha \, I \, 4.15 / \left(V_{LMAX} \, 10^6\right)\right]$$

where:

V_L = light dependent velocity (m/s)

V_{LMAX} = maximum light dependent velocity (m/s)

α = empirical constant (μm m^2/μEinst) (α = 10.0)

I = light intensity (watts/m^2)

Equation 5-16 includes conversion of the ambient light intensity from μEinst m^{-2} s^{-1} to watts/m^2 which is used in many models. The empirical constants (b, c, α), temperature constants (T_H, T_L), and swimming speed enhancement value (V_{off}) given above are from Kamylkowski and McCollum (1986) and Yamazaki and Kamykowski (1991).

The final equation for the swimming function is:

Equation 5-17

$$V_{TOTAL} = V_L + V_T + V_{\sin king}$$

where the velocities are vertical vectors.

The application logic of the swimming function (Equation 5-17) used to simulate the DVM in response to light splits between daytime and nighttime where the dinoflagellates swim up during daytime with both the light and temperature velocities and swim down at night with the temperature velocity. The sinking velocity is always acting downward.

Laboratory and field studies have shown the dinoflagellates will form subsurface maximum concentrations where the irradiance is optimal for photosynthesis (Cullen and Horrigan, 1981; Robinson and Brown, 1983; Harris et al., 1979; Eppley et al., 1968). In the case where nutrients are depleted throughout the water column, they will migrate to the surface (Cullen and Horrigan, 1981).

Dinoflagellate vertical migration has been incorporated in the JEEAI Generalized Environmental Model Surface Water System (GEMSS) water quality module. GEMSS is described in detail in www.jeeai.com.

Algal Blooms

An important component that is being examined for inclusion in eutrophication models is the production of algal blooms. Most of the studies of algal blooms are being performed under the harmful algal bloom (HAB) projects applied to coastal and estuarine waters, however, the basic questions of what generates algal blooms and how might they be formulated still apply to freshwater lakes and reservoirs.

The present hypothesis for the generation of algal blooms is that cysts and spores in sediments and contained within different types of aquatic vegetation germinate vascular cellular material under the right combination of light, temperature, nutrients and physical stirring can that develop into algae which in turn go through its growth and cellular division processes sometimes at very high rates. Different hypotheses for evaluating algal blooms are presented in Stock, et al (2005). A version of the Stock, et al model is presented in Edinger, et. al. 2003 that allows specifying the the time, duration and intensity of the germination of the vegetative cellular material as a source to phytoplankton resulting in a possible bloom given the right combination of nutrients and light.

The effects of the resulting algal growth and eventual decay on the distribution of the different water quality constituents is examined in Edinger, et. al, 2003 and can be compared to the usual post-bloom types of data available at times of bloom observations. The difficulty in studying algal blooms is having observations of all the antecedent conditions to the bloom, and it is here that modeling results might be helpful.

Sediment Source-Sinks

Figure 5.1 and 5.2 show that the main exchange process between the water column and the sediment in a eutrophication model are the sediment oxygen demand, SOD, and the release of nutrients NH3 and PO4 from the sediment to the water column. In it simplest form these rates are specified from field data. The phytoplankton densities are very sensitive to the release rates of NH3 and PO4. These rates are temperature dependent as indicated by the temperature coefficients in Table 5.2, and also can vary spatially throughout a water body.

The nitrogen and phosphorous is supplied to the sediment from the water column by settling of particulate ON and OP. These constituents mix within the sediment and undergo mineralization to form the NH3 and PO4 that eventually gets released. This process within the sediment is called diagenesis. Diagenesis is often incorporated as a module in an overall hydrodynamic and water quality model and has many different formulations depending on the processes incorporated. The diagenesis models are discussed in more detail in a separate section below.

Future Directions in Eutrophication Modeling

One relatively recent advent in eutrophication modeling is the linkage of models of eutrophication with models of hydrodynamics. In early eutrophication modeling, transport was often described rather than predicted. Recently, the linkage of water quality and hydrodynamic models has become the norm, and has greatly improved our predictive ability, particularly in hydrodynamically complex systems such as reservoirs. For example, Cole and Buchak (1995) indicated the two-dimensional model CE-QUAL-W2, in over 50 applications to reservoirs, produced remarkably accurate temperature predictions using default values of coefficients when provided adequate geometry and boundary conditions. Similar improvements have occurred in the ability to predict water quality variations, which are often directly associated with hydrodynamic transport.

A second recent advent is the inclusion of models of sediment diagenesis with models of water quality variations in the water column, as discussed in the following Section. While not as yet the norm, the use of sediment diagenesis models is becoming increasingly common.

Additional recent advances and future trends in eutrophication modeling include increasing model kinetic complexity, such as through the addition of model state variables and kinetic processes. This trend is somewhat “back to the future” in that many of the early eutrophication models were very complex. The complexity of early models often required inclusion of a large number of uncertain parameters and coefficients. Recent research is showing that many of these rates and parameters depend on different bacterial growth processes that will need to be considered (Philips, et al 2002). Since the uncertainty of model predictions increases non-linearly as the number of uncertain coefficients it contains increases, the usefulness of some of the early complex models was limited. However, increased knowledge and analytical capabilities have allowed modern state-of-the-art models to become more complex and to better represent processes occurring in natural waterbodies. Some of the recent advances and future trends include development of:

- Improved algorithms for predicting the cycling impact of labile and dissolved organic materials;
- Improved algorithms for phytoplankton, allowing for example simulation of seasonal succession;
- Improved algorithms for simulation of zooplankton;
- Improved algorithms for the simulation of periphyton and macrophytes; and

- Improved methods for coupling eutrophication models with those predicting non-point source loads from the watershed and from atmospheric deposition.

An additional recent and future trend is the development of improved methods for predicting higher trophic levels, such as fish. In some cases, prediction of higher trophic levels is necessary because of their impact on eutrophication, either direct or indirect. For example, predators may have a direct impact on zooplankton abundance and composition, thereby impacting algal abundance and composition. In addition, modern eutrophication models typically predict variations in the concentrations of specific chemicals, which are then related to an impact on fisheries through use of chemical-specific water quality criteria as discussed in Chapter 2. Future trends may involve development of aquatic life criteria, necessitating methods for predicting the impact of physical and chemical habitat changes on changes in species abundance and composition.

A major addition to the phytoplankton model would be the generation of algae from their source as spores on aquatic vegetation (Greer and Amsler, 2002), and cysts in the sediment (Bravo and Ramilo, 1998) Presently, the quantitative eutrophication models assume that there is always a very small density of plankton present as "seed", and kinetics related to the drifting plankton such as temperature parameters and light saturation values are varied to get growth to occur at the proper times of year. Godhe (2002a, 2002b, 2003) is examining the dynamics between the planktonic and benthic stages of cyst forming species of dinoflagellates. The studies are examining the factors triggering the planktonic dinoflagellates to produce resting spores and cysts, their length of survival in the sediment, the external environmental conditions causing the cyst to germinate and the role of benthic stages in inducing blooms of dinoflagellates. Investigation of sediment cores showed that dinoflagellate cysts can remain alive in the sediment for several decades. If cysts of toxic species can trigger blooms of toxic species, this is of ecological importance when considering the effects of trawling and dredging, or even storm stirring of shallow sediments. Inclusion of spores and cysts as source terms along with properties for release including temperature, bottom scour, and bottom water mineral chemistry, would be a major step forward in describing the onset of algal formation, and possibly estimates of occurrence of algal blooms.

5.3. <u>Sediment Exchange Processes</u>

One of the processes long been known to impact the water quality of surface waters is the oxygen demand by, and nutrient release from, sediments. Sediment oxygen demand (SOD), due to the mineralization (diagenesis) of organic materials in bottom sediments, can contribute to oxygen declines in waterbodies, such as seasonal declines in hypolimnetic oxygen observed in most reservoirs. Because of their importance, SOD and sediment nutrient releases have been included in essentially all modern numerical water quality models. However, most available models describe, rather than predict, SOD and sediment nutrient releases. For example, the most commonly used approach is to include SOD and sediment nutrient releases as temperature-corrected zeroth order rates (in units of g m^{-2} day^{-1}), so that the rate of change in concentrations in the overlying water are predicted from

Equation 5-18

$$\frac{dC}{dt} = -\frac{R}{H}\Theta^{(T-20)}$$

where

H	=	water depth,
R	=	rate (of SOD or nutrient release),
t	=	time
C	=	concentration in the overlying water,
T	=	temperature
Θ	=	temperature coefficient.

That is the rates of SOD and nutrient release vary only with temperature, and those rates are not impacted by the concentrations of materials in the sediments or the overlying water column. In some models, the rate is further modified by the dissolved oxygen concentration in the overlying water.

The zeroth order rates of SOD and nutrient release have commonly been estimated by model calibration. That is, the rates are adjusted until the observed oxygen and nutrient concentrations are close to those observed. While a commonly used method, this approach is flawed. The approach assumes that all other model processes impacting dissolved oxygen and nutrient concentrations are known with confidence, which is rarely the case. For example, inadequate predictions of reservoir hydrodynamics alone can result in rates determined by model calibration being highly uncertain.

A second commonly used, and preferable, method for the determination of the rates is through direct measurement. Methods developed, such as chamber and core methods, for measuring SOD and nutrient releases have been successfully applied in a variety of studies. However, the costs involved and difficulty in obtaining direct measurements often limits the number of such measurements available for use. A second limitation is often a lack of correlation of measured rates to other measurable sediment characteristics, making the interpolation between limited samples over a waterbody problematic.

The use of a zeroth order rate, or constant source term, in water quality model applications has a major additional limitation. That is it does not provide for a mechanistic link between sediment organic matter and its conversion into oxygen demand and nutrient release. In the absence of this missing link, one of two alternative approaches have commonly been employed. The first, and most commonly used approach has been to assume the rates of SOD and nutrient release are unchanged following waste load reductions, or implementation of other water quality management alternatives. Clearly, this should not be the case in that reduction in loads of organic matter, and commonly changes the characteristics of the organic materials, to a waterbody should consequently change organic loads to sediments and the resulting sediment oxygen demand and nutrient release. However, the "no change" approach has often been justified as being a "conservative" assumption. An alternative approach has been to assume linearity and lower SOD in direct proportion to a load reduction (Chapra 1997). However, there are rarely sufficient data available to estimate the magnitude and rate for the reduction.

The preferable approach would be the utilization of mechanistic models that provide a link between the influxes of organic materials to sediments, physical, biological and chemical processes occurring within the sediments, and consequent sediment oxygen demand and nutrient release. A number of approaches have been attempted. One limitation, as described by Chapra (1997) has been that some of the early and more simplified approaches have tended to overestimate the sediment oxygen demand and fail to capture the non-linear (e.g., Fair et al. 1941), or "square root" relationship, between loads and oxygen demand observed in natural waters.

In a landmark paper, Di Toro et al. (1990) developed a model of the SOD that mechanistically arrives at the observed non-linear relationships (Chapra 1997). Di Toro's approach calculates sediment oxygen demand and phosphorus and nitrogen release as functions of the downward flux of carbon, nitrogen, and phosphorus from the water column. This approach, well founded in diagenetic theory and supported by field and laboratory measurements, was an important advancement in the field of sediment-water interactions and has been integrated into the CE-QUAL-ICM model (Cerco and Cole 1995) and successfully applied to Chesapeake Bay (Cerco and Cole 1991, 1992, 1993; Cerco 1995). The CE-QUAL-ICM model, including the sediment diagenesis model, has subsequently been applied to a wide variety of other systems. The development and testing of the sediment diagenesis model was further described by Di Toro and Fitzpatrick (1993). A similar model for manganese and iron fluxes from the sediments was described by Di Toro et al. (1994). While other approaches to modeling sediment diagenesis have been reported and applied, this section will concentrate on the structure of the sediment diagenesis model as incorporated into CE-QUAL-ICM. A more detailed discussion of the processes and modeling approaches for sediment diagenesis can be found in Di Toro (2001).

Model Framework

The basic framework of the sediment model as described by Di Toro and Fitzpatrick (1993) and included in the CE-QUAL-ICM model (Cerco and Cole 1995) consists of two well-mixed sediment layers, underlying each surface water column segment: a thin upper sediment layer (the aerobic layer, on the order of 0.1 cm thick) and a thicker active layer (anaerobic, on the order of 10 cm thick; Di Toro 2001, Figure 5.3). Three major processes included in the sediment model are the (Figure 5.3):

- Fluxes of particulate organic matter from the water column to the sediments (note that since the upper sediment layer is assumed to have a negligible thickness, the fluxes are deposited directly into the second, or anaerobic layer),
- Mineralization (or diagenesis) of the particulate organic matter, and
- Reactions and transfers (between sediment layers, to the water column and deep inactive sediments) of the reaction products.

The sediments receive fluxes from the water column of particulate organic carbon (POC), nitrogen (PON), and phosphorus (POP), collectively referred to as particulate organic matter (POM). These forms of POM in the water column are represented by, in the CE-QUAL-ICM model (Cerco and Cole 1995), three forms of algae, labile and refractory POC, labile and refractory PON, and labile and refractory POP. The flux from the water column to the sediment is computed as the product of a net settling velocity and the water column particulate organic matter concentration.

Organic matter initially mineralizes rapidly in the sediments, but then slows down. In order to capture this process, the settled POM is fractionated to one of three "G classes" (Figure 5.3, Di Toro 2001). The three G classes represent a relatively rapidly mineralizing labile class (G1), a more refractory form (G2) and a relatively inert form (G3).

Diagenesis (mineralization) reactions occur in the second (anaerobic) sediment layer. The diagenesis equations solved for each form of particulate organic matter (N, P and C) and for each G class (1-3) are analogous. In order to compute the time-varying diagenesis for each modeled variable, a mass balance equation is written for each form and type of POM (e.g. POC, PON and POP in each G class) as

Equation 5-19

$$\frac{\partial C_{POM,2}}{\partial t} = \frac{J_2}{H_2} - \frac{W_2}{H_2} C_{POM,2}^{t+\Delta t} - K_2 \Theta^{T-20} C_{POM,2}^{t+\Delta t} \approx \frac{C_{POM,2}^{t+\Delta t} - C_{POM,2}^{t}}{\Delta t}$$

where

$C_{POM,2}^{t+\Delta t}$ = total concentration of POM type in layer 2 at time t+Δt

$C_{POM,2}^{t}$ = total concentration of POM type in layer 2 at time t (from initial conditions or computed value from previous time step)

Δt = time step

J_2 = flux from the water column,

H_2 = thickness of the active sediment layer,

K_2 = reaction velocity (specific to chemical and G class), and

W_2 = net sedimentation velocity

The mass balance equation is conveniently solved algebraically for the present concentration at the present time step, as

Equation 5-20

$$C_{POM,2}^{t+\Delta t} = \frac{J_2 \frac{\Delta t}{H_2} + C_{POM,2}^{t}}{1 + K_2 H_2 \frac{\Delta t}{H_2} + W_2 \frac{\Delta t}{H_2}}$$

Once the concentrations at the present time step are computed, the diagenesis source terms for reactions and transfers ($J_{T2}^{t+\Delta t}$) are computed. Diagenesis source terms are computed for C, N and P from the sum or the product of the chemical specific reaction velocities (K_2) and computed concentrations in each of the three G classes. For example,

Equation 5-21

$$J_{T2}^{t+\Delta t} = \sum_{i=1}^{3} K_{2.i} \Theta^{T-20} C_{POM,2,i}^{t+\Delta t}$$

where

$J_{T2}^{t+\Delta t}$ = source term for total chemical (C, N or P) in layer 2 at time t+Δt

K2,i = reaction velocity for total chemical in G class i

$C_{T2,i}^{t+\Delta t}$ = total chemical concentration for G class i

Once the sediment particulate organic matter (C, N and P) concentrations and source terms are computed for the present time step, as described above, the reactions and transfers are computed. Concentrations or ammonia, nitrates, methane, sulfates, sulfides, silica and phosphorus are computed, and then used to compute fluxes to the overlying water column, including sediment oxygen demand.

The total chemical concentrations are computed from mass balance relationships for each of the two sediment layers. Since the surface layer is thin compared to the active anaerobic layer, it can be assumed that layer 1 is at steady-state in comparison to the slower processes occurring in layer 2. Mass balance equations can then be written for the two layers as (DiToro 2001) as

Layer 1

Equation 5-22

$$\mathbf{0} = s\left(f_{d1}C_{T1}^{t+\Delta t} - C_{dO}^{t+\Delta t}\right) + \omega_{12}\left(f_{p2}C_{T2}^{t+\Delta t} - f_{p1}C_{T1}^{t+\Delta t}\right) +$$

$$K_{L12}\left(f_{d2}C_{T2}^{t+\Delta t} - f_{d1}C_{T1}^{t+\Delta t}\right) - \omega_2 C_{T1}^{t+\Delta t} - f_c f_o\left(f_d\frac{\kappa_{1,d}^2}{s} + f_p\frac{\kappa_{1,p}^2}{s}\right)C_{T1}^{t+\Delta t} + J_{T1}^{t+\Delta t}$$

Layer 2

Equation 5-23

$$\mathbf{0} = -\omega_{12}\left(f_{p2}C_{T2}^{t+\Delta t} - f_{p1}C_{T1}^{t+\Delta t}\right) - K_{L12}\left(f_{d2}C_{T2}^{t+\Delta t} - f_{d1}C_{T1}^{t+\Delta t}\right) - \kappa_2 C_{T2}^{t+\Delta t} + \omega_2(C_{T1}^{t+\Delta t} - C_{T2}^{t+\Delta t}) - \frac{H_2C}{\Delta}$$

$$+J_{T2}^{t+\Delta t} + \frac{H_2C_{T2}^{t}}{\Delta t}$$

where

s = surface transfer rate

f_c = reaction concentration dependence reduction (monod formulation)

f_o = reaction oxygen dependency (monod formulation)

f_{d1} = fraction dissolved in layer 1

f_{d2} = fraction dissolved in layer 2

f_{p1} = fraction particulate in layer 1

f_{p2} = fraction particulate in layer 2

$C_{T1}^{t+\Delta t}$ = total concentration in layer 1 at time t+Δt

$C_{T2}^{t+\Delta t}$ = total concentration in layer 2 at time t+Δt

C_{T2}^{t} = total concentration in layer 2 at time t

$C_{dO}^{t+\Delta t}$ = dissolved concentration in overlying water column

K_{L12} = dissolved phase transfer coefficient between layers 1 and 2

ω_{12} = particle mixing coefficient between layers 1 and 2

ω_2 = sedimentation velocity for layer 2

$J_{T1}^{t+\Delta t}$ = source term for total chemical in layer 1 at time t+Δt

$J_{T2}^{t+\Delta t}$ = source term for total chemical in layer 2 at time t+Δt

κ_1^2 = square of reaction velocity in layer 1 for dissolved (d) or particulate(p) phase

κ_2 = reaction velocity in layer 2

H_2 = thickness of layer 2

Δt = time step

where it is assumed that the thickness of the layers is constant and the fraction dissolved and particulate in the two layers are computed from:

Equation 5-24

$$f_{d,1} = \frac{1}{1+\pi_{C1}S_1};\ f_{p,1} = \frac{\pi_{C1}S_1}{1+\pi_{C1}S_1}$$

$$f_{d,2} = \frac{1}{1+\pi_{C2}S_2};\ f_{p,2} = \frac{\pi_{C2}S_2}{1+\pi_{C2}S_2}$$

where

π_{C1} = partition coefficient for total chemical in layer 1

π_{C2} = partition coefficient for total chemical in layer 2

S1 = solids concentration in layer 1

S2 = solids concentration in layer 2

The equations are conveniently solved for the new concentrations ($C_{T1}^{t+\Delta t}$ and $C_{T2}^{t+\Delta t}$) using a matrix solution. Once the concentrations have been updated, the flux of the material to the overlying water column (J) can be computed from:

Equation 5-25

$$J = s\left(f_{d1}C_{T1}^{t+\Delta t} - C_{dO}^{t+\Delta t}\right)$$

The surface transfer rate (s) quantifies the mixing between layer 1 and the overlying water column, which can be related to sediment oxygen demand by (Di Toro 2001)

Equation 5-26

$$SOD = D\left.\frac{d[O_2]}{dz}\right|_{z=0} \approx D\frac{[O_2(0) - O_2(H_1)]}{H_1} = \frac{D}{H_1}[O_2(0)]$$

where

D = rate of oxygen diffusion

$O_2(0)$ = oxygen concentration of the overlying water, and

$O_2(H_1)$= oxygen concentration at the depth H_1

assuming a straight line approximation of the derivative, so that the mass transfer coefficient ($K_{L,O2}$) may be estimated from (Di Toro 2001)

Equation 5-27

$$K_{L,O2} = \frac{D}{H_1} = \frac{SOD}{[O_2(o)]} = s$$

The reaction rate in the aerobic layer is formulated as a first order rate (K_1), where the term in the layer 1 equation is K_1H_1. The depth of the aerobic zone follows the definition of the surface mass transfer coefficient ($s=D/H_1$) so that $K_1H_1=K_1D_1/s$ so that

Equation 5-28

$$\kappa_1 = \sqrt{DK_1}$$

and

Equation 5-29

$$K_1 H_1 = \frac{\kappa_1^2}{s}$$

which is the term applied to the total chemical in the equation for layer 1.

The rate of mixing of the sediment by macrobenthos (bioturbation, w_{12}) is estimated by an apparent particle diffusion coefficient (D_p), temperature corrected that varies with the biomass of the benthos. Assuming that the mass of the benthos is proportional to the labile carbon in the sediment ($C^t_{POC,2,1}$, or POC in layer 2 in G class 1),

Equation 5-30

$$w^*_{12} = D_P \frac{\Theta^{(T-20)}}{H_2} C^t_{POC,2,1}$$

where w^*_{12} is a particle mixing coefficient that is further modified as discussed below.

An additional impact is that if anoxia occurs for periods of time, the benthic population is ultimately reduced or eliminated, so that bioturbuation is consequently reduced or eliminated. To include this effect, Di Toro (2001) computes the stress that low dissolved oxygen conditions (benthic stress, S) imposes on the population assuming that the stress accumulates as

Equation 5-31

$$\frac{\partial S}{\partial t} = -k_s S^{t+\Delta t} + \frac{K_{M,D_P}}{K_{M,D_P} + [O_2(0)]} \approx \frac{S^{t+\Delta t} - S^t}{\Delta t}$$

where

k_s = decay constant for benthic stress,

$K_{M,Dp}$ = particle mixing half-saturation concentration for oxygen

which can be solved for

Equation 5-32

$$S^{t+\Delta t} = \frac{S^t + \dfrac{K_{M,D_P}}{K_{M,D_P} + [O_2(0)]}}{1 + k_s \Delta t}$$

As $[O_2(0)]$ approaches zero, then $(1-k_s S)$ approaches zero, so that the particle mixing coefficient is similarly reduced, as

Equation 5-33

$$w_{12} = w_{12}^{*}\left(1 - k_s S^{t+\Delta t}\right) + \frac{D_{p,\min}}{H_2}$$

where a minimum rate of particle mixing ($D_{p,min}$) is maintained under anoxic conditions. The stress is continued at the minimum value for the year to conform with the observation that once the benthic population has been reduced by low dissolved oxygen, it does not recover until the next year (Di Toro 2001).

The dissolved phase mixing coefficient between layers 1 and 2 (K_{L12}) is due to passive molecular diffusion that is enhanced by the action of organisms (bio-irrigation). The mixing coefficient is computed from

Equation 5-34

$$K_{L12} = \frac{D_P}{H_2}\Theta^{(T-20)} + K_{L,B}w_{12}$$

where

D_P = pore-water diffusion coefficient

$K_{L,B}$ = ratio of bio-irrigation to bio-particle mixing.

<u>Computation of SOD and Sediment Release Rates</u>

Equation 5-22 and Equation 5-23, with chemical specific rate values, is applied to predict concentrations of ammonia, nitrate-nitrogen, sulfides, sulfate, methane, phosphates and dissolved silica. Given the chemical concentrations in layer 1, the SOD and release rates are then computed. However, with the exception of phosphates and silica, each of these constituents affects SOD, which in turn affects the surface transfer rate (s, Equation 5-27). Therefore, an iterative solution is required. The procedure employed in the sediment diagenesis model is as follows:

- Start with an initial estimate of the SOD.
- Solve layer 1 and 2 equations (for ammonia, nitrate, sulfate, sulfide, and methane).
- Refine the estimate of SOD. A root finding method is used to make the new estimate.

- Go to step (2) if no convergence.

Once the SOD is determined, then the layer 1 and 2 equations for phosphate and silica are solved and the flux rates determined.

Ammonia

For ammonia, Equation 5-22 and Equation 5-23 are solved with constituent specific rate terms and coefficients for the ammonia concentration in layers 1 and 2, with the source term for layer 2 being the production of organic nitrogen from the diagenesis reactions. Once the ammonia concentrations are known, the rate of flux to the overlying water column is computed from Equation 5-25. In order to compute the oxygen consumption due to the oxidation of ammonia in the aerobic layer, the reaction can be represented by (Chapra 1997, Di Toro 2001)

Equation 5-35

$$NH_3 + 2O_2 \rightarrow HNO_3 + H_2O$$

so that the consumption of oxygen during the process can be represented by

Equation 5-36

$$a_{no} = \frac{2(32)}{14} = 4.57 \quad gO \quad gN^{-1}$$

Therefore the contribution of the oxidation of ammonia to SOD can be estimated from

Equation 5-37

$$J_{O_2,NH_4} = a_{no} \frac{\kappa^2_{NH4,1}}{s} f_{d1} f_O f_{NH4} \, C^{t+\Delta t}_{NH_4,1}$$

The modification of the nitrification reaction for dissolved oxygen is computed from

Equation 5-38

$$f_O = \frac{O_{2,0}}{O_{2,0} + K_{NH4,O2}}$$

where

$O_{2,0}$ = dissolved oxygen concentration in the overlying water column (O20), and

$K_{NH4,O2}$= half-saturation concentration of dissolved oxygen in the nitrification reaction (KMNH4O2)

The concentration dependence of the reaction is given by

Equation 5-39

$$f_{NH4} = \frac{K_{NH4}}{C^{t}_{NH4,1} + K_{NH4}}$$

where

$C^{t}_{NH4,1}$ = ammonia concentration from the previous time step (C1TM1S), and

K_{NH4} = half-saturation concentration of ammonia in the nitrification reaction

Nitrate

For nitrate, Equation 5-22 and Equation 5-23 are solved with constituent specific rate terms and coefficients for the nitrate concentration in layers 1 and 2, with the source term for layer 1 being the oxidation of ammonia, and the source term for layer 2 being zero. Once the nitrate concentrations are known, the rate of flux to the overlying water column is computed from Equation 5-25.

The process of denitrfication requires a carbon source as indicated by (Di Toro 2001)

Equation 5-40

$$\frac{10}{8}CH_2O + H^{+} + NO_3^{-} \rightarrow \frac{10}{8}CO_2 + \frac{1}{2}N_2 + \frac{7}{4}H_2O$$

so that the carbon to nitrogen stoichiometric coefficient (a_{cn}) is 1.071 gC gN^{-1}. The oxidation of methane in the aerobic zone may be represented by

Equation 5-41

$$\frac{1}{2}CH_4 + O_2 \rightarrow \frac{1}{2}CO_2 + H_2O$$

so the oxygen to carbon stoichiometric coefficient (a_{oc}) is 2.67 g O g C^{-1}. If all of the carbon produced by the diagenesis reaction was converted to methane and fully oxidized, the maximum SOD that could be produced would be

Equation 5-42

$$J_{O_2,C} = a_{oc} J_{C,2}^{t+\Delta t}$$

However, this maximum is reduced by the carbon utilized during denitrification, so that the maximum oxygen utilization become

Equation 5-43

$$J_{O_2,C} = a_{o,c} J_{C,2}^{t+\Delta t} - a_{oc} a_{cn} \kappa_{NO3,2} \Theta^{T-20} C_{NO3,2}^{t+\Delta t}$$

where $\kappa_{NO3,2}$ is the product of the denitrification reaction rate (K_{NO3}) and the layer thickness (H_2), and $a_{oc}a_{cn}$ is 2.857.

Sulfides and Sulfates

The approach for modeling sulfide reactions varies between the steady-state model described by Di Toro (2001), the initial modeling approach incorporated in the CE-QUAL-ICM model (Di Toro and Fitzpatrick 1993, Cerco and Cole 1995) and that employed incorporated in code revisions during 1997. In the later, equations of the form of Equation 5-22 and Equation 5-23 are written for sulfate and sulfide in layers 1 and 2, for a total of four equations which are solved simultaneously. The source term for layer 2 is the carbon flux converted to oxygen equivalents and reduced by denitrification, as given by Equation 5-43. An additional refinement is that, based upon a continuous derivation, the depth of sulfate reduction (H_{SO4}) is computed which is then used to modify the rate of exchange between layers 1 and 2 ($K_{L12,SO4}=K_{L12}\,H_2/H_{SO4}$, Di Toro 2001). The loss term for the oxidation of sulfides in layer 1 is a source term for the sulfates. Once the sulfide concentrations are known, the rate of flux to the overlying water column is computed from Equation 5-25.

The sulfide and sulfate equations are written in terms of oxygen equivalenets, so the stoichiometric relationship between sulfide oxidation and oxygen consumption is unity. The oxygen demand due to sulfide oxidation can then be computed from

Equation 5-44

$$J_{O_2,HS} = f_O \left[\frac{\kappa_{D,H2S,1}^2}{s} f_{d1} + \frac{\kappa_{P,H2S,1}^2}{s} f_{p1} \right] C_{H2S,1}^{t+\Delta t}$$

Methane

As with sulfides, the modeling approach for methane varies between the steady-state model described by Di Toro (2001), the initial modeling approach incorporated in the CE-QUAL-ICM model (Di Toro and Fitzpatrick 1993,

Cerco and Cole 1995) and that incorporated in code revisions during 1997. In the later, equations of the form of Equation 5-22 and Equation 5-23 are written for methane in layers 1 and 2. The source term for methane in layer two is the carbon diagenesis flux converted to oxygen equivalents and corrected by denitrification, as given by Equation 5-43, and further reduced by a relationship for the carbon diagenesis flux available for methane formation. Once the methane concentrations are known, the rate of flux to the overlying water column is computed from Equation 5-25.

The methane equations are written in terms of oxygen equivalents, so the stoichiometric relationship between methane oxidation and oxygen consumption in the aerobic layer is unity. The oxygen demand due to methane oxidation can then be computed from

Equation 5-45

$$J_{O_2,CH4} = f_O \frac{\kappa^2_{CH4,1}}{s} C^{t+\Delta t}_{CH4,1}$$

SOD

Once the concentrations of materials affecting oxygen are computed, and the stoichiometric relationships described above applied, the SOD is computed from

Equation 5-46

$$SOD = J_{O_2,NH_4} + J_{O_2,HS} + J_{O_2,CH4}$$

this estimate is compared to that from the previous iteration, and as discussed above, if it differs by more than a specified amount, a new value of s is computed and the solution iterated.

Phosphate and Silica

Once the SOD has been determined, and the sediment of the surface exchange rate (s) made, then the concentrations of phosphate and silica in the two layers are determined. For ammonia and phosphate, equations of the form of Equation 5-22 and Equation 5-23 are solved with constituent specific rate terms and coefficients for the phosphate or silica concentrations in layers 1 and 2. The source term for layer 2 for phosphate is the phosphate source from diagenesis reactions plus the flux of particulate inorganic phosphorus from the water column. For silica, the source term for layer 2 is the dissolution reaction and the flux of silica from the water column. Once the silica and phosphate concentrations are known, the rates of flux to the overlying water column are computed from Equation 5-25.

Water Quality Linkage

The output from the sediment diagenesis model includes SOD and fluxes of ammonia, nitrate, phosphate and silica to the overlying water column. The equations in the water quality model for linkage to a model of sediment diagenesis are analogous to those for the zeroth order reactions (Equation 5-18). That is, using the zeroth order approach, the water quality model equations for oxygen, ammonia, nitrate, phosphate and silica contain source/sink terms for the affect of sediments. However, with the coupling of the water quality and sediment diagenesis models those terms are now predicted, rather than described. This coupling of water column and sediment models provides a long-needed predictive capability, a way to link external loads with sediments and their consequent impact on water quality.

5.4. Toxics modeling

Eutrophication modeling, as described in the preceding section, typically deals with the natural cycle of organic production and decomposition, and often the over-stimulation of these natural processes causing problems such as excessive plant growth or low dissolved oxygen concentrations (Chapra 1997). While specific models of the eutrophication may vary in complexity, typically the suite of organisms (e.g. algae), chemicals (e.g. nutrients, dissolved oxygen, etc.), and kinetic processes included in those models are somewhat similar. For example, models of eutrophication deal with on the order of 10 specific "pollutants" (Chapra 1997).

In contrast, the suite of chemicals regarded as toxic is large and the chemical characteristics and environmental impact of those chemicals varies widely. Chemicals regarded as toxic include a number of organic, inorganic and organometallic chemicals. The large number and variety of organic compounds include such major classes as pesticides, polychlorinated biphenyls, halogenated aliphatic hydrocarbons, halogenated ethers, monocyclic aromatics, phthalate esters, polycyclic aromatic hydrocarbons, and nitrosamines. As an example, the U.S. Environmental Protection Agency (EPA) lists over 100 "priority pollutants," including inorganic and organic toxicants and provides rankings of those chemicals based upon their potential exposure and effects on human health and aquatic life. The EPA ECOTOX (ECOTOXicology) database provides single chemical toxicity information for aquatic and terrestrial life for a wide variety of chemicals. The EPA's Toxics Release Inventory (TRI) contains information on over 650 toxic chemicals, their releases into the environment, and other waste management activities reported annually by certain covered industry groups as well as federal facilities. The EPA "Title III Consolidated List of Lists," a consolidated list of chemicals subject to reporting requirements, includes over 1700 chemicals. Many of these toxic pollutants occur naturally, but may result in toxic affects (chronic or acute) if in excess (e.g. nitrates and unionized ammonia). In modern technological societies, synthetic chemicals have been manufactured, used, and disposed of in large quantities. In contrast to "conventional pollutants" such as those included in models of eutrophication, many of these synthetic chemicals do not occur naturally in the environment.

As a result of the number and diversity in the chemicals involved, and factors affecting their fate and transport, models of toxic pollutants typically differ in structure from those designed for conventional pollutants (e.g. eutrophication). One typical difference is that models of toxic pollutants are often much more general in structure. That is these models (such as the Water Analysis Simulation Program, Ambrose et al. 1993a,b) typically include a variety of transformation and degradation processes. Then through selection of the appropriate processes and specification of chemical specific rates, the models may be applied to the specific chemical(s) of interest.

Toxic chemical models also typically differ between those designed to simulate organic and inorganic contaminants. The factors affecting many metal and organic toxicants are sufficiently different that a single model is rarely applicable to both unless a number of simplifying assumptions are made. For example, for toxic organics the processes impacting their fate and transport differ from those impacting metals. In addition, the degree of interaction between organic compounds is such that typical modeling studies of organic chemicals consider only one to a few chemicals. However, the environmental system behavior and toxicity of metals is governed by "simultaneous and interacting advancement of all possible chemical reactions, each controlled by its individual rate law (Morel, 1983)." The distinction between models of organic and inorganic toxicants is the basis for the organization of this Section of Chapter 6.

Organic chemicals and metals can enter the aquatic environment by various pathways, including point source waste discharges and nonpoint source runoff. Atmospheric deposition is a significant source of some toxic chemicals, such as mercury (U.S. EPA 1997). A number of organic compounds, such as some metals and organics strongly sorb to sediments. As a result, many hydrophobic organic materials such as Polychlorinated Biphenyls (PCBs) and PAH compounds are highly persistent in the environment. Modeling of these chemicals and the sediments to which they sorb over long time periods (decades) is typically required to evaluate their impact. Some organic compounds may bioconcentrate through the food chain where increased concentrations cause adverse impacts. Some organic and inorganic chemicals can cause toxicity to aquatic organisms, having a direct adverse effect on aquatic life by making certain areas uninhabitable. Organic compounds may bioconcentrate through the food chain where increased concentrations cause adverse impacts. For example, the presence of sediments contaminated with polycyclic aromatic hydrocarbons (PAHs) may result in the development of cancerous tumors in fish. Humans and wildlife may be affected by ingesting contaminated water or fish. The bioaccumulation of toxic materials in fish has resulted in consumption advisories in many areas. The prohibition or restriction of consumption of fish also has a negative economic impact on the effected areas (USEPA 1993).

A typical modeling goal for toxic chemicals is to evaluate the impacts of various management alternatives on the exposure of human and aquatic life to those chemicals and their resulting toxicity. To accomplish this and include the sources and impacts described above, complete models of toxic chemicals typically include: hydrodynamic fate and transport, sediment transport, chemical fate and transport, and bioaccumulation and toxicity. However, a number of these processes and relationships between them are poorly understood. In addition, modern models are typically designed to simulate chemical concentrations in the environment. But, environmental impact is often indicated by the presence of toxicity, where the specific causal factors for that toxicity are not fully known. Without specific relationships between chemical specific criterion and modeled concentrations, the application of toxic chemical models in the environmental management of toxic chemicals is problematic. Some of the methodologies commonly used in linking measured or predicted concentrations and toxic impact are discussed in the following section.

Risk and Hazard Assessments

Contamination by toxic materials is of concern primarily due to the potential risk it poses to humans and wildlife. Therefore, management of contamination by toxic chemicals includes risk management (USEPA 1993).

Humans may be exposed to contaminants through a variety of activities that result in the intake of contaminants through dermal, ingestion, and/or inhalation pathways. Risk may include non-carcinogenic (i.e. chronic or sub-chronic effects) and carcinogenic (i.e. probability of an individual developing cancer over a lifetime) effects over a range of exposure scenarios. For non-carcinogenic contaminants, typically a safe daily dose or reference dose is used to estimate risk. For carcinogenic chemicals, typically a cancer slope factor (e.g. mg/kg-day) is used along with estimates of uptake rates to determine risk. Reference doses and cancer slope factors have been estimated, and are available, for a wide variety of contaminants.

Guidance and methods for ecological risk assessments are not as well developed as methods of human health risk. Assessments of the impacts of contaminants often rely on toxicity testing and models to extrapolate data to different organisms or life stages. In some cases community impacts (i.e. benthic species diversity) or indirect population based impacts (such as toxicity tests on individual species) are used in an ecological risk assessment. The toxicity data may be site-specific or based upon experimental data. For example, experimental data has been used to estimate No Observable Adverse Effect Levels (NOAEL) or Lowest Observable Adverse Effect Levels (LOAEL) for some chemicals and organisms.

A general approach to risk assessment includes a) the identification of existing risks at sites with contamination, b) estimating the potential impact of various remedial alternatives on contaminant concentrations in various media and their associated risk, and c) the comparison of existing and potential risks in order to aid in the selection of remedial alternatives. General steps in the process include (EPA 1993):

- An initial screening of areas of concern to identify sites that may pose a potential threat to human or ecological health
- Development of a remedial action plan, which characterizes the risk, identifies potential remedial actions, and provides estimates of resources needed to pursue the remaining elements of the risk management process.
- Development of a baseline risk and hazard evaluation, which estimates potential human and ecological risks resulting from direct or indirect exposure to the contaminants for existing conditions
- Field sampling to provide for a detailed assessment of the site
- The ranking of information on hazards, risks and costs to prioritize specific sites.
- Selection of remedial alternatives to be investigated
- Analysis of contaminated fate and transport related to the remedial alternatives, which provides estimates of pollutant releases from the various components of a remedial alternative during and following the implementation of that alternative. Typically in this step, data and models describing the chemical/physical and biological characteristics of the area are combined to evaluate the effects of remediation and changes over time,
- A comparative risk assessment to estimate increases or decreases (relative to the baseline) that would result in implementation of the remedial alternatives,
- Selection and implementation of the remedial alternative.
- Post remediation monitoring.

Contamination by toxic metals occurs in a variety of aquatic habitats. Aquatic resource management strategies, including the assessment of waste allocation, potential remedial action (restoration) effectiveness or to assess the potential consequences of contaminant loadings (Willingham and Medine, 1989), requires an estimation of the impacts of processes affecting metals concentrations in the water, sediments and biota. These types of analyses are complicated by the fact that metal behavior is non-conservative in aquatic systems and that the transport, transformations, and attenuation depend upon the particular forms of metal present in the system.

Modeling Organic Chemicals

Several environmental processes can affect the transport and fate of organic chemicals in the aquatic environment. The most important include physical processes such as hydrophobic sorption, volatilization, and sedimentation; chemical processes such as ionization, precipitation, dissolution, hydrolysis, photolysis, oxidation and reduction; and biological processes such as biodegradation and bioconcentration (Ambrose et al. 1993). In organic chemical models typically sorption and ionization are treated as equilibrium reactions. All other processes are typically described by rate equations. Rate equations may be quantified by first-order constants or by second-order chemical specific constants and environment-specific parameters that may vary in space and time. Available organic chemical models have recently been reviewed by Imhoff et al. (2003).

Sorption

Sorption is the process whereby a dissolved substance is transferred to and becomes associated with a solids material (Chapra 1997). Sorption is an important process controlling the environmental fate and the toxicity of chemicals. For example, sorption may cause the chemical to accumulate in bed sediment or bioconcentrate in fish. Since sorption reduces the fraction of the chemical in the dissolved form, sorption may retard such reactions as volatilization and base hydrolysis (Ambrose et al. 1993).

Sorption to solids is usually described using a linear partition coefficient,

Equation 5-47

$$C_{s'} = K_{ps} \bullet C_{w'}$$

where $C_{s'}$ is the concentration in the sorbed form (e.g. mg/kg), K_{ps} is the partition coefficient (L/kg), and C_w the concentration in the dissolved form (mg/L). Typical assumptions on which the above formulation is based include:

- The assumption that sorption reactions are fast relative to other environmental processes so that the sorption reaction can be considered to be in equilibrium.

- The assumption that the concentrations are low so that the partitioning relationship is linear. This is typically the case for environmentally relevant concentrations (less than 10^{-5} M or one-half water solubility; (Karickhoff, 1984)

- There is no hysteresis, or the rates of sorption and desorption are equal. Laboratory data for very hydrophobic chemicals suggest, however, that a hysteresis exists, with desorption being a much slower process than adsorption. This phenomenon is not well understood and no quantitative modeling framework is available to characterize it (Ambrose et al. 1993a,b).

At equilibrium, then, the distribution among the phases is controlled by the partition coefficient K_{ps} so that the fraction of chemical in the dissolved (f_D) and particulate (f_s) form may be estimated from

Equation 5-48

$$f_D = \frac{n}{n + \sum_s K_{ps} \bullet M_s}$$

and

Equation 5-49

$$f_s = \frac{K_{ps} \bullet M_s}{n + \sum_s K_{ps} \bullet M_s}$$

where M_s is the mass of the solid and n the porosity of the sediment or water layer.

Given the total concentration (C_T) and the phase fractions of the chemical, the dissolved and sorbed concentrations are uniquely determined:

Equation 5-50

$$C_w = C_T \bullet f_D$$

Equation 5-51

$$C_s = C_T \bullet f_s$$

The processes impacting sorption can vary between chemicals, and may be associated with the surface characteristics of the solid, such as the surface charge. For many hydrophobic organic materials, the chemical often becomes associated with the organic matter in the particulate phase, such as such as benthic and suspended sediment, biological material, and sometimes dissolved or colloidal organic material. For non-polar organic chemicals, such as PCBs, the partition coefficient is commonly assumed a function of the organic content of the solids, as in (Chapra 1997)

Equation 5-52

$$K_{ps} = f_{oc} K_{oc}$$

where f_{oc} is the weight fraction of the total carbon in the particulate matter and K_{oc} is the organic carbon partition coefficient (L/kg). The f_{oc} is increasingly a commonly measured environmental parameter. The K_{oc} is commonly estimated from the octanol-water partition coefficient (K_{ow}, L_w/L_{oct}), such as from (Karickoff et al. 1984)

Equation 5-53

$$\log(K_{oc}) = a_o + a_1 \log(K_{ow})$$

where typical values for a_o and a_1 are 0.6 and 1.0, respectively (Ambrose et al. 1993). The K_{ow} may be obtained from field studies, published values, or published formulations (Chapra 1997).

Transformations and Daughter Products

Some organic chemicals may be treated individually. However, for some cases it may be necessary to simulate both the parent chemical and reaction (daughter) products in order to assess environmental impacts. Linked transformations are usually implemented by simulating two or three chemicals and by specifying appropriate yield coefficients for each process, such as give by for three chemicals (Ambrose et al. 1993):

Equation 5-54

$$S_{kc1} = \sum_c \sum_k K_{kc} C_c Y_{kc1}, \quad c = 2, 3$$

Equation 5-55

$$S_{kc2} = \sum_c \sum_k K_{kc} C_c Y_{kc2}, \quad c = 1, 3$$

Equation 5-56

$$S_{kc3} = \sum_c \sum_k K_{kc} C_c Y_{kc3}, \quad c = 1, 2$$

where:

S_{kci} = production of chemical "i" from chemical "c" undergoing reaction "k," mg_i/L-day

K_{kc} = effective rate coefficient for chemical "c," reaction "k," day^{-1}

Y_{kci} = yield coefficients for production of chemical "i" from chemical "c" undergoing reaction "k," mg_i/mg_c

An example where the formation of daughter products was important was in the simulation of the 1991 metam-sodium spill into the Sacramento River in Northern California in 1991. Metam-sodium is used as a soil fumigant, and its use depends upon its decomposition to give the toxic and volatile product MITC. Therefore, simulation of both the parent and daughter product were necessary to assess the impact of the spill (Wang et al. 1997).

Biodegradation

Bacterial degradation, sometimes referred to as microbial transformation, biodegradation or biotransformation, refers to the microbially mediated transformation of organic chemicals (Chapra 1997). Although these transformations can detoxify and mineralize toxins and defuse potential toxins, they can also activate potential toxins (Ambrose et al. 1993 a,b).

Typical degrading organisms include heterotrophic bacteria, actinomycetes, autotrophic bacteria, fungi and protozoa (Chapra 1997). Two general types of biodegradation are recognized--growth metabolism and cometabolism. Growth metabolism occurs when the organic compound serves as a food source for the bacteria. Adaptation times from 2 to 20 days were suggested in EPA (1985). Adaptation may not be required for some chemicals or in chronically exposed environments. Adaptation times may be lengthy in environments with a low initial density of degraders (EPA, 1985). For cases where biodegradation is limited by the degrader population size, adaptation is faster for high initial microbial populations and slower for low initial populations. Following adaptation, biodegradation proceeds at fast first-order rates. Cometabolism occurs when the organic compound is not a food source for the bacteria. Adaptation is seldom necessary, and the transformation rates are slow compared with growth metabolism (Ambrose et al. 1993a,b).

The growth kinetics of the bacterial population degrading a toxic chemical are not well understood. As a result, toxic chemical models assume a constant biological activity rather than modeling the bacteria directly (a pseudo-first order rate). Alternatively, measured first order biodegradation rate constants are used directly (Ambrose et al. 1993a,b). However, the water body's productivity tends to increase the effectiveness of biodegradation as a removal mechanism (Chapra 1997).

Volatilization

Volatilization is the movement of chemical across the air-water interface as the dissolved neutral concentration attempts to equilibrate with the gas phase concentration (Ambrose et al. 1993a,b). It is important both as a loss mechanism from the water column and as a source mechanism for the atmosphere.

The rate of exchange of a chemical with the atmosphere is proportional to the gradient between the dissolved concentration and the concentration in the overlying atmosphere, as in (Ambrose et al. 1993)

Equation 5-57

$$\frac{\partial C}{\partial t}\Big|_{volat} = \frac{K_v}{D}\left(f_d C - \frac{C_a}{\frac{H}{R T_K}} \right)$$

where

K_v = the transfer rate, m/day

D = segment depth, m

f_d = fraction of the total chemical that is dissolved,

C_a = atmospheric concentration, ug/L

R = universal gas constant, 8.206×10^{-5} atm-m^3/mole °K

T_K = water temperature, °K

H = Henry's law coefficient for the air-water partitioning of the chemical, atm-m^3/mole.

The rate of exchange becomes zero when the concentrations are in equilibrium. Equilibrium occurs when the partial pressure exerted by the chemical in solution equals the partial pressure of the chemical in the overlying atmosphere, which are related by the Henry's law constant, a characteristic of the chemical, and the water temperature, a function of the environmental conditions.

The rate of exchange (K_v) is influenced by environment specific properties, the conditions at the air-water interface (turbulence-controlled by wind speed, current velocity, and water depth). For quiescent conditions, such as lakes and reservoirs, the rate at which the dissolved concentration of a chemical can volatilize is commonly described by the two-layer resistance model (Whitman, 1923). The two-resistance method assumes that two "stagnant films (a liquid and gas film)" are bounded on either side by well-mixed compartments. For systems such as rivers, where the flow is turbulent, a surface renewal model applies (Chapra 1997). The total transfer rate is related to the liquid and gas-film rate by:

Equation 5-58

$$K_v = (R_L + R_G)^{-1} = \left[K_L^{-1} + \left(K_G \frac{H}{R T_K} \right)^{-1} \right]^{-1}$$

where

R_L = liquid phase resistance, day/m

K_L = liquid phase transfer coefficient, m/day

R_G = gas phase resistance, day/m

K_G = gas phase transfer coefficient, m/day.

As the Henry's Law coefficient increases, the conductivity tends to be increasingly influenced by the intensity of turbulence in water (liquid-film mass transfer coefficient). As the Henry's Law coefficient decreases, the value of the conductivity tends to be increasingly influenced by the intensity of atmospheric turbulence (gas-film mass transfer coefficient, Ambrose et al. 1993). Henry's Law coefficient generally increases with increasing vapor pressure of a compound and generally decreases with increasing solubility of a compound. As a result, highly volatile low solubility compounds are most likely to exhibit mass transfer limitations in water and relatively nonvolatile high solubility compounds are more likely to exhibit mass transfer limitations in the air (Ambrose et al. 1993).

A variety of expressions are available to compute the liquid and gas-film mass transfer rates, usually as a function of water depth, velocity and wind speed which are related to the specific chemical by its molecular weight. See Chapra (1997) and Ambrose et al. (1993a,b) for a discussion of the specific formulations. The volatilization rate constant is typically adjusted for temperature by:

Equation 5-59

$$K_{v,T} = K_{20} \Theta^{T-20}$$

where

Θ = temperature correction factor

T = water temperature, EC.

Ionization

Ionization is the dissociation of a chemical into multiple charged species. In an aquatic environment some chemicals may occur only in their neutral form while others may react with water molecules to form positively (cationic) or negatively (anionic) charged ions. These reactions are rapid and are generally assumed to be at (local) equilibrium. At equilibrium, the distribution of chemicals between the neutral and the ionized species is controlled by the pH and temperature of the water and the ionization constants (Ambrose et al. 1993).

Ionization can be important because of the different toxicological and chemical properties of the neutral and ionized species. For example, in some cases only the neutral form of the chemical may react or be transported through biotic membranes resulting in toxicity. As a result, it is often necessary to compute the distribution of chemicals among ionic forms as well as to allow them to react or transform at different rates. For example, in the WASP model different sorption and reaction constants (e.g. for hydrolysis, biodegradation, photolysis, etc.) may be specified for each ionic form of the chemical (Ambrose et al. 1993).

A chemical being modeled is typically presumed to exist as neutral molecules that may, or may not, react with water molecules to form singly and, possibly, doubly charged cations and anions. To illustrate, an organic acid (A^{--}) may react with water as described by (Ambrose et al. 1993):

Equation 5-60

$$AH_2 + H_2O \_ AH_3^+ + OH^-$$

Equation 5-61

$$AH_3^+ + H_2O \_ AH_4^{++} + OH^-$$

Equation 5-62

$$AH_2 + H_20 \_ AH^- + H_3O^+$$

Equation 5-63

$$AH^- + H_2O \_ A^{--} + H_3O^+$$

so that the chemical may exist in from one to a maximum of five species simultaneously (A^{--}, AH^-, AH_2, AH_3^+, AH_4^{++}). The law of mass action can be used to describe local chemical equilibrium for each of these reactions:

Equation 5-64

$$K_{b_1} = \frac{[AH_3^+][OH^-]}{[AH_2]}$$

Equation 5-65

$$K_{b2} = \frac{[AH_4^{++}][OH^-]}{[AH_3^+]}$$

Equation 5-66

$$K_{a2} = \frac{[A^{--}][H^+]}{[AH^-]}$$

Equation 5-67

$$K_{a1} = \frac{[AH^-][H^+]}{[AH_2]}$$

where K is the equilibrium constant for the formation of the acid (K_{ai}), or anionic species, or the base (K_{bi}) or cationic species.

The total concentration of the particular chemical is the sum of the concentration of each of these forms, as given by (Ambrose et al. 1993)

Equation 5-68

$$C = AH_2 + AH_3^+ + AH_4^{++} + AH^- + A^{--}$$

which may be combined with the law of mass action to form

Equation 5-69

$$C = AH_2\left[\frac{K_{b1}}{[OH^-]} + \frac{K_{b1}K_{b2}}{[OH^-]^2} + \frac{K_{a1}}{[H^+]} + \frac{K_{a1}K_{a2}}{[H^+]^2}\right]$$

By definition, $[H^+] = 10^{-pH}$ and $[OH^-] = 10^{14-pH}$; the bracketed term in Equation 5-69, denoted D, can be written

Equation 5-70

$$D = \left[\frac{K_b1}{10^{pH-14}} + \frac{K_b1\,K_b2}{(10^{pH-14})^2} + \frac{K_a1}{10^{-pH}} + \frac{K_a1\,K_a2}{(10^{-pH})^2}\right]$$

The preceding equations may be combined with equations and solved for the fraction of the total chemical f^k occurring in each of the chemical species k, given the total chemical concentration, the pH, and the equilibrium constants:

Equation 5-71

$$f^0 = \frac{1}{D}$$

Equation 5-72

$$f^{+} = \frac{K_b 1 / 10^{pH-14}}{D}$$

Equation 5-73

$$f^{++} = \frac{K_b 1\ K_b 2/(10^{pH-14})^2}{D}$$

Equation 5-74

$$f^{-} = \frac{K_a 1/10^{-pH}}{D}$$

Equation 5-75

$$f^{--} = \frac{K_a 1\ K_a 2/(10^{-pH})^2}{D}$$

The rates of chemical reactions may also vary with temperature so that the equilibrium constants are a function of temperature. The functional dependence of these constants on temperature may be described by the Van't Hoff equation:

Equation 5-76

$$\frac{d\,\ln K_i}{d\,T_K} = \frac{E_{ai}}{R\,T_K^2}$$

or in its integrated form:

Equation 5-77

$$\log K_i(T_K) = \log K_i(T_{Ri}) + \frac{E_{ai}}{2.303\,R} \bullet \left[\frac{T_K - T_{Ri}}{T_K T_{Ri}}\right] = -\,pK_{ai} + \frac{E_{ai}}{2.303\,R} \bullet \left[\frac{T_K - T_{Ri}}{T_k T_{Ri}}\right]$$

where

K_i	=	equilibrium constant
A_i	=	frequency factor
E_{ai}	=	standard enthalpy change for reaction, kcal/mole
R	=	the universal gas constant, kcal/mole EK
T_K	=	water temperature, EK.

T_{Ri} = reference temperature at which input ionization reaction constant was observed, EK.

Hydrolysis

Hydrolysis, or reaction of the chemical with water, is known to be a major pathway for degradation of many toxic organics (Ambrose et al. 1993a,b). Hydrolysis refers to reactions in which the bond of a molecule is cleaved and a new bond is formed with the hydrogen and hydroxyl components of a water molecule (Chapra 1997). Hydrolytic reactions are usually catalyzed by acid and/or base and the overriding factor affecting hydrolysis rates at a given temperature is generally hydrogen or hydroxide ion concentration (Wolfe, 1980, Ambrose et al. 1993a,b), as in (Chapra 1997)

Equation 5-78

$$k_h = k_b[OH^-] + k_n + k_a[H^+]$$

where k_b, k_n, and k_a are coefficients for the basic, neutral, and acidic dependencies of hydrolysis.

Photolysis

Photodegradation (photolysis) is the transformation or degradation of a compound that results directly from the adsorption of light energy. It is a function of the quantity and wavelength distribution of incident light, the light adsorption characteristics of the compound, and the efficiency at which absorbed light produces a chemical reaction. Photolysis is classified into two types that are defined by the mechanism of energy absorption. Direct photolysis is the result of direct absorption of photons by the toxic chemical molecule. Indirect or sensitized photolysis is the result of energy transfer to the toxic chemical from some other molecule that has absorbed the radiation (Ambrose et al. 1993a,b). Direct photolysis predominates in systems with little extraneous dissolved or particulate matter such as clear lakes, while indirect photolysis becomes more significant in more turbid or highly colored systems (Chapra 1997).

The first order rate coefficient for photolysis can be calculated from the absorption rate and the quantum yield for each ionic specie and phase (Ambrose et al. 1993):

Equation 5-79

$$K_{pG} = \sum_i \sum_j k_{ai} \phi_{ij} f_{ij}$$

where:

K_{pG} = first order photolysis rate coefficient at ambient light intensity, day^{-1}

k_{ai} = specific sunlight absorption rate for specie i, E/mole-day or (E/L)/(mole/L)/day

Φ_{ij} = reaction quantum yield for specie i in phase j, mole/E

f_{ij} = fraction of chemical as specie i in phase j

Modeling Metals

The common practice in modeling metals in surface waters has been to treat each metal as an independent state variable subject only to transport and sorption. Sorption is typically described using apparent partition coefficients, which reflect the combined effects of the aqueous chemistry. The apparent coefficients are sometimes based upon field measurements. A second approach has been to estimate the apparent partition coefficients for the modeling activity using geochemical equilibrium models, such as MINTEQA2 (Allison et al. 1991), PHREEQC (Parkhurst and Appelo, 1999), MINEQL and MINEQL+ (Westall et al. 1986, Schecher and McAvoy 1994), and incorporating site-specific water chemistry for the range of conditions expected. The former approach completely neglects interactive effects, and the latter approach, while preferable, has limited predictive capability in situations where the chemistry is variable, particularly under future loading and chemical conditions. For example, conditions may vary from complete sorption to complete desorption within as little as one-half a pH unit for some metals, making the range of the apparent partition coefficients large. Sorption is also dependent on the sorbate-sorbent ratio and may result in considerable error in the estimation of sorption using simple coefficients under varying conditions. (Dzombak and Morel, 1990).

Aquatic system toxicity is not generally governed by the total or total dissolved concentration of metals in water but is believed to be related to the aqueous species, in particular the free ion and possibly simple hydroxide complexes. Rather than to regulate metal concentrations in receiving waters based on the total metal, total recoverable metal or dissolved concentration, as presently implemented, a speciation-based approach may provide the more defensible and fair approach toward managing aquatic resources (Renner, 1997).

Geochemical equilibrium models, such as MINEQL and MINEQL+ (Westall et al. 1986, Schecher and McAvoy 1994), MINTEQA2 (Allison et al. 1991), META4 (Medine and Martin 2000a,b) and others are available, and have been widely applied to compute the aqueous speciation, adsorption-desorption, and precipitation-dissolution of metals in aqueous systems. Equilibrium speciation models, such as those mentioned above, are extremely useful for predictive interactive effects on ionic composition in static systems, including the evaluation of the potential water-sediment interactions and aquatic system toxicity. The equilibrium assumption appears valid for aqueous speciation (acid-base, complexation) and many divalent metal adsorption/desorption reactions relative to hydrologic processes in freshwater environments.

However, a need exists for models that consider the suite of both geochemical and physical processes (e.g. transport) affecting trace metals dynamics in receiving waters, 2 and 3 dimensional environments, and, steady as well as non-steady flow situations (Dzombak and Ali 1993). Field experiments and reactive transport models represent our most critical tools for progress toward a greater understanding of the complexity of natural systems (Broshears et al. 1996). Dzombak and Ali noted the relative lack of geochemical sophistication in most simulation tools. However, several authors have reported progress in coupling geochemical speciation and transport models. Chapman (1982) used a geochemical speciation code, based on MINEQL, coupled with a semi-analytical transport model to simulate the responses of aluminum, copper, iron and zinc to experimental manipulation of pH in a short stream reach. Runkel et al. (1996a) describe the development of an equilibrium based model, which couples a geochemical speciation model, based on MINTEQ, with a solute transport model for streams, which has been applied to predict conditions during pH modification experiments (Runkel et al. 1996b, Broshears et al. 1996). Medine and Martin (2000a,b) describe the coupling of an equilibrium simulation model, based upon MINEQL, with the generalized Water Analysis Simulation Program (WASP, Ambrose et al. 1993), which includes algorithms for predicting the dynamic impacts of transport and kinetics. The META4 combines these two complementary approaches in order to improve our capacity to predict trace metal behavior in aquatic systems.

As with organic chemical models, the rate of some of the processes impacting the speciation and fate of metals are assumed to be "fast" so that equilibrium can be assumed. These include reactions such as complexation, sorption, and precipitation. The equilibrium solution of equations for the fast reactions is the basis for models such as MINEQL and MINEQL+ (Westall et al. 1986, Schecher and McAvoy 1994) and MINTEQA2 (Allison et al. 1991). However, for some metals, slow reactions are also important, such as degradation and transformation reactions. Since the form or species of metal impacts the rates of the degradation and transformation reactions, predicting these impacts usually requires the coupling of equilibrium speciation models with models or algorithms for the slow reactions.

Complexation

Complexation refers to the aqueous reactions of a metal (e.g. As, Cd, Cu, Pb, Zn) with organic and inorganic ligands (e.g. OH^-, Cl^{2-}, S^{2-}, Cl^-, amino acids, humates, fulvates, etc.) to form a third species (the metal-ligand complex) as described by

Equation 5-80

$$a\,X(1) + b\,X(2) \leftrightarrow X(1)_a\,X(2)_b$$

which, at chemical equilibrium, can be written as

Equation 5-81

$$\frac{[\,X(1)_a\,X(2)_b\,]}{[\,X(1)\,]^a\,[\,X(2)\,]^b} = K(i)$$

or

Equation 5-82

$$C(i) = K(i)\,X(1)^a\,X(2)^b$$

where X are the components of the reaction (in units of activity, or molar concentration if the ionic strength is zero), C(i) is the concentration of the ith complex, a and b are stoichiometric coefficients for the complex, and K(i) is a stability constant for the ith complex. The reaction may be written in more general form as

Equation 5-83

$$C(i) = K(i) \prod_{k=1}^{N} X(k)^{a(i,k)}$$

where N is the total number of components and a(i,k) is the stoichiometric coefficient for the kth component of the ith complex. Each complex is defined by its stability constant and stoichiometric coefficients.

Soprtion

Sorption refers to the association of a component with a solid surface. Sorption is handled similarly to complexation reactions using the method used in geochemical models such as MINEQL or MINTEQA2. In the simplest description of sorption, the formation of a particular sorbed species, C(i), may be determined as before from

Equation 5-84

$$C(i) = N_s(i)\,K_p(i) \prod_{k=1}^{N} X(k)^{a(i,k)}$$

However, the stability constant, K(i), is replaced with the product of a partitioning coefficient, $K_p(i)$, and a variable indicating the maximum number of available sites on the solid for sorption, $N_s(i)$, which may vary with pH. The effect of pH on available sites may be included by specifying complexation reactions for the sorbing surfaces, the selected parameter and each of the solids types (e.g., HP, HP_2 and HP_3 where H is the hydrogen ion concentration and P is the solids or particulate type). If these chemical species are included in the model, they will be used to determine the amount of free ionic solids forms available for sorption. If they are not included then the concentration of the solids used in the sorption reactions will be the total molar concentration.

More complex sediment modeling requires a delineation of the importance of dominant mechanisms that affect accumulation of metals in sediments and potential releases of metals from sediments. The accumulation of metals in sediments generally includes five major mechanisms:

- Adsorption onto fine-grained materials, predominantly iron oxides
- Precipitation of trace element compounds
- Coprecipitation with hydrous Fe, Al and Mn oxides and carbonates
- Association with organic matter (adsorption or organometallic bonding)
- Incorporation in crystalline minerals.

In many surface waters, it is generally accepted that sorption of metals to iron oxyhydroxides is the dominant chemical process regulating the dissolved concentrations of metals in waters and sediments (Horowitz, 1991; Dzombak and Morel, 1990). Chemical precipitation of iron is typically very rapid under neutral pH conditions in surface waters and contributes to the sorption properties of suspended and bed sediment components. Desorption of metals from sediments may also be an important process regulating overlying water quality. Desorption depends on several factors including changing pH, interstitial and adsorbed metal concentration, sorbent concentration and redox potential of the aqueous microenvironment surrounding the sediment particles. It should also be mentioned that while adsorptive processes are very effective in removing metals from the water column, the equilibration of the sediment porewater with the contaminated sediments will generally result in the presence of a significant dissolved concentration of metals in porewaters. Stream flow changes that result in an increase in the sediment resuspension can lead to significant releases of dissolved metals to the overlying water column.

Due to the importance of iron oxides in regulating metal concentrations, the current approach for mathematically describing the mechanism of metal sorption to iron oxyhydroxides is based on surface complexation theory (Dzombak and Morel, 1990). While other simple, empirical approaches can be used (e.g. simple K_p approach), the two-layer surface complexation formulation, as developed by Dzombak and Morel (1990), forms the basis for the description of metal accumulations to hydrous ferric oxides in metal speciation models.

The two layer model is based on a charge balance, analogous to the mass balance equations, with the couloumbic factor as a component. Reactions are specified for two different binding sites (e.g. weak and strong surface hydoroxyl species) and the coulombic component, such as (Dzomback and Morel 1987, Schecher and McAvoy 1994)

Equation 5-85

$$S_{wk}OH + H_s = S_{wk}OH_2^+ \quad \mathbf{\log} K = \mathbf{3.52}$$

where the subscript "s" represents a site near the surface. The relevant expressions, for this example, are then

Equation 5-86

$$\mathbf{10^{3,52}} = \frac{S_{wk}OH_2^+}{\left[S_{wk}OH\right]\left[H^+\right]}\left[Coul.\right]^{\Delta Z}$$

$$\mathbf{10^{-7.99}} = \frac{\left[S_{wk}O^-\right]\left[H^+\right]}{\left[S_{wk}OH\right]}\left[Coul.\right]^{\Delta Z}$$

where the coulombic factor is given by

Equation 5-87

$$\left[Coul.\right] = \mathbf{exp}\left(\frac{-F\ \Psi_o}{R\ T}\right)$$

where ψ_o is the surface potential, R the ideal gas constant, T is absolute temperature and ΔZ the change in charge of the surface species due to the adsorption reaction. The relationship between the total surface charge (T_σ) and the surface potential is given by

Equation 5-88

$$T_\sigma = 0.1174 \quad I^{1/2} \sinh\left(\frac{Z\,\psi_o}{2RT}\right)$$

where Z is the valency of the electrolyte and I the ionic strength. The ionic strength is specified or computed. In addition to the reactions, the specific surface area of the solids and site density for the two surfaces are specified as well as initial concentrations for the surfaces and the coulombic factor.

Precipitation

Precipitation refers to the formation of a solid phase of a particular chemical species. Precipitation differs from sorption in that it removes components from the aqueous phase. The formation of a solid is normally computed based on solubility data, and where the solubility is exceeded the solid will then precipitate. In equilibrium models such as MINEQL (Westall et al. 1986) or MINTEQA2 (Allison et al. 1991), the solubility is checked for each species of interest. If the solubility product is exceeded, the component of the solids is essentially "removed" and the equilibrium solution obtained using a reduced number of components. This process is repeated until all the remaining species are calculated to be soluble and all specified solids are at equilibrium or undersaturated with respect to the allowed precipitation products.

For the case where solids formation is treated similarly to complexation and sorption, the components of the solid species are may be specified along with an equilibrium constant for the solids formation and the formation of the solid phase and is described by

Equation 5-89

$$C_s(i) = g_{chem,i}\, K_s(i) \prod_{k=1}^{N} X(k)^{a(i,k)}$$

in which the stability constant ($K_s(i)$) may be modified by a geochemical multiplier. The geochemical multiplier allows the formation of the solid to be modified by existing geochemical conditions as reflected by concentrations of some other component (e.g. dissolved oxygen or SO_4). However, once formed, the solid is treated as any other complex or sorbed species. Geochemical multipliers that may be used, which are computed from a dependence, and constant (K), where the dependence specifies the chemistry constituent (X) which controls the respective chemical include a power law

Equation 5-90

$$g_{chem} = [X]^K$$

saturation kinetics

Equation 5-91

$$g_{chem} = \frac{[X]}{[X] + K}$$

noncompetitive inhibition

Equation 5-92

$$g_{chem} = \frac{K}{[X] + K}$$

concentration limited function

Equation 5-93

$$g_{chem} = 1 \ \ for\ [X] \le K$$

$$g_{chem} = 0 \ \ for\ [X] > K$$

where the concentration of [X] is molar (Hudson et al. 1990, 1991; Porcella et al. 1992, Martin 1994).

In order to solve for the form of a metal, solids, or parameter present, two sets of equations are required. First, the concentration of each chemical species is a function of the available concentration of each of the components, so that

Equation 5-94

$$C(i) = f(X_1, X_2, \ldots X_N) = K(i) \prod_{j=1}^{N} X(j)^{a(i,j)}$$

where the conditional stability constant for sorbed or precipitated species has been adjusted by the number of sites or geochemical multiplier respectively. Each of the components may occur in its free ionic form or may be complexed with other components. However, it is only the free ionic form of the component that is available for reaction, which is an unknown.

A second set of equations may be written based on a mass balance of each of the components over all species. That is, the total concentration of a component distributed among M species is simply the sum of all of those species (i=1 to M) multiplied by the stoichiometric coefficient for the particular component, which may be written as

Equation 5-95

$$X_t(j) = \sum_{i=1}^{M} a(i,j)\, C(i) = \sum_{i=1}^{M} a(i,j)\, K(i) \prod_{k=1}^{N} X(k)^{a(i,k)}$$

where X_t is the total concentration (molar). A similar mass balance may be written for each of the N components, resulting in N mass balance equations. The unknowns are the free ionic concentrations, or the amount of each component available for chemical reaction (complexation, sorption or precipitation).

Once the free ionic concentrations of each of the components have been determined, the concentration of each of the species may be determined. The mass balance equations for all components are solved simultaneously in models such as MINEQL (Westall et al. 1986), MINTEQA2 (Allison et al. 1991) and META4 (Medine and Martin 2000a,b). Initial estimates of the free ionic concentrations are required for all components as well as convergence criteria. The solution procedure is to make an initial guess as to the activity (or concentration) of each of the components at the start-up time for the model. For subsequent iterations, the final free ionic concentrations from the previous time step are used as the initial estimates. The solution procedure is accomplished numerically using the Newton-Raphson method. The function that is solved in the method is

Equation 5-96

$$[Y(j)] = [X_t(j)] - T(j)$$

where Y(j) is the difference between the computed and known (T(j)) totals for the jth component. To solve the j simultaneous equations, the derivative of the function is taken with respect to each component and the derivatives structured in the Jacobian matrix. The matrix is augmented with the function. An initial guess is made and the matrix is solved iteratively using the Newton-Raphson technique for the simultaneous solution of non-linear equations. Solution is achieved when Y(j) is less than some given quantity (the convergence criteria specified in input) for all components. Westall et al. (1986) and Allison et al. (1991) provide a more complete description of the solution technique.

<u>Slow Reactions</u>

Slow reactions, such as degradation and transformation reactions are important for some metals, such as methylation of Hg(II), demethylation of CH_3Hg, and volatilization of mercury. Volatilization reactions are predicted similarly to methods described for organic chemicals.

<u>Bioaccumulaton Modeling</u>

The previously discussed components of toxic chemical models are designed to provide estimates of chemical concentrations in the water column or sediments. However, for some toxic chemicals, uptake through the water column, ingested sediments, or through the food chain can result harmful levels of concentrations. The levels may cause adverse impacts on the organisms themselves or those that eat them (humans or wildlife).

Bioconcentration refers to the direct uptake of a toxicant through the gills or epithelial tissues of an aquatic organism. Bioaccumulation refers to the direct contamination (bioconcentration) from the water column as well as the ingestion of contaminated food. Biomagnification refers to the process whereby bioaccumulation increases along the food chain (Chapra 1997).

One common approach historically has been to make direct measurements of bioconcentration factors (BCF) or bioaccumulation factors (BAF) based upon water column and tissue concentrations and then use these measurements to estimate chemical residues in biota from measured or predicted concentrations. More commonly, processes based bioaccumulation models have been used to determine how individual organisms take up and purge contaminants from the water column and then transfer the contaminants through the food chain. Bioaccumulation models consider both biological attributes of various levels of the food chain as well as uptake mechanisms such as ingestion, diffusive exchange across gill membranes and intestinal mucosa. Typically, the growth of individual fish is simulated using a bioenergetic model, which inlcudes the impacts of growth, ingestion, egestion, respiration, specific dynamic actions and excretion), in which a fish's realized ingestion is a function of both its maximum ingestion rate and the availability of prey of the appropriate size and taxonomy. The bioaccumulation model FGETS (Barber, et al., 1991) and BASS (Bioaccumulation and Aquatic System Simulator) models developed by EPA and the WASTOX food chain model (Connolly and Thomann, 1985) are good examples of models of this type.

Evaluating Sediment Transport and Stability

Many toxic organics of environmental importance are hydrophobic, and strongly absorbed to sediments. The result is that often the bulk of the mass of these contaminants is sequestered in sediments, which impacts the bioavailability and transport of these materials. As a consequence, the simulation of the sediments (in the water column and bed) to which these materials sorb is a critical component of studies designed to estimate the impact of management alternatives on chemical concentrations.

Including sediments in the water column and sediment bed is a common practice in modeling toxic organic chemicals. Historically, measured data have been used to estimate long-term average rates of sediment settling, resuspension, and burial (sedimentation) for use in contaminant modeling studies. These studies have proven useful in estimating the long-term fate of organic contaminants. However, an additional concern is the short-term (event) related factors affecting the sediments, the associated contaminants, and the bioavailability of those contaminants, or the stability of the sediments.

Sediment stability refers to the ability of the sediment bed to withstand erosion in response to natural or induced hydrodynamic stress (e.g., floods, waves, prop wash, etc.; Ziegler 1999, 2002). For example, if the contaminants are sequestered in deep sediments (such as due to natural processes or artificial capping) they are not of environmental concern. However, if these sediments may be exposed or transported due to some event that needs to be considered in the sediment/contaminant/risk management process.

Because of the uncertainties associated with sediment transport, the stability of sediments is often estimated from available data and modeling studies, using a "weight of evidence" approach. Typical types of analyses used in this process are combinations of bathymetric analysis, geochronological analyses, geomorphological analyses, contaminant bed concentration analyses, contaminant water column analyses, and sediment transport modeling (Connoly 2002).

Sediment transport modeling may be used not only to evaluate short term stability but to predict the long-term fate of organic chemicals. A variety of sediment transport models are available and commonly used in hydraulic design. However, many of the available models are based on the assumption that the sediments are non-cohesive. For non-cohesive particles, such as sands, the processes impacting the settling and resuspension are independent. That is, for example, the settling of a particle does not impact the rates of settling of the particles around it at normal concentrations. However, this is not the case for cohesive sediments, such as silts and clays, which for example may flocculate during settling. For organic chemicals, the rate of sorption to sediments is often directly related to the organic content of the sediments. For inorganic chemicals, the sorption is related to the surface charge. For both cases, the smaller cohesive sediments are of primary importance. Typically grain sizes greater than approximately 40 ③m are practically cohesionless, with cohesion becoming increasingly important as the particle size decreases from that point. For example, for particle sizes in the range of 0.1 to 2 ③m, cohesion is very important (Metha and McAnally 2003).

A detailed discussion of factors impacting sediment transport and sediment transport modeling is beyond the scope of this text. A general description of the processes and models is provided by McAnally and Mehta (2002), while more detailed descriptions are provided in Mehta and McAnally (2003).

5.5. Heat budget modeling

The net rate of heat exchange across the air-water interface consists of five individual terms:

Equation 5-97

$$R_n = R_s + R_a - R_b - R_e - R_c$$

where

R_s = net shortwave solar radiation, W m^{-2}

R_a = net longwave atmospheric radiation, W m^{-2}

R_b = back radiation, W m^{-2}

R_e = evaporative heat loss or condensation-related heat gain, W m^{-2}

R_c = conduction, W m^{-2}

Shortwave solar radiation passes directly from the sun through the earth's atmosphere to the surface of the waterbody. Its magnitude at a fixed location on the earth depends on the angle of the sun's altitude, which varies daily and seasonally, and on the amount of cloud cover. Atmospheric longwave radiation is shortwave solar radiation which the constituents of the atmosphere, notably water droplets and vapor, absorb and reradiate; its value depends primarily on the amount and altitude of cloud cover. A small fraction of both the shortwave and longwave radiation components are reflected and scattered by the water surface. The solar shortwave and atmospheric longwave radiation components are independent of the water temperature.

The three surface heat loss components are back radiation, evaporation, and conduction. Their magnitudes are all dependent on the temperature of the water surface. The rate of back radiation is proportional to the fourth power of the absolute temperature of the surface. The heat conducted from the surface is proportional to the difference between the water surface temperature and the air temperature. The amount of heat lost by evaporation (or gained by condensation) is proportional to the difference between the water surface temperature and the dew point temperature of the overlying air.

The five individual components that make up the net rate of heat exchange across the air-water interface can be computed from air and dew point temperature, wind speed, cloud cover, solar radiation, and atmospheric pressure (see, for example Wunderlich, 1972; Edinger, et al., 1974; and Ryan and Stolzenbach 1972).

Shortwave solar radiation

Shortwave solar radiation is the radiant energy which passes directly from the sun to the earth. Its intensity at the water surface is calculated from the "solar constant" (radiation at the earth's outer atmosphere) as modified to account for the location and time; atmospheric conditions, cloud cover, and shade; and, reflection at the water surface. Atmospheric conditions deplete solar radiation by absorption by ozone in the upper atmosphere; scattering by dry air; absorption, scattering, and diffuse reflection by suspended particulate matter; and absorption and scattering by water vapor. Shortwave solar radiation reaching the earth's surface is more reliably measured than computed. Relatively few observations are available, however, and the following methods may be employed.

As noted, solar radiation at the water surface is a function of the radiation at the top of the earth's atmosphere, attenuation in the atmosphere (clear sky), cloud cover, shade, and reflection

Equation 5-98

$$R_s = R_0 \, t_a t_c t_s t_r$$

where

R_0 = extraterrestrial radiation (i.e., at the top of the earth's atmosphere), W/m^2

t_a = atmospheric transmission through clear sky (fraction transmitted from 0 to 1)

t_c = transmission through clouds (fraction from 0 to 1)

t_s = transmission through topographic features and vegetative shading (fraction from 0 to 1)

t_r = transmission through reflection from the water surface albedo (fraction from 0 to 1)

Extraterrestrial radiation. The extraterrestrial radiation is computed as (Bras 1990)

Equation 5-99

$$R_0 = \frac{W_0}{r^2} \sin\alpha$$

where

W_0 = the solar constant, 1367 W/m^2

r = normalized radius of the earth's orbit (i.e., the ratio of actual earth-sun distance to mean earth-sun distance)

α = the sun's altitude, radians, which can be computed as

Equation 5-100

$$\sin\alpha = \sin\delta \sin L_{at} + \cos\delta \cos L_{at} \cos(\tau)$$

where

δ = solar declination, radians

L_{at} = local latitude, radians

τ = the local hour angle of the sun, radians.

The normalized radius can be approximated as (Bras 1990)

Equation 5-101

$$r = 1 + 0.017\cos\left(\frac{2\pi}{365}(186 - D)\right)$$

where

D = day of the year (e.g., D = 1 on Jan 1, = 365 on Dec 31 of a non-leap year).

The solar declination can be approximated as (Bras 1990)

Equation 5-102

$$\delta = 23.45\frac{\pi}{180}\cos\left(\frac{2\pi}{365}(172 - D)\right)$$

The local hour angle in radians is given by

Equation 5-103

$$\tau = ((trueSolarTime / 4) - 180)\frac{\pi}{180}$$

where

trueSolarTime = *localTime* + *eqtime* + (4 L_{lm}) - (60 *timezone*) (104)

where *localTime* is the local time in minutes (local standard time), L_{lm} is the local longitude (negative decimal degrees for the western hemisphere), *timezone* is the local time zone in hours relative to Greenwich Mean Time (e.g. -8 hours for Pacific Standard Time). The equation of time (*eqtime*) is an astronomical term that accounts for changes in the time of solar noon for a given location over the course of a year. The value of *eqtime* represents the difference between true solar time and mean solar time in minutes and is most accurately estimated using the algorithms of Meeus (1999). An example programming application of the Meeus algorithims is provided in a solar position calculator that is available from NOAA's Surface Radiation Research Branch (www.srrb.noaa.gov/highlights/sunrise/azel.html).

Atmospheric transmission. Various methods have been published to estimate the fraction of solar radiation attenuated by the clear sky (t_a). Four examples of alternative methods of estimating t_a are

- Bras (1990)
- Iqbal (1983) as implemented by Corripio (2002)
- Bird's model (Bird and Hulstrom, 1981)
- Ryan and Stolzenbach (1972)

The Ryan and Stolzenbach (1972) and Bras (1990) methods are the simplest of the four alternatives. Ryan and Stolzenbach (1972) compute t_a from ground surface elevation and solar altitude as

$$t_a = a_{tc}^{\,m\left(\frac{288-0.0065Z}{288}\right)^{5.256}}$$

where a_{tc} is the atmospheric transmission coefficient (0.70-0.91, typically approximately 0.8), Z is the ground surface elevation in meters, and m is the optical air mass, calculated as

Equation 5-105

$$m = \frac{1}{\sin\alpha_d + 0.15(\alpha_d + 3.885)^{-1.253}}$$

where $\alpha_d = \alpha\frac{180}{\pi}$ is the solar altitude in degrees from the horizon.

The Bras (1990) method computes t_a as:

$t_a = \exp(-n_{fac}\, a_1\, m)$ (5-106)

where n_{fac} is an atmospheric turbidity factor that varies from approximately 2 for clear skies to 4 or 5 for smoggy urban areas. The molecular scattering coefficient (a_1) is calculated as

$a_1 = 0.128 - 0.054 \log_{10} m$ (5-107)

The Bird model is a simple broadband model for direct and diffuse insolation under clear sky conditions that accounts for atmospheric transmission as a function of Rayleigh and aerosol scattering, and absorbance by ozone, uniformly mixed gases, water vapor, and aerosols (Bird and Hulstrom 1981). Atmospheric turbidity values at aerosol optical depth (AOD) for 380 and 500 nm wavelengths are required for input, as well as ozone and water vapor content of the atmosphere and barometric pressure. The Iqbal method (Iqbal,

1983; Corripio, 2002) is a simplification of the Bird model that estimates scattering and absorbance from atmospheric turbidity based on visibility measurements that are commonly reported at airport meteorological stations.

Direct measurements of solar radiation are available at some locations. The selection of the solar radiation model for a particular application should ideally be guided by a comparison of computed and measured values at a reference location.

Transmission through clouds. Transmission of solar radiation through cloud cover can be estimated from (TVA, 1972):

Equation 5-108

$$t_c = 1 - 0.65N^2$$

where N = fraction of the sky covered with clouds (0 to 1).

Transmission through topographic features and vegetative shading. Shade is an important parameter that modifies stream heating due to solar radiation. Human activities can degrade near-stream vegetation and/or channel morphology, and in turn, decrease shade. Reductions in stream surface shade have the potential to cause significant increases in heat delivery to a stream system. Stream shade may be measured or calculated using a variety of methods (Chen, 1996; Chen et al., 1998; Ice, 2001; OWEB, 1999; Teti, 2001; Ecology, 2003).

Transmission through shading features is calculated as

Equation 5-109

$$t_s = 1 - S_f$$

where S_f is the effective shade, which is defined as the fraction of the total possible solar radiation that is prevented from reaching the surface of the water by topography and riparian vegetation.

In the Northern Hemisphere, the earth tilts on its axis toward the sun during summer months, allowing longer day length and higher solar altitude, both of which are functions of solar declination (i.e., a measure of the earth's tilt toward the sun) (Figure 5.4). Geographic position (i.e., latitude and longitude) fixes the stream to a position on the globe, while aspect provides the stream/riparian orientation (direction of stream flow). Vegetation height, width and density describe the physical barriers between the stream and sun that can attenuate and scatter incoming solar radiation (i.e., produce shade) (Table 5-3). The solar position has a vertical component (i.e., solar altitude) and a horizontal component (i.e., solar azimuth) that are both functions of time/date (i.e., solar declination) and the earth's rotation.

While the interaction of these shade variables may seem complex, the mathematics that describes them is relatively straightforward geometry. Using solar tables or mathematical simulations, the potential solar heat flux can be quantified. The shade from riparian vegetation can be measured with a variety of methods, including (Ice, 2001, OWEB, 1999, Teti, 2001):

- hemispherical photography
- angular canopy densiometer
- solar pathfinder

Hemispherical photography is generally regarded as the most accurate method for measuring shade, although the equipment that is required is significantly more expensive compared with other methods. Angular canopy densiometers (ACD) provide a good balance of cost and accuracy for measuring the importance of riparian vegetation for preventing increases in stream temperature (Teti, 2001, Beschta et al., 1987.) Whereas canopy density is usually expressed as a vertical projection of the canopy onto a horizontal surface, the ACD is a projection of the canopy measured at an angle above the horizon at which direct beam solar radiation passes through the canopy. This angle is typically determined by the position of the sun above the horizon during that portion of the day (usually between 10 A.M. and 2 P.M. in mid to late summer) when the potential solar heat flux is most significant. Typical values of the ACD for old-growth stands in western Oregon have been reported to range from about 80% to 90%.

Computer programs for the mathematical simulation of shade may also be used to estimate shade from measurements or estimates of the key parameters listed in Table 5-3 (Ecology 2003, Chen, 1996, Chen et al., 1998, Boyd, 1996, Boyd and Park, 1998, Boyd and Kasper, 2003).

Transmission through surface albedo. Transmission through albedo or reflectivity can be estimated as (Anderson 1954)

$$t_r = A\alpha_d{}^B$$

where A and B are coefficients related to cloud cover (Table 5-4).

Longwave Atmospheric Radiation

The downward flux of longwave radiation depends primarily on air temperature and humidity and increases as the moisture content of the air increases. It constitutes the major thermal input to a waterbody at night and on warm, cloudy days. In some heat studies, its magnitude has been computed by using the fourth-power law and a constant emissivity of about 0.87. Atmospheric radiation does not, however, follow a simple law. It is a function of many variables, notably the distributions of moisture, temperature, ozone, carbon dioxide, and other atmospheric constituents.

The downward flux of longwave radiation from the atmosphere is one of the largest terms in the surface heat balance. This flux can be estimated using the Stefan-Boltzmann law

Equation 5-110

$$R_a = \sigma(T_{air} + 273)^4 \quad (\varepsilon_{sky}) \quad (1 - R_L)$$

where σ = the Stefan-Boltzmann constant = 5.67×10^{-8} Wm^{-2} K^{-4}, T_{air} = air temperature (°C), ε_{sky} = effective emissivity of the atmosphere (fraction from 0 to 1), and R_L = longwave reflection coefficient. The reflection coefficient is generally small and in typically is assumed to equal 0.03.

Brutsaert (1982) and Federer (2002) provide a summary of various methods for estimating downward longwave radiation from a clear sky based only on air temperature and humidity. The most commonly used methods differ in how they express the effective emissivity (ε_{clear}) of a clear sky:

- Brunt (1932) $\varepsilon_{clear} = c + d\, e_{kpa}^{0.5}$
- Swinbank (1963) $\varepsilon_{clear} = 0.0000092\, T_{air,k}^{2}$
- Idso and Jackson (1969) $\varepsilon_{clear} = 1 - 0.261 \exp(-0.000777\, T_{air}^2)$
- Satterlund (1979) $\varepsilon_{clear} = 1.08 \{1 - \exp[-(10\, e_{kpa}) T_{air,k} / 2016\}$
- Brutsaert (1982) $\varepsilon_{clear} = 1.24\, (10\, e_{kpa} / T_{air,k})^{1/7}$

where e_{kpa} is the air vapor pressure in kPa, T_{air} is the air temperature in °C, and $T_{air,k}$ is the air temperature in °K.

The Brunt method typically has values of c ranging from 0.5 to 0.7, and d = 0.134 kPa$^{-0.5}$ for vapor pressure in units of kPa or 0.031 mmHg$^{-0.5}$ for vapor pressure in units of mmHg. Koberg (1964) developed an empirical model to estimate the c constant for Brunt's equation based on air temperature and the ratio of actual to clear sky solar radiation (Figure 5.5) assuming that the value of d in Brunt's equation is 0.0263 millibars$^{-0.5}$ (approximately 0.031 mmHg$^{-0.5}$).

The Swinbank and Idso-Jackson curves have no dependence on humidity. Brutsaert (1982) suggests that Satterlund's equation is better below freezing, but the Satterlund model has a flat temperature response and produces the highest results for temperatures of 0 to 20°C and relative humidity above 60%. The Brutsaert method is the only method that is physically-based instead of empirically derived.

The air vapor pressure (in mm Hg) can be computed as (Chapra 1997):

Equation 5-111

$$e_{air} = 4.596e^{\frac{17.27T_d}{237.3+T_d}}$$

where Td = the dew point temperature, oC.

For cloudy conditions the atmospheric emissivity may increase as a result of increased water vapor content. High cirrus clouds may have a negligible affect on atmospheric emissivity, but lower stratus and cumulus clouds may have a significant affect. The effective atmospheric emissivity for cloudy skies (ε_{sky}) is often estimated from the clear sky emissivity using a nonlinear function of the fractional cloud cover (*N*) of the form:

Equation 5-112

$$\varepsilon_{sky} = \varepsilon_{clear}(1 + a_{cc}N^2)$$

Various values for the coefficient *acc* for the cloud function have been reported:

- TVA (1972) after Anderson (1954): a_{cc}=0.17 based on the average for various cloud types
- Bolz (1949) reported values for various cloud types (Brutsaert, 1982):
  - Cirrus: a_{cc}=0.04
  - Cirrostratus: a_{cc} =0.08
  - Altocumulus: a_{cc} =0.17
  - Altostratus: a_{cc} =0.20
  - Cumulonimbus: a_{cc} =0.20
  - Cumulus: a_{cc} =0.20
  - Stratocumulus: a_{cc} =0.22
  - Nimbostratus: a_{cc} =0.25
  - Fog: a_{cc} =0.25

The selection of the longwave model for a particular application should ideally be guided by a comparison of predicted results with measured values at a reference location.

Back radiation

Water returns energy back to the atmosphere in the form of longwave radiation. Since water radiates as an almost perfect black body, the rate at which heat is lost by back-radiation can be computed from the Stephan-Boltzman fourth power radiation law:

Equation 5-113

$$R_b = \varepsilon_{water}\sigma(T + 273)^4$$

where ε_{water} is the emissivity of water (= 0.97)

σ is the Stephan-Boltzman constant (= 5.67×10^{-8} W m^{-2} K^{-4})

T is the water surface temperature (°C)

Evaporative heat loss

Heat is lost from a body of water to the atmosphere through evaporation of the water. Each kilogram of water that leaves as water vapor carries away its latent heat of vaporization; heat is regained for each kilogram of water that condenses on the water surface, which happens rarely and usually at night when the water temperature falls below the dew point temperature.

A general evaporation formula gives the rate of evaporation as a function of the wind speed and the difference between the saturated water vapor pressure at water surface temperature and the water vapor pressure in the overlying air, thus (Chapra, 1997):

Equation 5-114

$$R_e = f(U_w)(e_s - e_{air})$$

where U_w is the wind speed (m/s), e_{air} is the air-vapor pressure (mm Hg) and e_s is the saturation vapor pressure of the air adjacent to the water surface (at temperature T), and $f(Uw)$, defines the dependence of the transfer on wind velocity over the water surface where Uw is the wind speed measured a fixed distance above the water surface.

Many relationships exist to define the wind dependence. Bras (1990), Edinger et al. (1974), and Shanahan (1984) provide reviews of various methods. Some researchers have proposed that conduction/convection and evaporation are negligible in the absence of wind (e.g. Marciano and

Harbeck, 1962), which is consistent with the assumption that only molecular processes contribute to the transfer of mass and heat without wind (Edinger et al. 1974). Others have shown that significant conduction/convection and evaporation can occur in the absence of wind (e.g. Brady Graves and Geyer 1969, Harbeck 1962, Ryan and Harleman 1971, Helfrich et al. 1982, and Adams et al. 1987). This latter opinion has gained favor (Edinger et al., 1974), especially for waterbodies that exhibit water temperatures that are greater than the air temperature.

Brady, Graves, and Geyer (1969) pointed out that if the water surface temperature is warmer than the air temperature, then "the air adjacent to the water surface would tend to become both warmer and more moist than that above it, thereby (due to both of these factors) becoming less dense. The resulting vertical convective air currents ... might be expected to achieve much higher rates of heat and mass transfer from the water surface, even in the absence of wind than would be possible by molecular diffusion alone" (Edinger et al., 1974). Water temperatures in natural waterbodies are frequently greater than the air temperature, especially at night.

Edinger et al. (1974) recommend that the relationship that was proposed by Brady, Graves and Geyer (1969) based on data from cooling ponds, could be representative of most environmental conditions. Shanahan (1984) recommends that the Lake Hefner equation (Marciano and Harbeck 1952) is appropriate for natural waters in which the water temperature is less than the air temperature. Shanahan also recommends that the Ryan and Harleman (1971) equation as recalibrated by Helfrich et al. (1982) is best suited for waterbodies that experience water temperatures that are greater than the air temperature. Adams et al. (1987) revisited the Ryan and Harleman and Helfrich et al. models and proposed another recalibration using additional data for waterbodies that exhibit water temperatures that are greater than the air temperature.

The following is a summary of various published methods to calculate $f(U_w)$, in units of $W/m^2/mmHg$:

- Marciano and Harbeck (1952): $f(U_w) = 0.1313 \times 17U_{w,mph}$, also known as the Lake Hefner method ($U_{w,mph}$ is the wind speed in miles/hour measured at a height of 7 m). The constant 0.1313 converts the units from BTU ft^{-2} day^{-1} $mmHg^{-1}$ to W m^{-2} $mmHg^{-1}$.

- Harbeck (1962): $f(U_w) = 0.1313 \times 24.2A_{acres,i}^{-0.05}U_{w,mph}$, is a function of surface area of the reach ($A_{acres,i}$ is surface area in acres).

- Brady, Graves, and Geyer (1969): $f(U_w) = 9.2 + 0.46U_w^2$.

- Ryan and Harleman (1971): $f(U_w) = 0.1313 \times (22.4\Delta\theta_v^{1/3} + 14U_{w,mph})$, derived an empirical model of the wind speed function for heated waters that accounts for the enhancement of convection currents when the virtual temperature difference between the water and air ($\Delta\theta_v$ in degrees F) is greater than zero. Wind speed in this equation is at a height of 2m.

- Helfrich et al. (1982): $f(U_w) = 0.1313 \times (17.7\Delta\theta_v^{1/3} + 11.1U_{w,mph})$ recalibrated the equation of Ryan and Harleman (1971) with data from several additional cooling ponds and recommended the recalibrated equation for heated waterbodies. Wind speed in this equation is at a height of 2m.

- Adams et al. (1987) updated the work of Helfrich et al. (1982). Two wind functions were proposed by Adams et al., also known as the East Mesa method (wind speed in these equations is at a height of 2m):

East Mesa 1: $f(U_w) = 0.1313 \times ((22.4\Delta\theta_v^{1/3})^2 + (24.2A_{acres,i}^{-0.05}U_{w,mph})^2)^{0.5}$, uses a recalibrated function to estimate effect of convection currents caused by virtual temperature differences between water and air, and the Harbeck (1962) equation is used for the contribution to conduction/convection and evaporation that is not due to convection currents caused by high virtual water temperature

East Mesa 2: $f(U_w) = 0.1313 \times ((22.4\Delta\theta_v^{1/3})^2 + (17U_{w,mph})^2)^{0.5}$ uses a recalibrated function of virtual temperature differences with the Marciano and Harbeck (1952) equation for the contribution to conduction/convection and evaporation that is not due to the high virtual water temperature

Virtual temperature is defined as the temperature of dry air that has the same density as air under the in situ conditions of humidity. The virtual temperature difference between the water and air ($\Delta\theta$ in °F) is estimated from water temperature ($T_{w,f}$ in °F), air temperature ($T_{air,f}$ in °F), vapor pressure of water and air (e_s and e_{air} in mmHg), and the atmospheric pressure (p_{atm} in mmHg) as:

Equation 5-115

$$\Delta\theta = (\frac{T_{w,f} + 460}{1 + 0.378e_s / p_{atm}} - 460) - (\frac{T_{air,f} + 460}{1 + 0.378e_{air} / p_{atm}} - 460)$$

The height of wind speed measurements is also an important consideration for estimating conduction/convection and evaporation. For example, to convert wind speed measurements take at a height z_1 ($U_{w,z1}$ in m/s) to the equivalent wind speed ($U_{w,z2}$ in m/s) at any other height (z_2 in meters), the exponential wind law equation may be used (TVA, 1972):

$$U_{w,z2} = U_{w,z1} \left(\frac{z_2}{z_1}\right)^{0.15}$$

For example, if wind speed data were collected from a height of 2 m, then the equivalent wind speed at 7 m above the water would be estimated by multiplying the measured wind speed by a factor of about 1.2.

Heat conduction and convection between the air and water

Heat can enter or leave water by conduction and convection if the air temperature is greater than or less than the water temperature. The rate at which heat is exchanged between the two media through conduction and convection is proportional to the temperature difference between the two media:

Equation 5-116

$$R_c = C_0(T_{res} - T_a)$$

where C_0 is a proportionality coefficient that depends on the wind speed. Similarity arguments relating heat transfer by conduction/convection and mass transfer by evaporation have been used by Bowen (1926) to show that C_0 must involve the same wind speed function as Equation 116, i.e.

Equation 5-117

$$C = C_c f(W)$$

where C_c is Bowen's coefficient, which has been empirically found for natural waters with negligible bulk flow to be approximately 0.47 mm Hg C^{-1}.

5.6. REFERENCES

Ambrose, B, Jr., T.A. Wool, and J.L. Martin, "The Water Quality Analysis Simulation Program, WASP5. Part B: the WASP5 Input Dataset, Version 5.00," Internal Report, Distributed by USEPA Center for Exposure Assessment Modeling, U.S. Environmental Protection Agency, Athens, GA, May 1993.

Ambrose, B, Jr., T.A. Wool, and J.L. Martin, "The Water Quality Analysis Simulation Program, WASP5; Part A: Model Documentation," Internal Report Distributed by USEPA Center for Exposure Assessment Modeling, U.S. Environmental Protection Agency, Athens, GA, June 1993.

Ambrose, R. (2005) Personal Communications with J.E.Edinger.

Barber, M.C., L.A. Suarez, and R.R. Lassiter. 1991. Modeling Bioaccumulation of Organic Pollutants in Fish with an Application to PCBs in Lake Ontario Salmonids. Canadian J. Fisheries and Aquatic Sciences. Vol. 48, pp. 318-337.

Bravo, I and I. Ramilo. 1998. Distribution of microreticulate dinoflagellate cysts from the Galician and Portuguese coast . Sci. Mar., 63(1): 45-50.

Broshears, R.E., R.L. Runkel, B.A. Kimball, D.M. McKnight, and K.E. Bencala. 1996. Reactive solute transport in an acidic stream: Experimental pH increase and simulation of controls on pH, aluminum and iron. Environmental Science and Technology, 30(10), pp. 3016-3024.

Cerco, C.F. 1995. "Response of Chesapeake Bay to nutrient load reductions," ASCE J. Environmental Engineering Div. 121(8), 549-557.

Cerco, C.F. and T. M. Cole. 1991. "Thirty year simulation of Chesapeake Bay Dissolved Oxygen," in Environmental Hydraulics, J. H. Lee and Y. K. Cheung, eds. Balkema, Rotterdam, 771-776.

Cerco, C.F. and T. M. Cole. 1992. "Thirty year simulation of Chesapeake Bay eutrophication," In Estuarine and Coastal Modeling, M. L. Spaulding, K. Bedford, A. Blumberg, R. Chen, and C. Swanson, eds. ASCE, 116-126.

Cerco, C.F. and T. M. Cole. 1995. "User documentation for Release Version 1.0 of the CE-QUAL-ICM Three Dimensional Eutrophication Model," Technical Report EL-95-1, U.S. Army Corps of Engineers, Waterways Experiment Station, Vicksburg, MS.

Cerco, C.F., and T. Cole. 1993. Three-dimensional eutrophication model of Chesapeake Bay. Journal of Environmental Engineering. 119:1006-1025.

Chapman, B.M. 1982. Numerical simulation of the transport and speciation of nonconservative chemical reactions in rivers, Water Resources Research, 18(1), pp. 155-167.

Chapra, S.C. 1997. Surface Water-Quality Modeling, McGraw-Hill, New York, New York, 844 pp.

Cole, T.M. and E.M. Buchak. 1995. "CE-QUAL-W2: A Two-Dimensional, Laterally Averaged, Hydrodynamic and Water Quality Model, Version 2. Instruction Report EL-95-1. U.S. Army Corps of Engineers, Waterways Experiment Station, Vicksburg, MS.

Connolly, J.P. and R.V. Thomann. 1985. WASTOX, A Framework for Modeling the Fate of Toxic Chemicals in Aquatic Environments. Part 2: Food Chain. U.S. Environmental Protection Agency, Gulf Breeze, FL and Duluth, MN.

Connoly, J.P. 2002. "Empirical Methods to Evaluate Stability, Overview," Presented at the Sediment Stability Workshop, January 22-24, New Orleans, LA.

Cullen, J.J. and S.G. Horrigan. 1981. Effects of nitrate on the diurnal vertical migration, carbon to nitrogen ratio, and the photosynthetic capacity of the dinoflagellate Gymnodinium splendens. Marine Biology. 62:81-89.

Di Toro, D. M. and J. F. Fitzpatrick. 1993. Chesapeake Bay sediment flux model. Tech. Report EL-93-2, U.S. Army Corps of Engineers, Waterways Experiment Station, Vicksburg, Mississippi, 316 pp.

Di Toro, D. M., J. F. Fitzpatrick and T. R. Isleib. 1994. "A model of manganese and iron fluxes from sediments," Contract Report W-94-1, U.S. Army Corps of Engineers, Waterways Experiment Station, Vicksburg, Mississippi, 76 pp.

Di Toro, D. M., Paquin, P. R., Subburamu, K. and Gruber, D.A. 1990. Sediment oxygen demand model: methane and ammonia oxidation. J. Environ. Engr. ASCE, 116:945-986.

DiToro, Dominic M. 2001. Sediment Flux Modelling. Wiley-Interscience. New York. ISBN 0-471-13535-6.

Dzomback, D.A. and F.M.M. Morel. 1987. Adsorption of Inorganic Pollutants in Aquatic Systems, ASCE J. Hydraulic Eng. 113, p. 430.

Dzombak, D.A. and M.A. Ali. 1993. Hydrochemical modeling of metal fate and transport in freshwater environments, Water Pollut. Res. J. Can., 28(1), pp. 7-50.

Dzombak, D.A. and Morel, F.M.M. 1990. Surface Complexation Modeling Hydrous Ferric Oxides. John Wiley and Sons Publishers, New York.

Edinger, J. E., D. K. Brady and J. C. Geyer. 1974. Heat Exchange and Transport in the Environment. Cooling Water Studies for the Electric Power Research Institute, Research Project RP-49, Report 14. Palo Alto, California. EPRI Publication Number 74-049-00-3. November.

Edinger, J.E. 2002. Waterbody Hydrodynamic and Water Quality Modeling. An introductory workbook and CD-ROM on three-dimensional waterbody modeling. ASCE Press. Reston, VA. ISBN 0-7844-0550-6.

Edinger, J.E., C.D.Boatman and V.S.Kolluru. 2003. Influence of Multi-Algal Groups in the Calibration of a Water Quality Model. ASCE Estuarine and Coastal Waters Proceedings from Fall 2003 Conference.

Edinger, J.E., Scott Dierks and Venkat Kolluru. 2003. Density Dependent Grazing in Estuarine Water Quality Models. Water Air and Soil Pollution 147: 163-182, 2003. Kluwer Academic Publishers, the Netherlands.

Eppley, R.W. O. Holm-Hansen and J.D.H. Strickland. 1968. Some observations on the vertical migration of dinoflagellates. J. Phycol. 4:333-340.

Fair, G.M., Moore, E.W., and Thomas, H. A., Jr. 1941. The natural purification of river muds and pollutional sediments. Sewage Works J. 13(6):1209-1228.

Gentleman, W., A. Leising, B. Frost, J.Murry and S.Strom. 2000. Dynamics of Food-web Models with Multiple Nutritional Resources: A critical review of implicit assumptions, JGOFS SMP Workshop. School of Oceanography, University of Washington, Seattle, Washington, July.

Godhe A, S. Svensson & A-S Rehnstam-Holm (2002b) Oceanographic settings explain fluctuations in Dinophysis spp. and concentrations of diarrhetic shellfish toxin in the plankton community within a mussel farm area on the Swedish west coast. Marine Ecology Progress Series, Vol. 240, 71-83.

Godhe A. & M.R. McQuoid (2003). Distribution of dinoflagellate cysts in recent marine sediments in relation to benthic and pelagic environmental factors along the Swedish west coast. Accepted for publication Aquatic Microbial Ecology.

Godhe A., A-S Rehnstam-Holm, I. Karunasagar & I. Karunasagar (2002a). PCR detection of dinoflagellate cysts in field sediment samples from tropic and temperate environments. Harmful Algae, Vol 1(4), 361-373.

Greer, S.P. and C.D. Amsler. 2002. Light boundaries and the coupled effects of surface hydrophobicity and light on spore settlement in the brown alga *Hincksia irregularis* (Phaeophyceae). Journal of Phycology 38: 116-124.

Harris, G.P. S.I. Heaney and J.F. Talling. 1979. Physiological and environmental constraints in the ecology of the phytoplankton dinoflagellate Ciratium hirundinella. Freshwater Biol. 9:413-428.

Horowitz, A.J. 1991. Sediment-Trace Element Chemistry, Lewis Publishers, Chelsea, Michigan.

Hudson, R.J.M., S. Gherini and D. Porecella. 1991. Instructions for Running the MCM Lake Mercury Cycling Model, Tetra Tech, Inc., Lafayette, CA.

Hudson, R.J.M., S.A. Gherini, and R.K. Munson. 1990. The MTL Mercury Model: A Description of the Model, Discussion of Scientific Issues, and Presentation of Preliminary Results, Section 8 In: Mercury in Temperate Lakes Project Annual Report for 1990 to Electric Power Research Institute, Palo Alto, California.

Imhoff, J.C., A. Stoddard, and E.M. Buchak. 2003. "Evaluation of Contaminated Sediment Fate And Transport Models, Final Report," Contract Report, Contract Number 68-C-01-037, prepared for the U.S. EPA National Exposure Research Laboratory, Athens, Ga.

Kamylkowski, D. and S.A. McCollum. 1986. The temperature acclimatized swimming speed of selected marine dinoflagellates. J. Plankton Res. 8:275-287.

Kamylkowski, D., S.A. McCollum and G.J. Kirkpatrick. 1988. Observations and a model concerning the translational velocity of a photosynthetic marine dynoflagellate under variable light conditions. Limnol. Oceangr. 33:55-78.

Karickhoff, S.W. 1984. Organic Pollutant Sorption in Aquatic Systems. J. Hydraul. Eng. Div. ASCE. Vol. 110, p. 707.

Mackay, D. and P.J. Leinonen. 1975. Rate of Evaporation of Low-Solubility Contaminants from Water Bodies to Atmospheres. Environ. Sci. Technology. 7:611-614.

Martin, J.L. 1992. "MERC4, A Mercury Transport and Kinetics Model, Beat Test Version 1.0: Model Theory and User's Guide," Prepared for the U.S. Environmental Protection Agency, Center for Exposure Assessment Modeling, Athens, GA, July 1992.

McAnally, W. H. and Mehta, A. J. 2002. "Sediment Transport in Estuaries," Chapter in Knowledge for Sustainable Development - An Insight into the Encyclopedia of Life Support Systems, United Nations Educational, Scientific, and Cultural Organization, Paris.

Medine, A. and Willingham, T. 1990. Water Resource Management Strategies for Restoring and Maintaining Aquatic Life Uses. Intern'l

Symposium on Fish Toxicology, Physiology and Water Quality Management, Joint EPA/PRC, Sacramento, Oct, 1990.

Medine, A.J. and J.L. Martin. 2000a. "Development of a Metal Exposure and Transformation Model for Use in Watershed Management, Restoration and TMDL Analysis, Watershed 2000, July 9-12, 2000, Vancouver, BC.

Medine, A.J. and J.L. Martin. 2000b. "Development of the Speciation-Based Exposure and Transformation Assessment Model (META4): Application to Copper and Zinc Problems in the Alamosa River, Colorado," Presented at the American Chemical Society, Environmental Chemistry Division Symposium, August, 2000, Boulder, CO.

Mehta, A.J. and W.H. McAnally. 2003. "Chapter 4, Fine-Grained Sediment Transport, " In Sedimentation Engineering, Volume II, American Society of Civil Engineers, New York.

Mill, T., W.R. Mabey, P.C. Bomberger, T.W. Chou, D.G. Herdry, and J.H. Smith. 1982. Laboratory Protocols for Evaluating the Fate of Organic Chemicals in Air and Water. U.S. Environmental Protection Agency, Athens, GA. EPA-600/3-82-0220.

Morel, F. 1983. Principles of Aquatic Chemistry, John Wiley and Sons Publishers, New York, p. 37.

Parkhurst, D.L. and C.A.J. Appelo. 1999. "User's Guide To PHREEQC (Version 2)—A Computer Program For Speciation,Batch-Reaction, One-Dimensional Transport, and Inverse Geochemical Calculations," U.G. Geological Survey, Water-Resources Investigations Report 99-4259, Washington, DC.

Philips, S., H.J. Laanbroek and W. Verstraete. 2002. Origin, causes and effects of increased nitrite concentrations in aquatic environments. Re/Views in Environmental Science and BioTechnology 1: 115-141, 2002. Kluwer Academic Publishers.

Renner, R., 1997. Rethinking Water Quality Standards for Metals Toxicity, Environmental Science and Technology, 31(10), pp. 466A-468A.

Renner, R., 1997. Rethinking Water Quality Standards for Metals Toxicity, Environmental Science and Technology, 31(10), pp. 466A-468A.

Robinson, M.G. and L.N. Brown. 1983. A recurrent red tide in a British Columbia coastal lagoon. Can. J. Fish. Aquat. Sci. 40:2135-2143.

Schecher, W.D. and D.C. McAvoy, 1994. MINEQL+: A Chemical Equilibrium Program for Personal Computers, User's Manual, Version 3.0, The Procter and Gamble Co., Ivorydale Technical Center, Cincinnati, Ohio.

Stock, C.A., D.J.McGillicuddy, A.R.Solow and D.M.Anderson. 2005. Evaluating Hypotheses for the Initiation and Development of *Alexandrium fundyense* blooms in the Western Gulf of Maine using a Coupled Physical-Biological Model. *Ecology and Oceanography of Harmful Algal Blooms in the Gulf of Maine.* Woods Hole Oceanographic Institution, Woods Hole, MA 02543. Accepted for publication in Deep Sea Research.

Tsivoglou, E.E. and J.R. Wallace. 1972. Characterization of Stream Reaeration Capacity. U.S. Environmental Protection Agency, Washington, DC EPA-R3-72-012.

US Environmental Protection Agency. 1985. Rates, Constants, and Kinetics Formulations in Surface Water Quality Modeling (Second Edition). EPA/800/3-85/040. Environmental Research Laboratory, Athens, GA, 30613. June.

US Environmental Protection Agency. 1993. Assessment and Remediation of Contaminated Sediments (ARCS) Program Risk Assessment and Modeling Overview Document, EPA 905-R93-007. Chicago, Ill.: Great Lakes National Program Office (GLNPO).

US Environmental Protection Agency. 1994. "Food and Gill Exchange of Toxic Substances, Version 3.0.18," Center for Exposure Assessment Modeling (CEAM), Athens, GA.

US Environmental Protection Agency. 1997. "Mercury Report to Congress," EPA 452/R-97-003, U.S. EPA Office of Research and Development, Washington, D.C.

US Environmental Protection Agency. 1999. Protocol for Developing Nutrient TMDLs; First Edition. EPA 841-B-99-007, Office of Water, Washington DC. November 1999

Wang, P.F., J.L. Martin and T. Wool. 1997. "Fate and Transport of Metal Spill in Sacramento River," ASCE J. Environmental Engineering, 123(7).

Westall, J.C., J.L. Zachary and F.M.M. Morel. 1986. MINEQL: A Computer Program for the Calculation of the Chemical Equilibrium Composition of Aqueous Systems, Report 86-01, Department of Chemistry, Oregon State University, Corvallis, OR.

Whitman, R.G. 1923. A Preliminary Experimental Confirmation of the Two-Film Theory of Gas Absorption, Chem. Metallurg. Eng. 29:146-148.

Wolfe, N.L. 1980. Determining the Role of Hydrolysis in the Fate of Organics in Natural Waters. pp. 163-178 in R. Haque (Ed.) Dynamics, Exposure, and Hazard Assessment of Toxic Substances. Ann Arbor Science Publishers, Ann Arbor.

Yamazaki, H. and D. Kamykowski. 1991. The vertical trajectories of motile phytoplankton in a wind-mixed water column. Deep-Sea Res. 38:219-241.

Ziegler, C. K. 2002. "Sediment Stability Analysis for Upper Hudson River," Presented at the Sediment Stability Workshop, January 22-24, New Orleans, LA.

Ziegler, C.K. 1999. "Sediment Stability at Contaminated Sediment Sites." Quantitative Environmental Analysis, LLC (QEA), Burlington, MA.

Figure 5.1. Process flow diagram for the WQDPM water quality model.

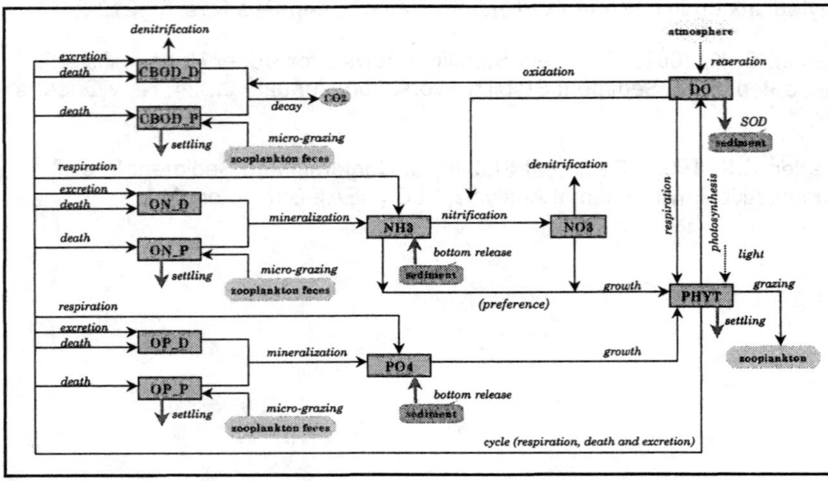

Figure 5.2-Source/Sink in transport relationship, Equation 5.1, for Processes in Figure 5.1.

Source/Sink	Constituent Dependency											External source/sinks	
	1	2	3	4	5	6	7	8	9	10	11		
	NH3	NO3	PHY	BODD	BODP	DO	OND	ONP	OPD	OPP	PO4		
SNH3	a1,1		a1,3				a1,7	a1,8				SEDnh3	
SNO3	a2,1	a2,2	a2,3										
SPHY			a3,3									-Settling	-grazing
SBODD		a4,2	a4,3	a4,4									
SBODP			a5,3		a5,5							-Settling	+grazing
SDO	a6,1		a6,3	a6,4	a6,5	a6,6						SOD	Rear.-Sat.
SOND			a7,3				a7,7						
SONP			a8,3					a8,8				-Settling	+grazing
SOPD			a9,3						a9,9				
SOPP			a10,3							a10,10		-Settling	+grazing
SPO4			a11,3						a11,9	a11,10		SEDpo4	

Table 5.1-Definition Source/Sink Rate Parameters in terms of Kinetic Coefficients

Source/Sink	Description of constituent related process		Coefficient Formulation
NH3 Ammonia			
a1,1	nitrification		-k12*{DO/(Knit+DO)}
a1,3	respiration minus growth		Dr*anc-Gp*anc*Pnh3
a1,7	mineralization	OND	k71
a1,8	mineralization	ONP	k71[PHY/(Kmnc+PHY)]
NO3 Nitrate			
a2,1	nitrification		k12{DO/(Knit+DO)}
a2,2	denitrification		-k2d[Kno3/(Kno3+DO)]
a2,3	for algal growth		-Gp*anc*(1-Pnh3)
PHY Phytoplankton			
a3,3	growth minus respiration minus death		(1-fe)*Gp-Dr-Dd
BODD Dissolved BOD			
a4,2	denitrification		-5/4*32/14*k2D*[Kno3/(Kno3+DO)]
a4,3	algal death + excretion		aoc*foc*Dd+aoc*fe*Gp
a4,4	oxidation		-kd*[DO/(Kbod+DO)]
BODP Particulate BOD			
a5,3	algal death		aoc*(1-foc)*Dd
a5,5	oxidation		-kd*[DO/(Kbod+DO)]
DO Dissolved Oxygen			
a6,1	nitrification		-64/14*k12*[DO/(Knit+DO)]
a6,3	growth		Gp*[32/12+48/14*14/12*(1-Pnh3)]
a6,4	heterotrophic respiration		-kd*[DO/(Kbod+DO)]
a6,5	heterotrophic respiration		-kd*[DO/(Kbod+DO)]
a6,6	De-rearation		k2
OND Dissolved Organic Nitrogen			
a7,3	algal death + excretion		Dd*anc*fon+anc*fe*Gp
a7,7	mineralization		k71
ONP Particulate Organic Nitrogen			
a8,3	death		Dd*anc*(1-fon)
a8,8	mineralization		-k71*[PHY/(Kmnc+PHY)]
OPD Dissolved Organic Phosphorus			
a9,3	death + excretion		Dd*apc*fop+apc*fe*Gp
a9,9	mineralization		-k83
OPP Particulate Organic Phosphorus			
a10,3	death		Dd*apc*(1-fop)
a10,10	mineralization		-k83*[PHY/(Kmpc+PHY)]
PO4 Orthophosphate			
a11,3	respiration		Dg*apc
a11,9	mineralization		k83
a11,10	mineralization		k83*[PHY/(Kmpc+PHY)]

Table 5.2-Definition of Water Quality Modeling Kinetic Coefficients with Ranges of Values

Parameter	Default	Range	Definition	Units	Reference
Θ_{2D}	1.08	1.08	Temperature coefficient	None	Ambrose (1993)
Θ_{12}	1.08	1.08	Temperature coefficient	None	Ambrose (1993)
Θ_{1c}	1.068	1.068	Temperature coefficient	None	Ambrose (1993)
Θ_{1R}	1.045	1.045	Temperature coefficient	None	Ambrose (1993)
Θ_{1R}	1.045	1.045	Temperature coefficient	None	Ambrose (1993)
Θ_{2D}	1.045	1.045	Temperature coefficient	None	Ambrose (1993)
Θ_{71}	1.08	1.08	Temperature coefficient	None	Ambrose (1993)
Θ_{83}	1.08	1.08	Temperature coefficient	None	Ambrose (1993)
Θ_{d}	1.047	1.047	Temperature coefficient	None	Ambrose (1993)
Θ_{KT}	1.045	1.045	Temperature coefficient	None	Ambrose (1993)
Θ_{s}	1.08	1.08	Temperature coefficient	None	Ambrose (1993)
a_{nc}	0.25	0.25	Nitrogen to carbon ratio	g N/g C	Ambrose (1993)
a_{ncp}	0.25	0.25	Particulate nitrogen to carbon ratio	None	Ambrose (1993)
a_{oc}	32/12	32/12	Oxygen uptake per unit of algae respired	g O_2/g C	Ambrose (1993)
a_{oc}	32/12	32/12	Oxygen to carbon ratio	g O_2/g C	Ambrose (1993)
a_{pc}	0.025	0.025	Phosphorus to carbon ratio	g P/g C	Ambrose (1993)
a_{pcp}	0.75	0.70 ~ 0.80	Particulate organic phosphorus to carbon ratio	None	Ambrose (1993)
as	0.5	0.5 - 0.8	Assimilation efficiency of zooplankton grazing	None	EPA (1985)
c2chla	30	10 - 100	Ratio of carbon to chlorophyll a	None	EPA (1985)
C_s	Equation (6.3)	Equation (6.3)	Dissolved oxygen saturation	gO_2/ m^3	Ambrose (1993)
f_e Specify	0.1	0.1~0.8	Excretion fraction of phytoplankton	None	EPA (1985)
f_{oc}	0.5	0.5	Organic carbon from dead algae	None	Ambrose (1993)
f_{on}	0.5	0.5	Organic nitrogen from dead algae	None	Ambrose (1993)
f_{op}	0.5	0.5	Organic phosphorus from dead algae	None	Ambrose (1993)
I_s Specify	127	110-150	Saturating light intensity	watt/m^2	WDOE (1984)
Gzoo	1.0	0.0-2.5	Multiplier on graze rate	ND	
k_{12} Specify	0.02	(0.09~0.13)	Nitrification rate	day^{-1}	WDOE (1984)
k_{1c} Specify	2.0	2.0	Maximum growth rate	day^{-1}	Ambrose (1993)
k_{1D} Specify	0.015	0.2	Death rate	day^{-1}	WDOE (1984)
k_{1R} Specify	0.125	0.125	Phytoplankton respiration rate, 20°C	day^{-1}	Ambrose (1993)
k_{1R} Specify	0.015	(.05-.20)	Endogenous respiration rate at 20°C	day^{-1}	WDOE (1984)
k_2		Equation (13.6.2)	Reaeration rate @ 20°C	day^{-1}	Ambrose (1993)
k_{2D} Specify	0.09	0.09	Denitrification rate at 20°C	day^{-1}	Ambrose (1993)
k_{71} Specify	.0075	0.01-0.15	Organic nitrogen mineralization rate	day^{-1}	WDOE (1984)
k_{83} Specify	0.22	0.10-0.30	Dissolved organic phosphorus mineralization at 20°C	day^{-1}	Ambrose (1993)

Parameter	Default	Range	Definition	Units	Reference
K_{BOD}	0.5	0.5	Half-saturation constant for oxygen limitation	$g\ O_2/\ m^3$	Ambrose (1993)
k_d Specify	0.15	(0.02-0.20)	Deoxygenation rate @ 20°C	day^{-1}	WDOE (1984)
Kg_{macro}	0.101	0.101	Grazing rate due to macrozooplankton	day^{-1}	Boatman (1997)
Kg_{micro}	0.081	0.081	Grazing rate due to microzooplankton	day^{-1}	Boatman (1997)
K_{mN}	10/1000	25/1000	Half-saturation constant for nitrogen	$g\ N/\ m^3$	WDOE (1984)
K_{mNc}	1.0	1.0	Half-saturation constant for nitrogen mineralization	$g\ C/\ m^3$	Boatman (1997)
K_{mP}	1.0/1000	1.0/1000	Half-saturation constant for phosphorus	$g\ P/\ m^3$	Ambrose (1993)
K_{mPc}	5.0	1.- 10.0	Half-saturation constant for phosphorus mineralization	$g\ C/\ m^3$	Boatman (1997)
K_{NIT}	2.0	2.0	Half-saturation constant for oxygen limitation of nitrification	$g\ 0_2/\ m^3$	Ambrose (1993)
K_{NO3}	0.1	0.1	Michaelis constant for denitrification	$g\ O_2/\ m^3$	Ambrose (1993)
P_{NH3}	Eq. (13-6.2)		Preference for ammonia uptake term	None	Ambrose (1993)
S4 Specify	0.02	0.01-0.09	Algae contribution to sediment oxygen demand	None	Pamamat (1971)
SED_{NH3}	Formula	Data	Sediment release of ammonia	$g\ N/m^2/d$	WDOE (1984)
SED_{NH3m} Specify	0.0	0.25-2.50	Measured. sediment release of NH3	$g\ N/\ m^2/d$	Measurement
$SED_{PO4}m$ Specify	1.5	Field data	Measured sediment release of phosphorus	$g\ P/m^2/d$	Measurement
SODm Specify	0.0	0.5-5.0	Background sediment oxygen release	$gO_2/m^2\ /d$	WDOE (1984)
Sp5 Specify	0.07	0.01-0.09	CBOD_P contribution to sediment oxygen demand	None	Pamamat(1971)
Sp7 Specify	0.07	0.01-0.09	ON_P contribution to sediment oxygen demand	None	Pamamat(1971)
SpNH3	0.45	0.20-0.60	Fraction of sediment oxygen demand released as NH3		Panamat(1971)
v_{s4} Specify	0.09	(0.05~0.5)	Settling velocity Phytoplankton .	m/day	WDOE (1984)
v_{s5} Specify	0.08	(0.05 0.2)	Settling velocity of CBOD_P	m/day	EPA (1985)
v_{s7} Specify	0.08	(0.05~0.5)	Settling velocity of ON_P	m/day	EPA (1985)
v_{s8} Specify	0.08	(0.05~0.2)	Settling velocity OP_P	m/day	EPA (1985)
Wad Specify	Field data	Field data	Reaeration Wind speed	m/s	Measurement

Figure 5.3-Different layers describing sediment exchange processes (after DiToro, 2001)

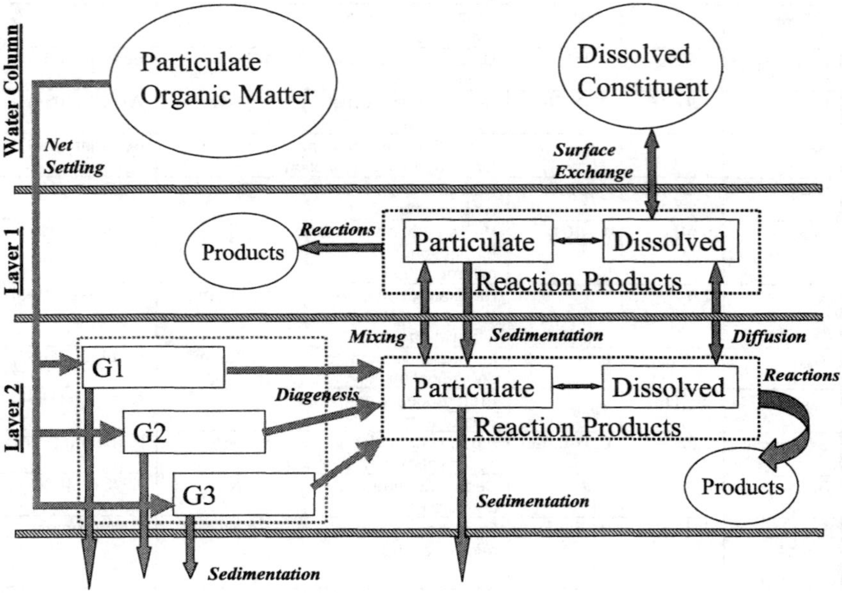

Figure 5-4 Parameters and geometric relationships that affect shade (after Boyd and Kasper, 2003).

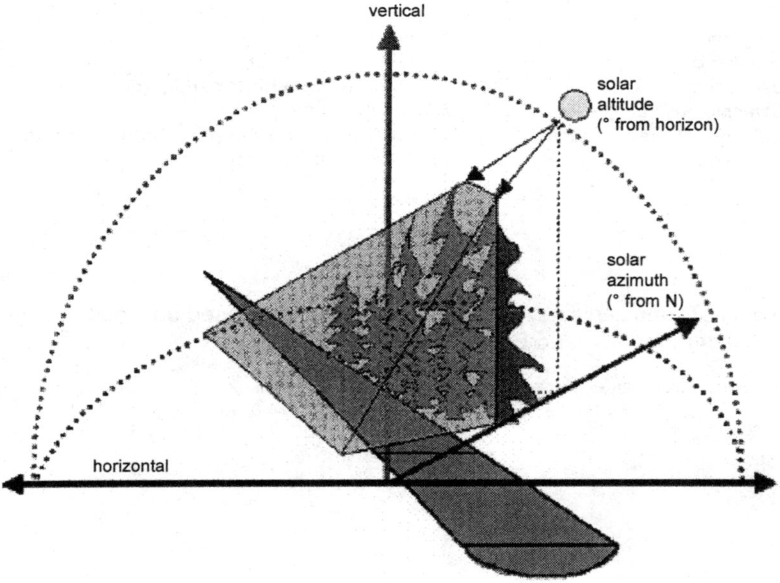

Table 5.3. Factors that influence stream shade (bold indicates those influenced by human activities).

Description	Parameter
Season/time	Date/time
Stream characteristics	Aspect, **channel width, depth, velocity**
Geographic position	Latitude, longitude
Vegetative characteristics	**Riparian vegetation height, width, and density**
Solar position	Solar altitude, solar azimuth

Table 5.4. Coefficients used to calculate reflectivity based on cloud cover (after Anderson 1954 and Brown and Barnwell 1987).

Cloudiness	Clear		Scattered		Broken		Overcast	
N	0-0.1		0.1-0.5		0.5-0.9		0.9-1	
Coefficients	A	B	A	B	A	B	A	B
	1.18	−0.77	2.2	−0.97	0.95	−0.75	0.35	−0.45

Figure 5-5 Koberg's family of curves for determining the value of the c constant in Brunt's equation for atmospheric infrared radiation (after Koberg, 1964).

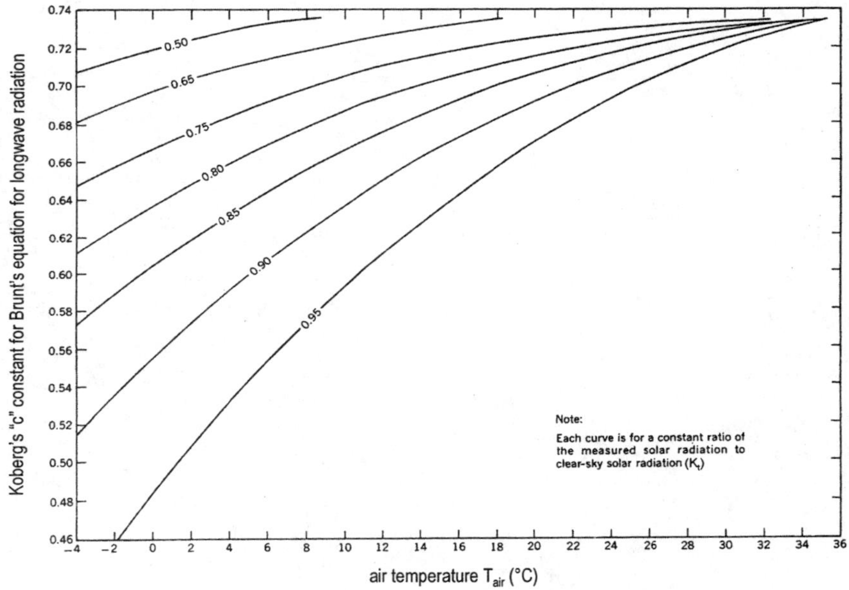

6. MODELING SYSTEMS AND THEIR APPLICATION

Edward M. Buchak

Jerad D. Bales

James L. Martin

In this chapter, we provide an overview of reservoir hydrodynamic and water quality models and the modeling systems in which they are embedded, with an emphasis on the most elaborate type of model likely to be encountered: the multi-dimensional, time varying hydrodynamic and water quality model. Guidance on model selection is included in the discussion on assessment of prototype behavior and development of the modeling plan. The value and importance of reliable data are emphasized throughout the sections on model grid construction, boundary conditions, and model calibration.

The primary purpose of this chapter is to provide a guide to water-resources managers who are supervising model applications and are using model results to guide decisions concerning use and allocation of water resources. Inexperienced model users should also find the discussion of the model application process to be a useful guide for project planning and implementation.

6.1. Models and modeling systems

Reservoir computer models are based on the partial differential equations presented and described in Chapter 4, "Theoretical basis of the hydrodynamic computations," and Chapter 5, "Theoretical basis of the water quality computations." The complexity of the fundamental relations is such, however, that models are often constructed by using a simplified form of the equations, with these simplifications based on reasonable assumptions about the reservoir system. These simplified equations may be based on assumptions about the importance of spatial variability (e.g., two-dimensional laterally-averaged instead of three-dimensional), time scale (e.g., steady-state instead of time varying), or various physical or chemical processes (e.g. simple empirical representations of turbulence instead of three-dimensional turbulence closure models).

The partial differential equations are written in finite difference (or finite volume or finite element) form so that the equations can be solved algebraically. The finite difference equations are augmented with algorithms that (1) map reservoir bathymetry to the finite difference grids; (2) allow the equations to step through time by accessing time varying boundary condition data; and (3) represent auxiliary functions such as wind shear, sub-grid scale turbulence, surface heat exchange, and chemical kinetics. The combination of the finite difference equations and associated algorithms, written in a scientific programming language such as FORTRAN, constitute a reservoir computer model. Newer coding techniques along with the advent of dynamic memory allocation allow models to be applied in different settings without recompilation of the computer code. Moreover, in recent years the models themselves have become more general so that the same computer code can be applied to a wide range of waterbodies in a variety of environmental settings.

The increasing sophistication of hydrodynamic and water quality models paralleled the improvements in computer hardware with sufficient speed, memory, and storage to make simulations within practical time frames, generally acknowledged to be on the order of several hours to several days. The accelerating capabilities of hardware and operating system software, especially with the advent of personal computers in the 1980's, allowed models to be embedded in modeling systems. In addition to the model itself, modeling systems include pre- and post-processors to aid in the creation of the input files and to summarize and display the model output.

Input files typically include a file that contains the parameters necessary to control the model execution, a file that provides the spatial domain information (model grid), and a set of time series boundary condition data. The time series boundary condition data may include flow rates into and out of the reservoir, as well as the temperatures and water quality properties of the inflows. Model output files generally include a file that repeats the input parameters for documentation and quality assurance purposes. Model results are typically (1) time series of velocities, temperatures and water quality parameters at one or more locations in the reservoir and (2) snapshots of these variables to show the state of the system at a particular time throughout the reservoir. These two output file types correspond to the two basic field observational methods: temporally continuous measurements at a point and synoptically collected data throughout the system at a specific time.

A prominent feature of modern modeling systems is the graphical user interface (GUI). GUI's allow use of descriptive parameter names rather than arcane computer variable names, avoid the need to precisely format input files, provide the user with immediate access to model documentation, and perform comprehensive error checking. The GUI's often incorporate elaborate post-processors that automatically generate time series graphs, animations of velocities and water quality parameters through time, and graphical comparisons of observed and computed values with statistical summaries of these comparisons.

GUI's can directly access model results stored in a relational database and thus take advantage of software developed specifically to store and retrieve millions of data points. An advantage of the relational database approach is that field observations can be stored and accessed in the same manner as model results so that that graphical, statistical, and visualization tools used to display model results can also be used to display field data. GUI's also have allowed modeling systems to methodically address spatial referencing issues, mostly in response to the development of software for processing geospatial data. Many modeling system read, for example, reservoir bathymetry directly from geographic information system (GIS) files and write results to shape files. As a result, all locations within the computational domain, or model grid, are geo-referenced with latitude and longitude (or other coordinate system such as UTM or state plane) assigned to each grid cell or node. Other geo-referenced files (e.g., biological communities, pipeline routes, industrial facilities and aerial photographs) can then be directly overlain on the model grid so that results can be clearly scaled to recognizable landmarks.

All of these advances have somewhat removed the typical model user from the details of the mathematics, physics, and biochemistry represented in the model. As a result, users now have the ability to apply models without a basic understanding of the computations they embody and may blindly accept model results without careful consideration of model limitations, assumptions, and unique characteristics of the system to which the model is applied. Nevertheless, it remains the responsibility of the user to demonstrate that the computer model is an adequate representation of the physical and chemical systems under consideration.

Reservoir water quality models are often used to assist in the design of new facilities. For example, the model can be used to identify the level of treatment for a new sewage treatment plant that is required to meet water quality standards in the reservoir, and to determine the best location of the outfall in order to minimize impacts on recreational areas of a reservoir. These design parameters cannot be readily determined from field experiments because the facility does not yet exist, so a model provides the means to conduct numerical experiments with various levels of treatment and alternate locations of outfalls. Reservoir water quality models also are used to assess the impact of changes for conditions that cannot be easily reproduced in the field. For example, a reservoir model can be applied to determine the impact of an existing power plant thermal discharge for a permit-specified extreme low-flow event occurring during maximum plant load. The extreme low-flow may not occur for many years, and when it does occur, may not coincide with a maximum plant load. Application of a model provides an immediate answer to the question, as well as the opportunity to modify the permit. In addition to these types of engineering applications, models can provide insight into important physical, chemical, and biological processes in a structured way, providing unforeseen benefits to water-resources managers.

The use of the term "model" may suggest to the public a level of certainty does not exist, given the complex and dynamic character of natural waterbodies, the assumptions in model code, and the limitations of data. As a result, perhaps the term "computation," the same term that engineers use when performing a hand calculation, is preferable to "model" when communicating model results to the general public. In this way, the basic notion that models are not perfect representations of the reservoir system remains intact.

6.2. <u>The model application procedure</u>

The use of a model for a particular reservoir is called a model application. Model applications are usually done in a sequence of steps (table 6-1) and are typically performed in an iterative manner. For example, the model grid is often adjusted during model calibration in order to achieve more satisfactory results.

<u>Step 1: Identify the Problem</u>

The first step in a modeling application is to clearly identify the problems to be solved, or to state the questions to be answered. Each party involved in an issue may have a particular set of concerns, and although each concern may be somewhat similar, distinctly different approaches could be required to address each of the concerns.

For example, in the case of a proposed new thermal discharge to a reservoir, issues might be:

- How does the thermal discharge affect fishing? (public)
- How will changes in water temperature affect phytoplankton growth? (state water quality agency)

Table 6-1. Reservoir model application procedure

Step	Typical tasks	Typical products
Identify the problem	Review issues with clients, stakeholders, and regulatory agencies to reach consensus on problem	Clear statement of questions to be answered.
Assess prototype condition	Review all available data, literature, and anecdotal evidence; visit site.	Data inventory; citation list; written summary of prototype features and processes
Develop the modeling plan	Define modeling objectives; assess time and financial resource limits; select model; select study period; authorize and design field program	Modeling plan including schedule, budget; QA plan, SOP and QAPP for field program
Build the model grid	Acquire bathymetry maps, elevation-area-volume tables, and engineering drawings; develop grid	Functioning model grid and control file; steady-state simulations; QA report
Assemble Boundary Conditions	Acquire, review, and format time series boundary data; review, format, and upload all datasets to the project database	Documentation of methods used to acquire and/or synthesize boundary data; QA report
Assemble Initial Condition Data	Acquire or estimate data for initial conditions; review, format, and upload all datasets to the project database	Documentation of methods used to acquire and/or synthesize initial conditions; QA report
Assemble Calibration and Evaluation Data	Acquire or estimate data calibration and confirmation data; review, format, and upload all datasets to the project database	Documentation of methods used to acquire and/or synthesize data; QA report
Develop Model Input	Convert or modify data as necessary to conform to model input requirements, upload data to model database or incorporate into input fiies; debug model setup	Documentation of methods used and testing of input; QA report
Calibrate the model; test model performance	Compare model output to observations: water surface elevations, in-reservoir temperature and water quality profiles, release temperature and water quality time series; perform model performance analysis	Run log; calibrated model; QA report
Use the model for water resources management applications	Develop additional input files to represent management scenarios; model runs for each scenario	Report including archive of model with all input files, run log, and significant runs; report

- How much will the temperature in the lake be increased by the thermal discharge? (permitting agency)

- Is there sufficient cold water to operate a downstream power plant efficiently? (power company)

Each of these questions, while all related to the impacts of a thermal discharge, could require a slightly different type of data-collection and modeling program. In addition, those involved in the application of the model may have a set of preconceptions based on previous studies, so it is important that the unique conditions of the problem at hand be carefully identified. Consensus among all parties on the specific questions to be answered by the model application should ultimately lead to a well designed modeling program and to results that meet the needs of all stakeholders.

Step 2: Assess Prototype Conditions

After the problem(s) has been clearly stated, the next critically important task in the model application process is to systematically assess the prototype conditions. The objectives of this step are to (1) articulate hypotheses on dominant physical processes that control reservoir behavior (see, for example Fischer, et al., 1979 and Martin and McCutcheon 1999) and (2) identify important aspects of the system that should be included in the model. This step includes an analysis of available field data, review of previous pertinent reports including previous modeling studies, and identification of key physical attributes of the system.

Some questions that should be answered in this step include:

- What are the general physical characteristics of the lake?
- What are the dominant thermal and water quality gradients: with depth, across the reservoir, or down the longitudinal axis of the reservoir?
- When and to what extent does the reservoir thermally stratify?
- Is there ice cover in the winter?
- Are density currents (thermal or sediment) present?
- Is wind a dominant factor in mixing and transport?
- Where and what magnitude are the thermal discharges to the reservoir?
- What are the characteristics (flow, constituent loading) of other point-source discharges to the reservoir?
- What is the trophic state of the reservoir, and how does the trophic status vary throughout the system?
- Are macrophytes a major part of the nutrient cycle of the reservoir?
- How does the clarity of the water vary with time and through the system?

- What are the major inflows to the reservoir?
- What are the inflow constituents of concern?
- Is groundwater a major contributor to the reservoir?
- What are the characteristics of the withdrawals at the dam and elsewhere?
- Is there an ongoing data-collection program in the reservoir, and if so, what are the characteristics of the data-collection program?
- What data are collected on streams and rivers draining to the reservoir?
- What kind of meteorological data are collected nearby?

Products from this step include an inventory of previously collected data, a summary of ongoing data collection, a citation list, and a written summary of salient features and processes specific to the prototype. This summary can be updated as the model application proceeds, informed by insights developed as the modeler uses both the datasets and the model to gain further insight into the underlying physical, chemical, and biological processes in the structured way made possible by the model application process.

<u>Step 3: Develop the Modeling Plan</u>

The objectives of the study and the approach to be used to achieve the objectives are described in the modeling plan. Most aspects of the study plan depend on a clear statement of modeling objectives, so it is important that the objectives are clearly defined. The approach should describe the model selection process; identify the study period; describe data collection, compilation, and analysis; describe methods used to synthesize required model input data in the absence of measurements; define the model calibration process, including acceptable tolerances between simulations and measurements; and present a summary of expected products

Model selection

Any one modeling code is generally not applicable to all types of reservoir problems because most models are developed with certain limiting assumptions (or simplifications) and typically include a subset of all of the physical and biogeochemical processes present in the reservoir. Selection of the appropriate model for the study is primarily dependent on the objectives identified in the modeling plan. Model selection should include consideration of technical factors, such as the scale (temporal and spatial) of processes to be simulated, the types of physical, chemical, geomorphic, and biological processes to be considered, and conditions in the study area. Institutional factors, such as familiarity of the user with the model, model support, and ease of model application, also should be considered.

Generally, the most sophisticated model that can be supported by available datasets, financial resources, and schedule is chosen. In some cases, a second, more simplified model is used in conjunction with the primary model. The role of the secondary model is to provide quality assurance in much the same way a simple hand calculation is used to check the results of a more complex calculation.

There are many models available for use, and there are numerous reviews that provide comparisons of selected models. The USGS Surface water quality and flow Modeling Interest Group (SMIG) web site contains a description of 38 models with links to model web sites (U.S. Geological Survey, 2005a). The SMIG web pages also contain links to various model archives maintained by the USGS, EPA, Army Corps of Engineers, National Weather Service, US Department of Agriculture, US Bureau of Reclamation, and others. The Water Environment Research Foundation Model Selection Tool is an application that allows users to identify characteristics of the problem to which the model is to be applied and to define desired features of the model (Water Environment Federation, 2005). The tool then automatically identifies candidate models from a database of more than 150 available options. Other reports that provide reviews of models include Walton and others (1998), who reviewed 13 three-dimensional models for potential application to the Duwamish River-Elliott Bay system in Seattle, Washington {see also WEST Consultants, Inc (1996); Imhoff et al. (2003); and HGL and Aqua Terra (1999)}.

There are two basic classification schemes that can be useful in the model selection process. The first classification scheme is based on the technical characteristics of the models and coincides with the "features and processes" approach to evaluating the study area. The second classification scheme addresses the institutional properties of the model, which includes such user-oriented issues as model availability and technical support.

Model classification scheme – technical characteristics

Four important technical characteristics of reservoir models are (1) spatial dimensions, (2) time scale, (3) approach used to simulate water movement, and (4) constituents and methodology used in the water quality simulations. Technical properties of several commonly used reservoir models are summarized in Table 6-2.

Table 6-2. Characteristics of selected reservoir models
[1D, one-dimensional; 2D, two-dimensional; 3D, three-dimensional; phen., phenomenological; hyd., hydrodynamic]

Model	Technical Properties				Institutional Properties			Reference
	Spatial averaging	Temporal averaging	Water Movement Simulation Method	Coupling of Water Quality to Water Movement	Availability	Applications	Support	
CE-QUAL-R1	1D layer averaged	time varying	phen.	coupled	Public, no new updates expected.	Wolsinski and Collins, 1985.	Very limited support.	U.S. Army Corps of Engineers, 1986
DYRESM-CAEDYM	1D layer averaged	time varying	phen.	uncoupled	Public.	DYRESM User Group, 2005	Available through user group	Antenucc and Imerito, 2002
MINLAKE	1-D layer averaged	time varying	phen.	coupled	Public	Stefan and Fang, 1994	Limited	Riley and Stefan, 1988
BETTER	2D laterally-averaged	time varying	phen.	coupled	Public; new updates not expected.	TVA reservoirs	Limited	Bender, et al., 1990
CE-QUAL-W2	2D laterally-averaged	time varying	hyd.	coupled	Public	Water Quality Research Group, 2005	Training classes and web page	Cole and Wells (2004)
GEMSS	3D	time varying	hyd.	coupled	Public	J.E. Edinger Associates, 2005b	Training classes and other	J.E. Edinger Associates, 2005a
EFDC/WASP or EFDC/CE-QUAL-ICM	3D	time varying	hyd.	uncoupled	Public	Typically used in estuaries and large lakes	Training classes ocassionally available	US EPA, 2005a (EFDC); US EPA, 2005b (WASP); Cerco and Cole, 1995 (CE-QUAL-ICM)

Reservoir models based on the assumption that the entire reservoir is fully mixed and are called box or zero-dimensional models. Properties (velocity, temperature, dissolved-oxygen concentration, etc.) measured at one location in the reservoir are assumed to be representative of that property at any location in the reservoir. Zero-dimensional models generally have limited applications, but these models are capable of performing simulations for very long time periods (decades, for example) because of the rapid computation times associated the simple mathematical formulation of the model. Such models can incorporate a limited number of key parameters, such as surface heat exchange and inflows rates for temperature-related problems, and can often be operated by using information readily obtained from existing databases (for example, Buchak and Krallis 2002).

One-dimensional models can be used to represent either cross-sectionally averaged conditions or layer-averaged conditions. Cross-sectionally averaged models are based on the assumption that the important gradients in the reservoir are in the longitudinal direction, and that conditions measured at any point laterally or vertically in the cross section are representative of conditions throughout the cross section. These types of models are most typically applied to unsteady flows in rivers (for example, the U.S. Army Corps of Engineers (2005) HEC River Analysis System), but might be appropriate for shallow (limited thermal stratification) and relatively narrow run-of-river reservoirs.

One-dimensional layer-averaged models are based on the assumption that the primary gradients are in the vertical direction, and that conditions measured at a particular depth in the reservoir are representative of conditions measured at that same depth throughout the reservoir. Layer-averaged models would be appropriate for reservoirs that are rather small, have relatively low inflow, and do not have extensive coves, such as thermal power facility cooling ponds. The Dynamic Reservoir Simulation Model (DYRESM; Antenucci and Imerito, 2002) is one of the most widely used one-dimensional layer-averaged models, although most two-dimensional lateral-averaged and three-dimensional models can be configured to operate in the one-dimensional mode.

Two-dimensional models can be vertically averaged, in which vertical gradients at any point in the reservoir are assumed to be negligibly small, although variations in the depth-averaged property can vary from location to location within the waterbody. Laterally averaged models are based on the simplifying assumption that lateral gradients are small relative to vertical and longitudinal gradients in the waterbody. Depth averaged models have been applied to shallow, wind-driven lakes, but are seldom applied to reservoirs. Laterally averaged models, however, are perhaps the most commonly used approach for reservoir water quality modeling. Most three-dimensional models can be configured to operate as a two-dimensional vertically averaged or a laterally averaged model.

Three-dimensional models are applied when vertical, longitudinal, and lateral gradients must all be resolved in order to properly meet the objectives of the modeling study. An example of an application for a three-dimensional model is a long and deep reservoir, in which several tributaries enter the reservoir laterally, and the mixing of the tributary inflow through the reservoir must be resolved.

It should be noted that because the hydrodynamics and water quality transport properties of reservoirs are dependent on boundary conditions – meteorological parameters, inflow and outflow rates and constituent concentrations, and effects of complex bathymetries – models of the fully-mixed, 1D, 2D and 3D types necessarily require essentially the same boundary condition datasets. Furthermore, as computers become faster and less expensive, applications in the future will trend to using 3D models with very fine spatial and temporal resolution. For example, the boundary-fitted coordinates currently used to provide accurate domain definition while allowing coarse grids could be supplanted by simple orthogonal grids using 1 m x 1 m horizontal spacing and 0.1 m vertical spacing.

Models can be applied in a deterministic or a stochastic temporal mode. Historical periods are reconstructed in deterministic applications by using concurrently collected boundary condition data, allowing for hindcast simulations. Historical conditions can then be compared with possible future scenarios in which future boundary conditions are synthesized, or in which other possible future conditions (changes in point-source discharges or withdrawals, revised reservoir rule curve, etc.) are incorporated into the model. A simplified deterministic approach can also be used by assuming steady state boundary conditions. Such an approach is often used to determine if the model is performing properly or to compare model simulations with results obtained from analytical solutions of very simple forms of the governing equations. Stochastic applications are performed by reducing time varying forcing function data into basic statistical properties (for example, mean and standard deviation). The model is then repeatedly applied by using randomly generated boundary data that have the same statistical properties as the measured conditions.

Models compute water movement by using either numerical solutions of the hydrodynamic relations presented in Chapter 4 or by using a phenomenological approach. One-dimensional, layer-averaged models often use the phenomenological approach, in which water is introduced into a specific layer by matching the density of the water in the reservoir with the density of the inflow. The volume of the layer is increased by this introduced volume of water, and the temperature and other properties of the layer are estimated by assuming instantaneous mixing in the layer. Wind energy is introduced at the surface and allowed to mix the water column as it decays with depth. This method strings together established empirical computations in a quasi-finite difference format. The advantages of the method are that it is easy to understand, simple to code into a model, relatively easy to calibrate (few parameters to adjust), and the computer applications process rapidly. The approach, however, does not have an overall framework that corresponds directly to the fundamental hydrodynamic relations, and complex flow systems generally are not simulated accurately using this approach.

The transport of water quality constituents is performed in one of two general ways: (1) all hydrodynamic and water quality computations for the entire reservoir are performed simultaneously for each computational time step (coupled approach) or (2) the hydrodynamic computations are performed and the flow fields stored; water quality transport computations are subsequently performed using the stored velocities (uncoupled approach). An exception to separating the hydrodynamic and water quality transport is made for those constituents that affect density. Flow and density are physically coupled, because flows move properties (heat, dissolved solids, and sediment) that affect density and density in turn affects the flow field; consequently the transport of these constituents needs to be part of the hydrodynamic computation. Most, although not all, reservoir models include this coupling between flow and density by solving the hydrodynamic and heat transport equations simultaneously. Some reservoir models, primarily those derived from estuarine or ocean models, retain the coupling between dissolved solids and flow. Few reservoir models include the ability to simulate density differences resulting from extremely high suspended sediment concentrations. A final distinguishing feature of the water quality component is the number and types of water quality processes included in the model. As an example, oxygen demand from bottom sediments can be treated as a zero-order process (a constant demand rate is specified); a first order process (demand is a function of some other parameter, such as water temperature); or as a complex process that involves chemical reactions among a large number of constituents.

A model can be readily selected for application to a particular problem if existing datasets are available for the study site, if the reservoir to which the model is to be applied is reasonably well understood, and if the model user is experienced. If the existing datasets are inadequate, it is reasonable to initially assume that the study reservoir has characteristics similar to other reservoirs of similar morphological and flow characteristics. For example, it is reasonable to assume that flow processes in a deep and narrow reservoir with relatively small annual inflow and outflow rates located in a temperate climate are dominated by solar radiation and wind mixing, and that the reservoir is seasonally stratified with little or no lateral variation. In this case a two-dimensional, laterally averaged model would be appropriate.

Model classification scheme – institutional properties

Four important institutional properties of reservoir models are (1) model availability, (2) model complexity, usability, and required resources for application, (3) number of applications of the model to similar conditions, and (4) the level of technical support available to the model user.

Models may be available in the public domain or proprietary, and for either case, the model source code may or may not be available. The initial choice of many modelers is to use public domain models for which source code is routinely included in the distribution package. Cost, a large user community, and credibility are reasons often cited in making this choice. Many experienced modelers will not work with models for which they do not have access to the source code. The use of proprietary models can sometimes be justified because many of these codes have advanced computations and include GUI's that save substantial time in completing the application. Version control and professional assistance can also be important features of proprietary codes.

Model complexity and usability can be important issues, particularly for first-time users and for applications that require strict adherence to project schedules. Resources required for model application include the hardware and software requirements (cost of compilers and the associated training costs, for example), as well as the level of expertise and experience required of the user. When selecting a model, potential users should be aware of previous applications done for reservoirs similar to the one for which the study is to be performed. The professional literature is an important source for assessing model success in previous applications, as are the model developers and general user community.

Study period selection

Deterministic models are typically applied to periods of at least one year in order to simulate the seasonal variations in thermal structure and inflows. These models require datasets without significant gaps in boundary condition data. For example, daily reservoir inflow records missing more than a few days cannot be used without reconstructing the missing data. A good way to quantify the availability of data and select the most appropriate study period is to list the start and end times for the available datasets and the longest period(s) in which all required data are available.

If multiple years of data are available, then some consideration can be given to selecting years in which some form of extreme conditions (high inflow or low inflow, for example) are represented. Extreme conditions should be specified relative to the problem to be addressed. For example, biological criteria are often related to weekly average water temperature, whereas thermal discharge requirements are often related to an instantaneous maximum water temperature.

Step 4: Build the Model Grid

For purposes of this discussion, the reservoir dimensional data are referred to as bathymetry which can be in the form of soundings, transects, cross-sections, or contours displayed on a map. Digital elevation models, formatted for viewing with GIS software, can be used to view and manipulate bathymetric data. The finite difference (or finite element or finite volume) form of the bathymetry is called the computational grid. After the grid has been constructed, engineering drawings can be used to identify grid cells and their indices for various structures, such as bridge, intakes, weirs, etc.

The complex reservoir bathymetry is represented by the computational, or model, grid (fig. 6-1). The two major classes of grids are structured and unstructured.

In the simplest case, a structured grid is a rectangular array of points with a well-defined and regular relation between each point on the grid and its neighbor (fig. 6-2A). Irregular structured grids can be thought of as regular orthogonal grids, but with all of the connections between grid points made of rubber bands so the grid can be stretched to a complicated shape as long as none of the connections are broken and none of the grid elements are turned inside out. The accuracy of numerical schemes for solving equations on irregular structured grids typically diminishes as the stretched grid becomes increasingly non-orthogonal. Unstructured grids (fig 6-2B), often used in finite volume or finite element computations, offer more flexibility in representing complex geometries than do structured grids. Grids can be constructed from a variety of elements (triangles, quadrilaterals, tetrahedral, hexahedra). Unlike structured grids, however, the spatial relation of one element to another must be explicitly stated because there is no implied or logical relation among grid elements.

In addition to structured and unstructured, computational grids can be classified based on the number of spatial dimensions represented in the grid. A zero-dimension model would be one in which the entire reservoir is represented as a single box having a volume equal to the volume of the reservoir. As previously discussed in Chapter 4 and Section 6.2 Step 3, one-dimensional models can be either layer-averaged, in which the primary gradient of important conditions (hydraulics, temperature, water chemistry, etc.) is in the vertical direction, or cross-sectionally averaged, in which the primary gradient is from upstream to downstream. For a layer-averaged model, the model user specifies the thickness of each layer, the length of the layer, and the width of the layer. The length of each segment and a relation between water depth and cross-sectional area for the segment is specified for a one-dimensional cross-sectionally averaged model. Two-dimensional

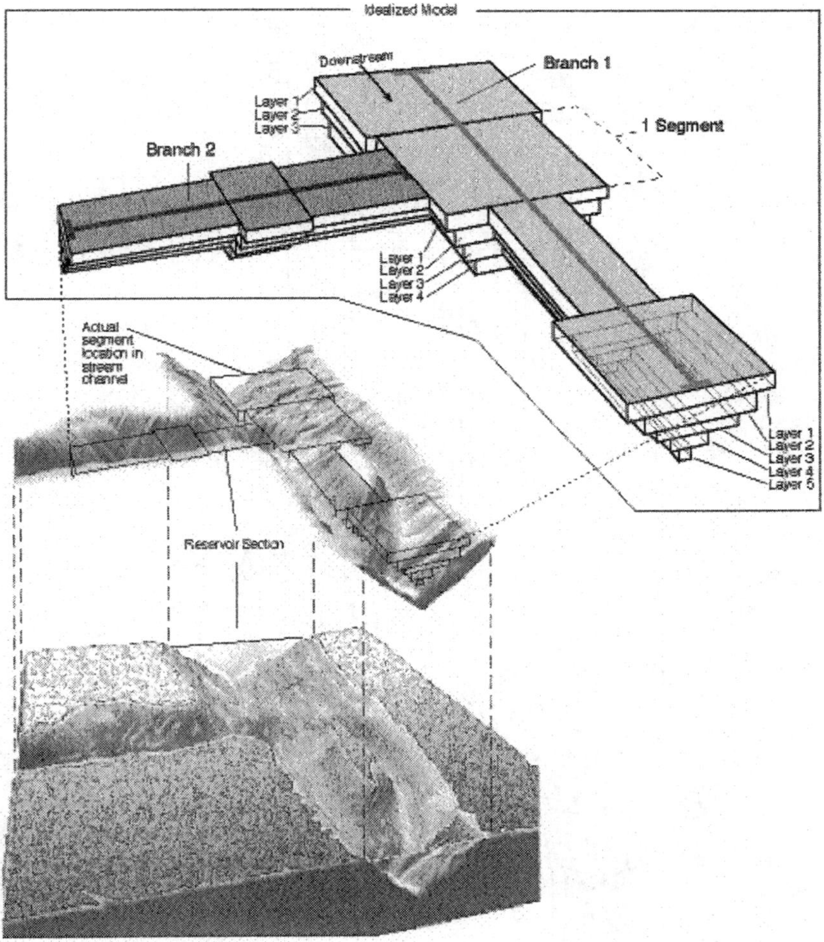

Figure 6-1. Transformation from natural bathymetry to an idealized, two-dimensional laterally averaged model grid.

A

B

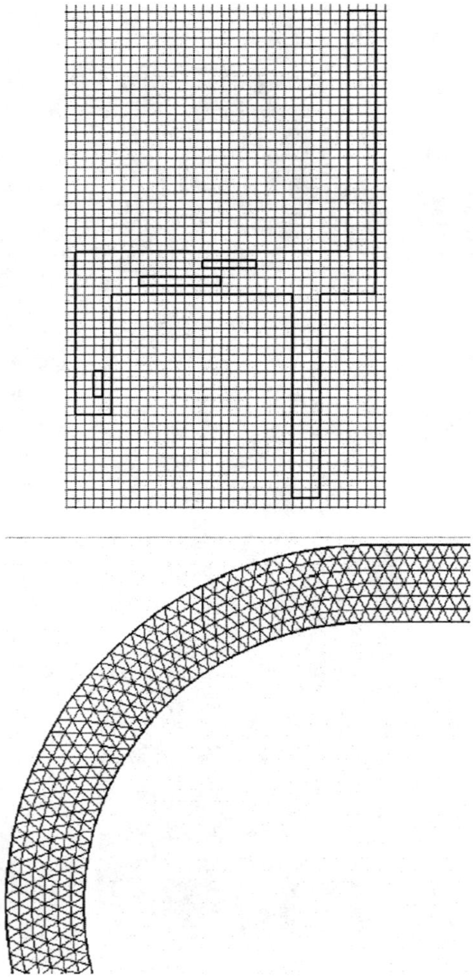

Figure 6-2. Examples of (A) structured and (B) unstructured grids, showing plan views of two-dimensional depth averaged model grids.

models can be represented as laterally-averaged (Chapter 4, fig. 6-1), in which conditions in a computational cell are assumed to be relatively uniform across the entire width of the reservoir, or depth-averaged (see Chapter 4), in which conditions in a cell are assumed to be uniform over the full water column. Model grids provide the fundamental reservoir dimensions of length, width and depth to the model. The width, thickness, length, and horizontal position of each computational cell must be specified for 2D and 3D models.

Because laterally averaged models can have infinitely variable cell widths, these models represent the bathymetry of long, narrow reservoirs in a fashion that mimics the hydrodynamics of such systems. For 3D models to represent reservoir bathymetries as well as laterally averaged models do, they typically must have detailed lateral resolution, something like 10 or more cells in the lateral.

Several factors should be considered in the development and refinement of the model grid, but perhaps the most important for reservoir models is to ensure that the volume of water represented in the model grid at a given water-surface elevation is the same as in the prototype. Accurate representation of the prototype surface area in the model grid also is important for models in which heat transport and wind mixing are important processes.

There are a wide variety of sources for bathymetric data (table 6-3). The reservoir elevation-area-volume information can usually be obtained from the project report, which is the basic engineering description of the design phase of reservoir construction. The elevation-area-volume table in the project report for older reservoirs was usually developed from planimetered topographic maps on which elevation contours were drawn. These tables

Table 6-3. Use and sources of bathymetric and structural data

Category/Type	Use	Sources
Reservoir shoreline at maximum water-surface elevation	Develop model grid	Topographic maps; GIS digitized data; digital elevation models.
Reservoir cross-sections, depth measurements, or bathymetric contours	Develop model grid	Sediment range surveys; project contour maps; or project surveys.
Elevation-area-volume table or curves	Verify model grid	Reservoir project report; bathymetric data collected during or prior to the modeling study
Engineering drawings of release structures, weirs, flumes, and dams	Locate structures within the grid	Reservoir design report
Reservoir mile points	Identify locations of tributaries, withdrawals, discharges, and other features.	Reservoir project report, or developed during modeling study.

may reflect inaccuracies and lack of precision in both the topographic maps and the planimetering procedure. In addition, sedimentation that has occurred during the life of the reservoir may have resulted in significant loss of storage. Any available data from post-construction range surveys (bottom elevation traces at regular intervals along the reservoir centerline or at benchmarked locations typically used to measure sedimentation rates over the life of the reservoir project) should be compared with information from the project report.

The most accurate and reliable bathymetric data for constructing model grids are obtained with modern geo-referenced acoustic sounding systems that record hundreds of thousands of depths simultaneously with horizontal position (for example, fig. 6-3). The shoreline can be accurately defined by

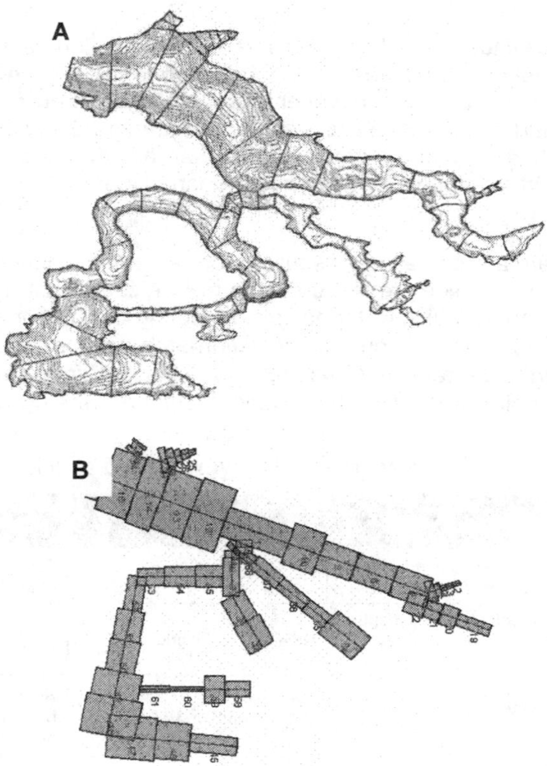

Figure 6-3. (A) Bathymetry of Kensico Reservoir, New York, showing contours and model segments, and (B) corresponding CE-QUAL-W2 model grid.

using digital orthophotoquads to complete the geo-referenced datasets. The grid should be constructed to include the highest water surface elevation anticipated, so the bathymetry at higher water-surface elevations may have to be obtained from existing digital elevation models (for example, U.S. Geological Survey, 2005b) or from topographic maps.

Another issue to consider as the model grid is developed is the need to balance adequate spatial resolution of the reservoir in the computational grid with the requirement for reasonable computational times. Most models also have numerical limitations such that the computational time step must be less then the volume of any cell divided by the flow into or out of the cell during the time step, or

$$\text{Computational time step} < (\text{Cell volume})/(\text{Flow into or out of the cell})$$

which is a form of the Courant condition for mass flux. There are two consequences of using small grid sizes: the number of computations increases with the increased spatial resolution (i.e. increased number of cells) and, for most models, the computational time step decreases when the resolution increases, resulting in increased computer run times. Horizontal resolutions of 0.5 to 1 km along the reservoir centerline and lateral directions and 0.5 to 1 m in the vertical are common.

Step 5: Assemble Boundary Condition Data

Boundary conditions for reservoir models can generally be described as a time series of data that are used to define the inflows of momentum, mass, and heat to the model domain, and the outflows of water from the model domain. Outflows of heat and chemical constituents from the model domain typically do not need to be pre-defined, but are computed by the model. Since the models described here are generally of the deterministic type, all boundary condition data must be obtained for a concurrent period in order for the model to run and represent hindcast conditions. In other words, meteorological data from 2004 cannot be used with inflow rates measured in 1999 to properly simulate observed conditions.

The creation of boundary data files for model applications can be a time consuming process because of all of the various types of data that must be assembled (table 6-4). The quality of the model simulations, however, depends directly on the quality of the boundary condition data so it is important to ensure that all inflows and outflows from the study area are properly represented. Uncertainty in boundary conditions translates directly to uncertainty in model results.

Essentially all deterministic models, regardless of the degree of temporal and spatial averaging, require the same general types of boundary condition data. Hence, the notion that, for example, a three-dimensional model requires more

Table 6-4. Typical types of boundary condition data

Category	Types of Data (data required depends on model)	Typical Measurement Frequency	Use of Data in Model	Sources of Data
Meteorological data	• Dry bulb and dew point temperatures •Wind speed and direction •Cloud cover and solar radiation • Atmospheric pressure •Precipitation	Hourly	Calculation of surface heat exchange, evaporation, wind shear stress and water balance	•National Weather Service •Local weather stations •Project observations
Water inflow and outflow rates	• All river and stream inflows • Estimates of diffuse (or nonpoint) inflows for large systems • All point source discharges • Releases at the dam • All withdrawals	Hourly to daily	Calculation of water balance; also used in calculation of heat in constituent transport	•U.S. Geological Survey •U.S. Army Corps of Engineers •U.S. Bureau of Reclamation •Hydropower companies •Dam operators •Utilities •State NPDES records •Project observations
Water temperature	Temperature of all inflows, including streams, rivers, nonpoint sources, and point sources (if inflows are thermally stratified, then vertical distribution of water temperature may be required	Typically same frequency as water inflow rates	Calculation of heat balance	•Project observations •Estimates of water temperature derived from meteorological data •State NPDES records
Chemical constituent inflow rates	Concentration (or sometimes mass) of all relevant constituents (dependent on model and model application) from all inflow sources	Monthly to weekly	Calculation of mass balance of chemical constituents	•U.S. Geological Survey •State and local monitoring programs •Project observations

boundary data than a two-dimensional model is not necessarily correct, although a three-dimensional model may require more data for a good calibration than a two-dimensional model.

Initial model applications are often made by using steady (temporally invariant) boundary conditions. The purpose for using steady boundary conditions is to test the computational grid and other features of the model without introducing the complexity of time varying conditions. Steady boundary conditions might include estimates of mean seasonal inflow rates, mean seasonal temperature, and so on. These steady-state simulations can be used to confirm that the model is properly constructed, and could be followed by simulations in which important daily cycles (meteorological conditions, for example) are repeated for many days until a quasi – steady state is developed, providing insight into important reservoir time scales and characteristics of the numerical scheme used in the model. Finally, the full set of measured time varying boundary conditions can be applied in order to calibrate and further test the model.

Boundary conditions can be assembled by using previously collected data or data collected specifically for the study. When time and resources permit, the collection of new data is preferred because data collection strategies can be optimized to reduce model uncertainty and because data will represent current reservoir conditions, rather than some past state that may differ from the present. One challenge associated with monitoring programs is to collect data over a relatively short period that consistently represents the range of conditions that occur in the reservoir over the longer term.

Historical data should be carefully reviewed to ensure acceptable data quality (see section 6.2.10). If data were collected by more than one group, or if data were collected over a long period, differences in sample locations, collection methods, sample handling, and laboratory analyses may suggest trends or variability which are, in fact, an artifact of variability in methods. Another difficulty in using previously collected data is that all the required data for model application may not be available from a concurrent period. The decision must then be made to either synthesize the needed data or to design and implement a new data collection program. If data are synthesized, the sensitivity of model results to changes in these synthesized boundary conditions should be carefully documented during model testing.

The general types of required boundary condition data for reservoir models includes (1) meteorological data, (2) water inflows and outflows, (3) water temperature of inflows, and (4) chemical concentrations (or loads) of inflows (table 6-4). Boundary condition data requirements are discussed in the following sections. As noted, one or more supporting models (sometimes referred to as "sub-models") must be run to complete boundary condition datasets which contain periods of missing data. For example, watershed runoff rates and loadings are sometime generated with sophisticated hydrologic models like HSPF. Even proportioning ungaged inflows to nearby gaged subwatersheds must be regarded as the application of a supporting model.

Meteorological data

Meteorological data are required to compute the reservoir water mass balance, heat balance, and wind stress at the water surface, which is calculated from wind speed and direction. Data generally are required at hourly intervals to account for the daily variation in temperature, radiation, and wind, but daily mean values may be adequate for some applications.

Precipitation and evaporation rates are required for the water mass balance. Precipitation is measured. Evaporation can be calculated by the model, which is the typical approach, or can be based on measurements. The type of boundary condition data required for the calculation of evaporation depends on the formulation used by the model to compute evaporation, but may include wind speed, barometric pressure, air temperature, dew point temperature, and relative humidity.

Surface heat exchange rates also are dependent on meteorological boundary condition data. Surface heat exchange can be calculated by either summing the various terms contributing to heat exchange or by using the equilibrium temperature and an estimate of the surface heat exchange coefficient. Boundary condition data required for the first approach (summation of terms) include measured or estimated short-wave solar radiation (estimation typically requires observed cloud cover), measured or estimated long-wave radiation (estimation typically requires air temperature and cloud cover), and data required for the computation of evaporation (previously described), which also are used to compute evaporative heat flux. The equilibrium temperature approach requires a time series of equilibrium temperatures and coefficients of surface heat exchange (see Chapter 5). However, computation of these two parameters requires the availability of the same meteorological data used in the summation of terms approach.

Water inflows and outflows

An accurate water mass balance for the reservoir over the simulation period is fundamentally important to the development and application of a reliable reservoir model. The water mass balance states that the change in reservoir contents is equal to the difference between reservoir inflows and outflows. The mass balance depends on good measures of inflows and outflows, as well as an accurate relation between reservoir water-surface elevation and reservoir volume. The setting and hydrology of the study reservoir determine the relative importance of the various inflows and outflows and the time increment at which data must be supplied.

All inflows to the reservoir must be measured or estimated, with data required at hourly to daily time increments for the entire simulation period. Inflows generally include all rivers and streams, point-source discharges, precipitation, diffuse runoff directly into the reservoir, and ground-water discharge to the reservoir.

Streamflow records, both current and historical, are available for many locations throughout the nation from the U.S. Geological Survey (2005c). Streamflow can be estimated at ungaged sites by using a rainfall – runoff model or by estimating streamflow from nearby gaging station records. Rainfall-runoff models typically are not used to estimate flows into a reservoir unless there is a need to simulate the effects of changes in basin land use and chemical export on reservoir water quality.

Two methods are available for using gaging station records to estimate streamflow at nearby ungaged sites. The recorded streamflow can be multiplied by the ratio of the drainage area at the gaged site to the drainage area at the ungaged site to provide an estimate of streamflow at the ungaged site. This procedure may be adequate if the two drainage areas are relatively small, have similar characteristics, and experience similar weather patterns. A series of instantaneous discharge measurements at the ungaged site can also be correlated with streamflow at the gaged site. This relation can then be applied to the entire measured streamflow record to provide a time series of streamflow at the ungaged site. In general, however, measured streamflow boundary conditions are preferred to estimated streamflow, particularly for the larger inflows to the reservoir.

Point-source discharge records are typically available from the discharger or the permitting agency. In most cases, a daily mean flow is adequate. Diffuse runoff to the reservoir can be estimated from the contributing land area and an estimate of the average annual runoff. In many cases, estimates of diffuse runoff are used to close the water balance after all other sources have been determined. Ground-water discharge directly to the reservoir is almost always negligibly small, although this is not the case for many natural lakes.

Outflows from the reservoir include releases at the dam, leakage through the dam, evaporation, withdrawals for water supply or industrial uses, and ground-water recharge, which is typically quite small. Evaporation is often determined as part of the heat balance calculations. If evaporation is not calculated by the model, long-term daily mean evaporation rates are typically adequate except for very large reservoirs in arid areas. For those situations, more sophisticated calculations may be required (for example, Wiche, 1992; Winter et al., 1995).

Releases at the dam should be readily available, as should registered water withdrawal rates. The location of the release, release structure, timing of the release and release magnitude will impact the reservoir hydrodynamics and the zones from which water is withdrawn, impacting release water quality (Martin and McCutcheon, 1999). Distinctions must be made between, for example, release through hydropower ports or from flood release structures.

Inflow water temperatures

The statement of the reservoir heat balance, similar to the water balance, is given as the change in heat content in the reservoir is equal to the difference between heat inputs and heat losses. Boundary conditions for the heat balance include the various meteorological data time series, previously described, and a time series of water temperature for all inflows.

Water temperature boundary conditions for all major inflowing streams should be based on measurements, if at all possible. Water temperature sensors are quite inexpensive, can be deployed easily for long periods of time with little or no maintenance, and produce reliable results. In most cases, the water temperature in free-flowing streams is generally uniform throughout the cross-section, except perhaps along the stream bank where water is pooled or has a very low velocity. Hence, a single measurement point located near the center of flow should provide a representative measure of temperature for the entire stream. Methods for collecting and processing water temperature data are given by Wagner et al. (2000) and Whitfield and Wade (1993). Statistical models also have been used to estimate daily mean stream temperature from various measures such as air temperature, elevation, and geologic formation (for example, Donato, 2002; Moheni and Stefan, 1999; Sinokrot and Stefan, 1993; Ford and Stein, 1986).

The water temperature of the diffuse runoff can be reasonably approximated as equal to water temperature in one of the measured streams. Ground-water discharge temperature can usually be assumed as equal to the mean annual air temperature at the study area.

The temperature of point-source discharges, with the exception of thermal discharges, may not be available, but these discharges are typically small relative to the total inflow to the reservoir so the added heat content is small. Monthly mean water temperatures of non-thermal point-source discharges, estimated from air temperature, are adequate for most applications. Water temperature of point sources may be important, however, to ensure that the discharge rises or sinks to the appropriate density level within the reservoir. In that case, accurate measures and more frequent data on point-source water temperature are required.

Outflow water temperature boundary conditions are not required, because the model calculates outflowing water temperatures. The temperature of water withdrawn from the reservoir at a specific location in the reservoir is equal to the simulated water temperature within the reservoir at that location.

Inflow chemical and biological conditions

A time series record of each key chemical and biological constituent is required at all inflow boundaries. Constituent concentrations are required for all inflowing rivers and streams, diffuse inflows, point-source discharges, ground-water discharge, and at the water surface.

The particular constituents required for boundary conditions depend on (1) the problem to be addressed by the reservoir model and (2) the specific formulations used by the model to simulate the process under consideration (table 6-5). The user's manual for the reservoir model should provide guidance on the constituents required for chemical boundary conditions to simulate important processes. Even with this guidance, however, the user will be required to make some decisions on the constituents to include in the model. For example, there are several methods for performing simulations that address reservoir DO issues:

- Prescribe temperature dependent photosynthesis and respirations rates within the reservoir and specify inflow DO and CBOD (or organic carbon) concentrations;

- Simulate time dependent algal growth and mortality, along with nutrient and DO balance;

- Simulate time dependent algal growth and mortality for multiple algal groups, along with nutrient and DO balance; and

- Simulate time dependent algal growth and mortality for multiple algal groups, along with zooplankton grazing, nutrient balance (including zooplankton nutrient release), and DO balance.

Each of these four approaches results in a simulation of the DO balance in the study reservoir, but the four approaches require successively more boundary condition data. However, the approaches also could successively provide more insight into processes affecting the DO balance. The two most important factors to consider when developing an approach for addressing a particular reservoir problem are system characteristics and project resources.

Step 6: Assemble Initial Condition Data

All dynamic models require initial conditions at all points in the computational grid for dependent variables as a starting point for simulations. Initial conditions may include the initial water surface elevations, ice thickness, water temperatures and concentrations of state variables (table 6.5).

Determination of initial conditions from limited field data for all points in a computational grid is problematic where those data are spatially variable. As a result, a common practice is to initiate dynamic simulations during periods

Table 6-5. Typical chemical and biological boundary condition data for selected reservoir model applications

Model application	Typical chemical and biological boundary condition data
Dissolved oxygen	• Dissolved oxygen • Carbonaceous oxygen demand (or organic carbon)
Eutrophication: Basic	• Dissolved oxygen • Carbonaceous oxygen demand (including ultimate demand and either 5-day or 20-day demand) and/or dissolved and particulate organic carbon • Total suspended solids • Volatile suspended solids • Dissolved nitrate • Dissolved ammonium • Total phosphorus • Dissolved inorganic phosphorus • Biomass (or chlorophyll a) • Total iron • Dissolved silica
Eutrophication: Multiple phytoplankton groups	• Same as above, except biomass (or chlorophyll *a*) for each algal group.
Eutrophication: Zooplankton	• Same as above, except zooplankton populations for each zooplankton group specified (inflow zooplankton populations often are assumed to be negligible relative to resident populations).
pH	• pH • Alkalinity • Total dissolved solids • Total inorganic carbon
Suspended sediment	• Suspended sediment concentrations for desired number of grain size categories
Metals and organics	• Concentration of each metal and organic compound under consideration • Boundary conditions required for pH simulations

of complete mixing, where spatial differences may be at a minimum. The CE-QUAL-W2 model (Cole and Wells, 2004), as an alternative, allows specification of a vertical distribution for model state variables where horizontal variations are minimal. The pre-processor for GEMSS, a 3D model (J.E. Edinger Associates, 2005a), allows randomly spaced observations to be automatically interpolated throughout the model domain to develop initial conditions. However, these elaborate initialization schemes are only important in long-residence time systems.

The effort required, and resources allocated, to accurately establish initial conditions should be based in part upon the impact those initial conditions have upon the model predictions and the intended use of the model. For example, in hydrodynamic model applications it is not common practice to provide initial conditions for all hydraulic variables. Rather, because the simulated flow fields are established relatively quickly, selected initial conditions (flows and/or water surface elevations) are estimated and then the model run until an equilibrium is established with boundary conditions. Typically predictions prior to that point are discarded. The final predicted conditions are then commonly used as the initial conditions for subsequent simulations (referred to as a restart or "hot start"). A similar approach may be taken for water quality constituents for reservoirs with relatively short retention times, where the initial conditions may be quickly flushed out of the system. For reservoirs (or portions of the reservoir) with long retention times, and particularly for models that include bottom sediments (i.e., for prediction of contaminants or sediment diagenesis), errors resulting from inappropriate or incomplete initial conditions may persist throughout the simulation period and affect the usefulness of the model.

Step 7: Assemble calibration and evaluation data

In addition to data to drive the model (bathymetry and boundary condition data), data are required to estimate site-specific values of model kinetic coefficients and parameters and to evaluate model performance. Many kinetic coefficients (for example, sediment oxygen demand rate) and model parameters (for example, longitudinal dispersion coefficient) can be measured in the field. These measurements, however, are generally expensive and often are performed for only one set of field conditions, whereas some coefficients vary as a function of various reservoir conditions. If measurements of kinetic coefficients and model parameters are not available, the usual process is to "tune" or adjust model parameters within scientifically defensible ranges until model simulations agree with observations. This process is referred to as model calibration (Chapra 1997, Pascual et al., 2003). Calibration is an iterative process whereby model coefficients are adjusted until an adequate fit of observed versus predicted data is obtained (Cole and Wells, 2004). The definition of 'adequate' agreement between simulations and observations should be agreed upon by all stakeholders early in the modeling process.

Before "tuning" coefficients, however, it is important to compare initial model results and field data to search for the possibility of missing processes not included within the model. Examples in water quality models are accounting for respiration and oxygen production due to vertical migration of dynoflagellates, and rapid releases of vegetative cellular material from sediments that can undergo transformation to algae resulting in blooms. Tuning of coefficients can only adjust the processes they are related to, not compensate for missing processes.

In addition to calibration data, data are required to evaluate the robustness (model performance under a wide range of prototype conditions) of the model by application to one or more independent datasets. In historical modeling literature this step was often referred to as "validation" or "verification" which implied some establishment of "truth" or that the model was an accurate representation of reality. More appropriate and preferable terms for this process are model "confirmation" (Chapra 1997) or corroboration (Pascual et al. 2003), which imply a claim of usefulness and not truth. Corroboration and confirmation involve the use of quantitative and qualitative methods for evaluating the degree to which model results correspond to observations and a general understanding of prototype behavior.

All available data should be used in the calibration and confirmation (corroboration) process. These data should include measurements for variables predicted by the model, such as water surface elevations, current velocities, and concentrations for water quality state variables. These data may also include process data, such as rates of productivity and respiration, for comparison with model predicted rates. Mass balances of important variables can provide useful insight into model performance and identify processes which may not be properly represented in the model. Subjective, or qualitative, information about how the reservoir functions may also be useful in the calibration and confirmation of a model.

Available data for calibration and confirmation should be sufficient to determine the expected quality of the model application as measured by the integrity, utility, and objectivity of the model. Model objectivity is determined from accuracy and bias of model predictions, and from the reliability of the model. The accuracy of model predictions can only be assessed by comparing simulations to observations and is typically stated as some statistical measure of the differences in simulated values and observed data. Bias refers to a systematic deviation between predictions and the independently observed values (Pascual et al., 2003). Reliability may be estimated from the performance history of a model (in this and other applications) and performance of the during the confirmation phase of a particular application.

The types of data required for model calibration and evaluation depend on the objectives of the study. The temporal and spatial resolution of these data depend on the variability in the prototype and on the areas (either a location in the reservoir or a particular time period) of interest in the modeling study. Finally, the data required for calibration and evaluation will be governed by the acceptance criteria for the model, as discussed in a following section.

The time scale (frequency and duration) of the model calibration and evaluation data will vary with the processes simulated and objectives of the study. For example, reservoir eutrophication processes (see Chapter 5) typically vary both over short-term and seasonal to yearly time scales. Hence, a model designed to assess changes in the degree of eutrophication must typically be run dynamically over periods of from seasons to years. Therefore, data for model evaluation are required at some reasonable frequency and for a duration of from seasons to years. For example, a single measurement within a year would not be sufficient for model evaluation, nor would data from an intensive set of measurements made within a relatively short time period, such as a single week or month. A long period of record of reasonably frequent (weekly to monthly) measurements would more closely correspond to the time scale of the problem and model application.

The spatial scale at which measurements are taken will also vary with the study objectives and spatial variability of the data. For reservoirs with appreciable seasonal variations in stratification, vertical profiles of simulated water quality constituents are typically required. Vertical profiles are typically taken near the dam and at locations up-reservoir within the old river channel or thalweg. Data may also be needed in embayments in order to assess mixing patterns and the influence of those embayments on the reservoir water quality. Depending upon the objectives of the study, there may also be other particular areas of importance (recreational beaches, water-supply intakes, etc.) that may affect the spatial extent of data collection. Frequently, the quality of reservoir releases is addressed in modeling studies, so data are needed both within and below the reservoir in order to evaluate model predictions.

Ideally, data would be available for to test model performance in predicting the reservoir water budget, water temperature, and all water quality constituents simulated. However, in some cases constituents may be simulated for which there are no directly comparable analytical measurements. There may also be differences in model state variables from those data routinely collected as required by a regulatory agency. For example, five-day Carbonaceous Biochemical Oxygen Demand ($CBOD_5$) is a common measurement for regulatory compliance, but many reservoir models are based upon state variables for organic carbon and do not directly predict CBOD. Therefore, some processing is typically required in order to compare model predictions to some common measurements. In addition, models are continually evolving as the body of knowledge increases. An iterative and evolving approach for sample design in support of models is then often optimal.

As with all aspects in a modeling study, there is uncertainty associated with data collected for comparison to model predictions. Hence, some estimate of the uncertainty in data is required. Data uncertainty is associated with variability in field conditions, sampling techniques, and laboratory analyses. Field replicates, split samples and laboratory quality assurance information can help quantify some of the uncertainty in field data. Specific data quality objectives should be assessed as part of the modeling study and be included in project quality assurance and quality control plans, as discussed in a following section.

Data can be compared to model simulations using a variety of tools. The type of graphical and/or statistical tools to be used for making these comparisons, as well as the often model-specific structure of the databases, may dictate the format of the data compiled for calibration and evaluation. Some models are designed so that calibration and evaluation data can housed in databases that are designed to facilitate comparison to model predictions. Comparisons can be, for example, time series at particular locations for comparison to time series data, snapshots at particular locations for comparison with synoptically collected data, or a series of measured vertical profiles compared to either static or animated contours of model predictions using graphical post-processors.

Step 8: Develop Model Input

Once the data required to run the model are assembled, the data must then be placed in a format that the model can recognize. Input may be in the form of ASCII files or databases from which data are extracted (e.g., boundary data) during model execution. The development of input usually requires manipulation of the original data in order to convert it to a “model ready” format. For example, boundary conditions are required for all state variables. There may be model state variables for which field data are not available, so that they must be estimated. The units of the boundary data or format of dates may also differ from that required by the model and must be converted. All steps in the process of converting original data to “model ready” data should be carefully planned, executed, and documented.

A critical component of input analysis is checking input files for errors. A number of modern models include input error checking routines, but these are usually limited to detecting format errors and it should not be assumed that all is necessarily well if no warnings or errors are reported (Cole and Wells 2004). The modeling study should include a check of all inputs to ensure they are correct. All time varying input data should be plotted and screened for errors. These plots will need to be included in any final report and can eliminate a number of problems early on in the project (Cole and Wells 2004).

Step 9: Test and Calibrate the model

The generalized modeling framework, including the computer code that implements the governing equations, does not become a model of a natural waterbody until site-specific values for boundary and initial conditions are determined. Even then, until calibrated and evaluated for reliability, the model has limited utility as a site-specific tool. Model parameters must be selected and tested, model formulations and assumptions must be assessed for validity, and the consistency of the initial and boundary conditions must be determined from the ability of the model to simulate site-specific conditions, as determined by comparisons to data (Martin and McCutcheon 1999). The overall process may be referred to as model evaluation, which includes model testing, calibration, and confirmation (or corroboration).

Testing involves determining whether the model is correctly implemented and/or identifying and correcting errors in that implementation. Conceptual errors or misunderstandings may lead to errors in the way a model is configured. Errors may occur in the implementation of model input, as discussed above. Numerical instabilities or mass imbalances may occur. With the increasing complexity of modern models, it often becomes increasingly difficult to identify and correct errors in model implementation.

Model Testing

A well-designed quality control process includes a systematic approach to test various aspects of the model. The testing stage of the application should include not only an evaluation of the model and model input, but the data used to drive and calibrate the model. An often overlooked step in model applications is plotting and analyzing observed data for all stations and times for which data are available (Cole and Wells, 2004). Data should also be checked for reasonableness to ensure that concentration variations in space and time make limnological sense and are reasonably consistent with observations in similar systems. For example, low DO near the water surface and high DO near the bottom probably indicates that vertical profile data were incorrectly recorded. Models will be unable to reproduce these unreasonable conditions, and this should not be viewed as a shortcoming of the model (Cole and Wells 2004). Checks for reasonableness must also extend to model predictions. Models can produce unreasonable results for a number of reasons, including model or user errors or excluding some critical process from model formulations. For example, initial model runs may include simulations of an arbitrary conservative material using time invariant input to ensure that mass is conserved in the model, that predictions are numerically stable, and that transport predictions are reasonable. Only after a carefully designed and staged series of model test simulations should fully dynamic simulations of complicated interactive processes (such as eutrophication) be attempted. Testing should be a continuous process throughout the model application.

Development of Assessment and Acceptance Criteria

The application and evaluation process includes model calibration and confirmation. Assessment criteria are required to determine if the application was successful and the model acceptable for use to make management decisions. Both subjective and objective (qualitative and quantitative) criteria may be used in assessing model performance (Chapra 1997). The criteria may vary with the specific purpose for which the model will be used. For example, the degree and rigor of acceptance criteria may vary between models used for system analysis, design, regulatory purposes, or those which may be used in litigation.

Subjective assessments typically involve expert judgment and tests of intuitive behavior (Pascual 2003). Subjective assessments may include, for example, review of the animation of model results to determine if predictions appear reasonable. The ability of a model to capture trends may also be subjectively evaluated using graphical comparisons of predictions to observed data. For example, vertical profiles may be used to determine if the model adequately captures the timing and magnitude of stratification, onset and location of hypolimnetic anoxia, or development of metalimnetic oxygen minima. In some cases a relatively small difference in timing of the events between predictions and observations may not result in the predictions being viewed as unsuccessful, when viewed subjectively, while an objective (e.g. statistical) comparison may imply a poor comparison. In some cases the determination of whether model-predicted behavior of a system agrees with best-available understanding of internal processes and functions can only be determined subjectively. Modelers often make this determination. However, another common practice to subjectively assess the credibility of a model application is through consensus. For example, an expert panel consisting of model developers, other parties and stakeholders could be convened to determine whether there is agreement that the methods and outputs of a model are consistent with processes, standards and results used in other models (Pascual 2003).

Objective approaches are typically based on some quantitative measure of goodness of fit between model predictions and observed data. Theoretical statistical "norms" can be derived using the mean error and its variance, and the goodness of fit determined from the normal distribution based on the number of observations. The other quantities; ie, the mean square error and the absolute mean error do not have a similar theoretical statistical test of mornalicy. All one can do with these number is compare them to what others obtained in other similar cases for similar numbers of comparisons. Some of the methods commonly used include (Pascual 2003):

- Graphical methods: These methods are typically used to compare the distribution of model predictions and observed data typically using some goodness of fit criteria. Alternatively they may include an evaluation of the distribution of the deviations between model predictions and observations.

- Deviation measures: used to compute some averaged deviation between model predictions and observations, such as the mean error, mean square error, or as defined below, absolute mean error

$$AME=\frac{\sum\left|\text{Predicted-Observed}\right|}{\text{number of observations}}$$

- Statistical tests: include formal methods for hypothesis testing such as whether model predictions are significantly different from observed data

Unfortunately, there is no universally accepted set of methods or criteria for evaluating a model's performance or determining an optimal model calibration. All of the above methods should be used, both subjective and objective

A series of tests and metrics should be determined at the onset of the study that will be used to determine the acceptability of the model. These metrics should be based upon knowledge of the goals and objectives of a particular study and the capabilities of the model.

The model acceptance will also depend upon the model output as compared to the required output. The model output should be evaluated during the model selection process because it may be necessary to modify model output to meet a particular study goal. For example, if the model were to be used to make regulatory decisions with respect to some water quality criterion, then the output must be suitable for comparison with that criterion or the study will not be accepted.

A systematic planning process to establish quality objectives and performance criteria are critical steps that are typically implemented in a model quality assurance (QA) plan. The possible negative consequences of making inappropriate decisions due to poor model prediction ability are a key concern. Additional details on the steps for developing a modeling QA plan are provided in the section on Quality Assurance Planning (p. 6-39) and in U.S. EPA (2002).

Calibration Sequence

The sequence of model calibration is typically from the simplest process to the most complex. Typically the model calibration involves first evaluating the water balance, followed by hydrodynamics and temperature, and finally by water quality constituents. The degree of calibration required depends, in part, upon the number of approximations and/or parameters included in model formulations. Model parameters are terms in the model that are fixed during a model run or simulation but can be changed in different runs as a method for conducting sensitivity analysis or to achieve calibration.

Water budget

The first step in the evaluation/calibration process typically involves running the model and then checking predicted water surface elevations with observed elevations. Errors in the water budget generally result from errors in input rather than uncertain model parameters. Errors in the water budget may be generated by the following (Cole and Wells, 2004): incorrect bathymetry, missing data from storm events, incomplete inflow data, inaccurate estimates of evaporation or seepage, and inaccurate outflow measurements.

It should be understood that water budgets for typical reservoirs are not well known because all inflows and outflows are rarely measured, and parameters such as evaporation and seepage and nearly impossible to measure. Unknown inflows are often estimated by difference from available data for inflows and outflows and estimated changes in reservoir volume. Changes in volume are typically estimated from elevation/volume relationships for the reservoir and measured changes in water surface elevations. However, this method should be used with caution. In practice, both the unknown flows required to compute observed elevations and the closeness of computed and observed elevations are part of the water budget calibration.

Hydrodynamics and Temperature

The hydrodynamics (water level and velocity) and water temperature component of the model are calibrated after the water budget is satisfactorily simulated. Typically hydrodynamic data for calibration are limited due to the relatively flat water surface profile and low velocities in reservoirs. If available, simulated velocities can be compared with velocity and flow measurements obtained from an acoustic Doppler current profiler (ADCP).

Water temperatures affect the reservoir hydrodynamics through changes in water density, as discussed in Chapter 4. Because of this linkage between water temperature and the flow field, the performance of the water temperature simulations gives some insight into hydrodynamic simulation performance. However, temperature is not an ideal variable because it is non-conservative and is usually not sufficient alone to test the hydrodynamic calibration in the absence of velocity data. Frequently, the hydrodynamic calibration may only be assessed after evaluating the transport of all water quality constituents. Cole and Wells (2004) indicate that each simulated constituent can contribute knowledge about the system and can have an impact on the hydrodynamic calibration. Their experiences have shown that dissolved oxygen and phytoplankton are often much better indicators of proper hydrodynamic calibration than water temperatures. For example, errors in the model grid and hydrodynamics will propagate through, and may only become apparent during, the water quality calibration. Therefore, the hydrodynamic and water quality calibration cannot be decoupled. The hydrodynamics, while it may initially be calibrated to flow and/or temperature data, should be reassessed during the calibration of water quality constituents.

Modern models that include mechanistic representations of both the reservoir hydrodynamics and heat budget, such as CE-QUAL-W2 (Cole and Wells 2004), include relatively few parameters that are typically varied during the hydrodynamic and temperature calibration. These may include, for example, bottom roughness coefficients and coefficients affecting wind stress at the water surface. More commonly the hydrodynamic and temperature calibration of modern models involves the testing and evaluation of input data. Problems in the calibration of hydrodynamics and temperature typically result from errors in, or inadequate, inflow data, meteorological data, outflow data or bathymetry (Cole and Wells 2004).

Water Quality

The water quality calibration is typically the most difficult aspect of model calibration because of the interactions and feedback between state variables for dissolved oxygen nutrients, and phytoplankton (see Chapter 5), and because of the uncertainty in the numerous kinetic coefficients and rate parameters that must be tuned during calibration. Calibration should involve a careful review of predictions for all model state variables. The review may be through a quantitative comparison to available data or, at a minimum where field data are not available, a qualitative determination of the reasonableness of the model predictions. Particular attention should be given to the timing and duration of events involving changes in phytoplankton, epiphyton, and dissolved oxygen. If the model does not represent the onset, extent, and duration of anoxic conditions, then nutrient dynamics also will not be represented adequately (Cole and Wells, 2004).

Some of the coefficients and parameters that must be adjusted for calibration of the water quality model can be measured in the field, although these measurements are seldom performed because of the expense and difficulty of making the measurements. The bounds on many of the water quality model parameters, however, are well known from previous field, laboratory, and modeling studies. Adjustments to model parameters during the calibration process, then, should be made with a good understanding of reasonable values of the model parameters and the sensitivity of model results to changes in these parameters (see following section). Models can be 'overtuned' so that simulations agree with observations for one set of conditions, but the adjustments to model parameters required to obtain these results are unrealistic with parameter values outside the range of expected values. As a result, the model likely will not perform well for another set of conditions that differs substantially from the calibration dataset.

Sensitivity Analyses

A sensitivity analyses is a common step in a model evaluation and provides a means of assessing the effect of uncertainty in model parameters and model input on simulations. The sensitivity analysis measures the effects of changes in input values or model assumptions on model predictions (Pascual et al., 2003).

There are a variety of ways to perform a sensitivity analysis. Some methods do not require running a model and may be based upon statistical evaluations of the uncertainty of model inputs. However, these methods are related to model output only for simple models or assuming a linear relationship between model input and output. More commonly sensitivity analyses require running the model for a range of conditions.

One of the more common methods used in sensitivity analyses is parameter perturbation (Chapra 1997). Parameter perturbation involves varying each model parameter in succession by some predetermined amount (e.g. 50 percent), while holding all other factors constant, in order to estimate the impact of that perturbation on model predictions (e.g., predicted concentration of dissolved oxygen). A variety of other methods are available, and often the selection of the method is based in part upon the complexity of the model. For example, complex models may make the application of Monte Carlo methods (see following section) impractical. Choosing the most appropriate sensitivity analysis method will often entail a trade-off between computational complexity, model assumptions, and the amount of information needed from the sensitivity analysis (Pascual et al., 2003).

Uncertainty Analyses

Model users typically are interested in the degree of uncertainty in model predictions, but it is difficult to quantify uncertainty for complex multi-dimensional reservoir models. The quantitative comparisons of model predictions and observed data (difference or statistical methods) described previously provide a measurement of model error. The error estimates are based on a "best fit" of calibrated model parameters. Differences in values between model calibration and confirmation (corroboration) may provide some further indication of the uncertainty in model predictions. However, those predictions are based upon model input and model parameters which are themselves uncertain. To determine the certainty of model predictions, it is necessary to estimate the effect of uncertainty in both model inputs and model parameters on those predictions.

An uncertainty analysis can be illustrated using the Monte Carlo technique. With this technique, the statistical distribution (uniform, normal, log-normal, etc.) of all model input (inflows, meteorological data, etc.) and parameters (phytoplankton growth rates, nitrification rates, etc.) are identified, along with the statistics necessary to define those distributions (e.g., mean and standard deviation for a normal distribution). Each distribution is then randomly sampled, new sets of input generated, and the model run for those inputs. The reader is referred to Chapra (1997) for a more detailed discussion of the Monte Carlo technique. This process is repeated multiple times, providing ultimately, a statistical distribution of model output. This distribution can then be used to identify, for example, the probability that the dissolved-oxygen concentration at a particular location and time will be within a specified range.

Quantifying model uncertainty using the Monte Carlo technique, even for a relatively simple model, may require tens of thousands of iterations or model runs. Historically, uncertainty techniques have been applied to simple analytical or steady-state models, such as QUAL2-UNCAS (Brown and Barnwell 1987), where the implementation may be practical. For modern reservoir models, the computational burden is typically prohibitive. In addition, in modern dynamic models not only "what" the variations in inputs are but "when" they vary is important, making quantification of uncertainty even more difficult. The development of new and more practical means of uncertainty analyses, however, is an area of active research.

Step 10: Apply the model

The final step in the modeling study is the application of the model for an intended use, typically model-aided decision-making. Common applications include hind-casting to determine what may have cased some event, now-casting of existing conditions, or forecasting to some either estimated future or rare condition or some change in design or operation. The success of the model application may depend upon how clearly the objectives of the model application were originally stated and how effectively those objectives were translated into the modeling plan and model data quality objectives.

Decision makers should be well informed as to the basis for a model, the steps used in the model application, and limitations in the model and/or modeling study. Model studies are limited by available resources (e.g., financial and/or time), a lack of theoretical understanding of some of the processes impacting the natural system, limitations of the model construct (e.g., model assumptions, numerical implementation, etc.), and the data available to apply the model (and uncertainties in those data). As a result no model or modeling study is a perfect representation of reality, which is true of all engineering computations. While deviations in predictions from observations require scrutiny, some deviation from observed data is to be expected due to both limitations in the model and data. The level of deviation considered acceptable should be included in project model quality objectives.

However, if these objectives are not clearly defined, they may limit the applicability or acceptance of the model by decision makers.

Models are not perfect representations of reality, which should be understood by both modelers and decision makers. In modern water quality models, we attempt to incorporate the underlying mechanisms that cause change, allowing us to establish cause-and-effect relationships that are useful in, for example, forecasting the impact of environmental controls (Martin and McCutcheon 1999). The theoretical understanding of the processes impacting those cause and effect relationships is constantly evolving, so perhaps an optimal use of models in decision-making is in the context of an iterative, ever-improving process of continuous model refinement (Pascual et al., 2003). However, a lack of complete understanding of a natural system, such as a reservoir, does not preclude the use of models in decision making. Disclosure about the state of science used in a model and future plans to update the model can be used to establish the record of reasoned, evidence-based application to inform decisions (Pascual 2003). The EPA, for example, has successfully defended a challenge to a model used in its TMDL program when it explained that it was basing its decision on the best available scientific information and that it intended to refine its model as better information surfaced (NRDC v. Muszynski, 268 F.3s 91, 2d Cir., 2001 cited by Pascaul et al., 2003).

<u>Data quality issues</u>

Issues regarding data quality should be considered during all aspects of the planning and application of a water quality model. "Garbage in, garbage out" also applies to reservoir models, so the quality of the model will not be better that the data used to drive the models (meteorological data, inflows, outflows, etc.). In addition, errors in data used to calibrate the model can lead to an inaccurate calibration, and errors in data used to determine model accuracy may result in a poor understanding of model performance. As a result, the quality of all data to be used in the modeling study should be scrutinized and documented (see Quality Assurance Planning, next page). Some of the considerations for data used in a modeling study should include (Pascaul 2003):

- Representativeness: Were the data representative of the conditions analyzed and were sampling and analytical methods used to generate the collected data acceptable to this project? How will potentially confounding effects in the data (e.g., season, time of day, location, and scale incompatibilities) be addressed so that these effects do not unduly impact the model output?

- Bias: Would any characteristics of the dataset directly impact the model output (e.g., unduly high or low process rates)?

- Precision: Is the statistical variability of the data sufficiently small to meet the uncertainty objectives of the modeling project?

- Qualifiers: Have the data been evaluated in a way to qualify the data quality and allow comparison of data from difference sources (e.g., analytical techniques, detection limits)?

- Summarization: Does the data summarization process allow distinction between averages or statistically transformed values from unaltered measurement values?

Quality Assurance (QA) Planning

Historically, quality assurance planning was a relatively ill-defined step in the modeling process. Quality assurance (QA) plans were frequently required as part of data collection efforts by various agencies, but it was not clear how the plans should be extended to data used in or produced by environmental models.

In recent years (Order 5360.1 A2, EPA 2000d) the definition of environmental data has been extended to those data produced from models, where the U.S. Environmental Protection Agency considered a model to be "something that creates a prediction." As such, QA plans for modeling are now typically required. Guidance on the development of a QA plan for data collection is provided in EPA Requirements for Quality Assurance Project Plans (QA/R-5; EPA, 2001) and EPA Manual 5360 A1 (EPA, 2000a) and for modeling in Guidance for Quality Assurance Project Plans for Modeling (QA/G-5M, EPA 2002).

The degree of QA will vary among projects, and a graded approach to QA development is generally recommended (EPA 2002), where the level of QA required may vary with the consequences of errors in data quality. The degree of rigor in the QA required may be higher where the results of the modeling study would be used in litigation or to determine regulatory compliance. For example, for nuclear safety-related model applications (for example, a reservoir used as the ultimate heat sink for reactor shutdown in the event of a loss of coolant accident), the QA requirements are stringent. Features of such QA plans include certification of datasets, use of a second model of another type applied by a team working independently of the primary modeling team, and documented reconciliation of any differences in the two model results. In any case, a well designed and implemented QA plan will enhance the credibility of any modeling study. QA typically adds between 10 and 15% to the overall modeling effort.

The elements of a QA project plan will include: project management; measurement and data acquisition; assessment and oversight; and data validation and usability. The reader is referred to EPA (2001, 2002) for a detailed description of the components of the QA plan. Many of the components of the QA plan were previously discussed in this Chapter. However, the QA plan itself provides a formal record of how each of these components are to be implemented, some of which are briefly discussed below. .

Project Managament

As discussed previously, the first step in any model application is problem identification. As such, the QA plan should define the problem and describe the scope of the project, including the project objectives. The QA plan would also typically include the project organization, tasks, project personnel, and support facilities and/or organizations. Once the problem and objectives are identified, specific management questions can be determined and the appropriate modeling framework needed to address those questions defined.

Modeling Quality Objectives and Acceptance

A second critical component to the QA plan is the identification of modeling and data quality objectives. Identification of quality objectives will aid in identifying the data that are needed, the use of the data, and the acceptable uncertainty in the data. Data from all sources, whether collected in the field as part of the study, model generated, or taken from the literature, should be included.

The modeling and data quality objectives must be related to the study objectives. Model acceptance criteria, for example, may depend upon a regulatory criterion to which the results will be compared.

The model acceptance criteria may, in turn, influence the data quality objectives, including the analytical and quality control methods required to meet those objectives. For example, acceptance criteria may be based upon differences or statistical relationships between model predictions and observed data, as discussed previously (see Section 6.2 Step 9). Selection of an inappropriate analytical method may result in the majority of measurements being below the limit of detection, which would make these data of limited use for comparisons to model predictions. Similarly, non-detects are difficult to use for establishing model boundary conditions. Failure to establish and follow data quality control measures such as sample preservation techniques, holding times, custody documentation, and other measures may make data and model comparisons suspect.

Quality Control

Quality control is the process by which a QA plan is implemented (EPA 1999). The degree and rigor of quality control checks for a model may vary with the model's history of use and degree of documentation. For example, for a relatively new model or one developed for the particular study, code verification may be required. During code verification the final set of computer code is scrutinized to assure that the equations are programmed correctly and that sources of error, such as rounding, are minimal. For a code with an established record of use, the verification may be minimal. However, errors may exist even in established codes and some degree of quality control testing is required for any application. It is also not uncommon for a computer code to be altered as part of a specific application. The QA plan may provide for any such modifications and identify requirements such as documentation, code testing, and peer review. It is also not uncommon for changes to occur due to computer architecture, such as changes made to operating systems to protect computers from viruses which may necessitate changes in, or in some cases result in disabling, some models.

The quality control process must also include checks of the model implementation, such as the development and implementation of input data. Specific tests should be identified, for example, to test continuity and mass conservation and numerical stability and convergence properties of model code algorithms. Model results should also be checked for reasonableness, and compared to other models or known analytical solutions. Periodic audits and/or peer review are components of the quality control process.

Data Quality and Record Keeping

As defined by EPA's Information Quality Guidelines (EPA 2002b), quality is a broad-term that includes notions of integrity, utility, and objectivity. Integrity refers to the protection of information from unauthorized access or revision to ensure that it is not compromised through corruption or falsification. For environmental models, integrity may be most relevant in the protection of the model computer code and input files from unauthorized or inappropriate manipulation (Pascual 2003). Utility refers to the usefulness of the information to the intended users. The utility should for example be reflected in the data and modeling quality objectives and may be the basis for the model acceptance criteria. Objectivity involves two distinct elements, presentation (information presented in an accurate, clear, complete and unbiased manner) and substance (accurate, reliable and unbiased information relevant to the issues at hand).

Data used in modeling studies may be derived from a variety of sources. The quality of those data (integrity, utility, and objectivity) may directly impact the resulting quality of the modeling study. Therefore an integral part of a QA plan for modeling is documentation of methods and procedures identifying the source and determining the quality of data used in the modeling study. There is also considerable manipulation required to transform data from the original source to a form that can be use for model input or for comparison with model predictions. The quality of these manipulated data should also be the subject of review and documentation. Databases and database tracking systems are often well suited for that purpose.

During the process of model calibration and model application, a large number of electronic files, such as for model input or output, are typically generated. In addition, the model itself may change over the course of the application so that version control is needed. Some type of model tracking system should be a part of the QA plan so that all model runs and files can be recalled, rerun, or examined as necessary (EPA 1999)

Record keeping and archival of records should also be an integral part of the QA/QC process. Records maintained should include modeler's logs and all electronic files (databases, input, output, model files, etc). The specific records and files that must be maintained should be identified and agreed upon early in the QA process. These files should be place in an acceptable format for archiving, transferred to and maintained by the designated QA officer. Record keeping standards are provided by EPA (2002b) and other agencies.

6.3. REFERENCES

Antenucci, J., and Imerito, A., 2002, The CWR dynamic reservoir simulation model, DYRESM. Science manual: Perth, Australia, Center for Water Research, University of Western Australia, accessed at http://www2.cwr.uwa.edu.au/~ttfadmin/cwrsoft/doc/dyresm_science/ dyresm_science.pdf on April 18, 2005.

Bender, M. D., Hauser, G. E., Shiao, M. C. and Proctor, W. D., 1990, Technical Reference Manual and User's Guide BETTER: A Two–Dimensional Reservoir Water Quality Model: Norris, Tennessee, Tennessee Valley Authority Engineering Laboratory, Report WR28-2-590-152, 160 pp

Brown, L. C. and T. O. Barnwell. 1987. The enhanced stream water quality models QUAL2E and QUAL2E-UNCAS: documentation and user manual. Env. Res. Laboratory. US EPA, EPA /600/3-87/007, Athens, GA. 189 pp.

Buchak, E. M. and G. A. Krallis. 2002. An Investigation to Increase the Cooling Pond Efficiency at Gerald Gentleman Station, Phase I - Develop, Calibrate and Test the Fully-Mixed System Model, Phase II – Scenario Development for Extreme Conditions. Prepared for Nebraska Public Power District, Columbus, Nebraska. Prepared by J. E. Edinger Associates, Inc., Wayne, Pennsylvania. April 17, 2002.

Chapra, S. C. 1997. Surface Water Quality Modeling. McGraw-Hill. 844pp.

Chapra, S.C. and Martin, J.L. 2004. LAKE2K: A Modeling Framework for Simulating Lake Water Quality (Version 1.2): Documentation and Users Manual. Civil and Environmental Engineering Dept., Tufts University, Medford, MA.

Cerco, C.F., and Cole, T., 1995, User's guide to the CE-QUAL-ICM three-dimensional eutrophication model: Vicksburg, Miss., U.S. Army Corps of Engineers Waterways Experiment Station, Technical Report EL-95-15.

Chesapeake Bay Program, 2005, Chesapeake Bay Quality Assurance Program, accessed at http://www.chesapeakebay.net/info/quality_assurance.cfm on March 23, 2005.

Cole, T.M., and Wells, S.A., 2004, CE-Qual-W2: A two-dimensional, laterally averaged, hydrodynamic and water quality model, version 3.2: Vicksburg, Miss., U.S. Army Corps of Engineers Waterways Experiment Station, Instruction Report EL-03-01.

Donato, M.M., 2002, A statistical model for estimating stream temperatures in the Salmon and Clearwater River basins, central Idaho: U.S. Geological Survey Water-Resources Investigations Report 2002-4195, 46 p.

DYRESM User Group, 2005, DYRESM References, accessed at http://www2.cwr.uwa.edu.au/~ttfadmin/usergroups/dyresm/index.html on July 20, 2005.

Fischer, H.G., List, E.J., Koh, R.C.Y., and Imberger, J., and Brooks, N.H., 1979, Mixing in inland and coastal waters: San Diego, CA, Academic Press, 483 p.

Ford, B. and Stein, A. B., 1984, The hydrometeorology of DeGray Lake, Arkansas": Vicksburg, Miss., U.S. Army Engineer Waterways Experiment Station, Miscellaneous Paper E-84-3.

HGL and Aqua Terra. 1999. Selection of Water Quality Components for Eutrophication-Related Total Maximum Daily Load Assessments. Task 4: Documentation of Review and Evaluation of Eutrophication Models and Components EPA Contract Number 68-C6 0020 Work Assignment No. 2-04. Prepared by HydroGeoLogic, Inc. Herndon, VA 20170 and AQUA TERRA Consultants, Mountain View, CA. June.

Imhoff, J. C., A. Stoddard, and E. M. Buchak. 2003. Evaluation of Contaminated Sediment Fate and Transport Models. Final Report. Prepared for National Exposure Research Laboratory, Office of Research and Development, U.S. Environmental Protection Agency. Athens, GA. Contract Number 68-C-01-037, Work Assignment No. 1-10. June.

J.E. Edinger Associates, Inc. 2005a, Models and Software—GEMSS: accessed at http://www.jeeai.com/Software-GEMSS.htm on July 22, 2005.

_____2005b, Publications: accessed at http://www.jeeai.com/publications.htm on July 22, 2005.

Martin, J.L. and S.C. McCutcheon. 1999. Hydrodynamics and Transport for Water Quality Modeling. CRC Lewis Publishers, Boca Raton, FL.

Mohseni, O., and Stefan, H.G., 1999, Stream temperature/air temperature relationship: a physical interpretation: J. of Hydrology, 218(3-4):128-141.

Pascual, P., Stiber, N. and Sunderland, E. 2003. Draft Guidance on the Development, Evaluation, and Application of Regulatory Environmental Models, Prepared by The Council for Regulatory Environmental

Modeling, Office of Science Policy, Office of Research and Development. U.S. EPA, Washington, D.C.

Riley, M.J., and H.G. Stefan. 1988. MINLAKE: A Lake Water Quality Simulation Model: Ecological Modeling. 43, no. 3-4: 155-182.

Sinokrot, B.A., and Stefan, H.G., 1993, Stream temperature dynamics: measurements and modeling: Water Resources Research, 29(7):2299-2312.

U.S. Army Corps of Engineers, 2005, Hydrologic Engineering Center River Analysis System: accessed at http://www.hec.usace.army.mil/software/hec-ras/hecras-hecras.html on April 18, 2005.

U.S. Army Corps of Engineers, 1986, CE-QUAL-R1: A Numerical One-Dimensional Model of Reservoir Water Quality: Users Manual: Vicksburg, Mississippi, Instruction Report E-82-1 (Revised Edition), Environmental Laboratory, US Army Engineer Waterways Experiment Station.

U.S. Environmental Protection Agency, 1999, Quality Assurance Plan for Mathematical Modeling, Lake Michigan Mass Balance Project. Great Lakes National Program Office, Chicago, IL.

_____, 2000a, EPA Quality Manual for Environmental Programs (Order 5360 A1), Washington, DC.

_____, 2000b. Guidance for the Data Quality Objectives Process (QA/G-4) (EPA/600/R-96/055). Washington, DC: Office of Environmental Information.

_____, 2000c, Guidance for Data Quality Assessment: Practical Methods for Data Analysis (QA/G-9) (EPA/600/R-96/084, QA00 Update). Washington, DC: Office of Environmental Information.

_____, 2000d, Policy and Program Requirements for the Mandatory Agency-wide Quality System (EPA Order 5360.1 A2). Washington, DC.

_____, 2001, EPA Requirements for Quality Assurance Project Plans (QA/R-5) (EPA/240/B-01/003), Office of Environmental Information, Washington, DC.

_____, 2002, Guidance for Quality Assurance Project Plans for Modeling, EPA QA/G-5M EPA/240/R-02/007, Office of Environmental Protection Information, Washington, DC.

_____, 2002b, Information Quality Guidelines. U.S. EPA Office of Environmental Information, Washington, D.C.

_____, 2005a, EFDC: Watershed and Water Quality Modeling Technical Support Center, accessed at http://www.epa.gov/athens/wwqtsc/html/efdc.html on July 22, 2005.

_____, 2005b, WASP, Watershed and Water Quality Modeling Technical Support Center, accessed at http://www.epa.gov/athens/wwqtsc/html/wasp.html on July 22, 2005.

U.S. Geological Survey, 2005a, Surface water quality and flow Modeling Interest Group: accessed at http://smig.usgs.gov/SMIG/model_archives.html on April 6, 2005.

_______2005b, National Elevation Dataset, accessed at http://gisdata.usgs.net/ned/ on March 25, 2005.

_______2005c, Surface Water Data for the Nation, accessed at http://waterdata.usgs.gov/nwis/sw on August 23, 2005.

U.S. Geological Survey, 2003, Quality-assurance plan for water quality activities of the U.S. Geological Survey in Miami, Florida: U.S. Geological Survey Open-File Report 03-195, 72 p.

Walton, R., Shock, K., and Swarmer, R., 1998, Review of 3-D hydrodynamic and contaminant transport models in Water Resources Engineering 98: New York, American Society of Civil Engineers, p. 1120-1125.

Warwick, J.J. 1997. Use of First-Order Uncertainty Analysis to Optimize Successful Stream Water Quality Simulation. Journal of the American Water Resources Association. 33(6) 1173-1185.

Water Environment Research Foundation, 2005, Product catalog. Water quality models: a survey and assessment: accessed at https://www.werf.org/acb/showdetl.cfm?Product_ID=203&DID=7 on April 18, 2005.

Water Quality Research Group, 2005, CE-QUAL-W2 hydrodynamic and water quality model applications history: accessed at http://www.ce.pdx.edu/~scott/w2/ on July 21, 2005.

WEST Consultants, Inc. 1996. "Evaluating Hydrodynamic and Contaminant Transport Computer Models - For Application to Duwamish River and Elliott Bay". Prepared for King County Department of Natural Resources, Seattle, Washington. Prepared by WEST Consultants, Inc. 2101 Fourth Avenue. Suite 1050. Seattle, Washington.

Whitfield, P.H. and N.L. Wade. 1993. Quality Assurance Techniques for Electronic Data Acquisition. Water Resources Bulletin. 29(2): 301-308.

Wiche, G. J., 1992, Evaporation computed by energy-budget and mass-transfer methos and water-balance estimates for Devils Lake, North Dakota, 1986-88: North Dakota State Water Commission Water Resources Investigation 11, 52 p.

Winter, T.C., D.O. Rosenberry, and A.M. Sturrock. 1995. Evaluation of eleven equations for determining evaporation for a small lake in the north central United States, Water Resources Research, v. 31, no. 4, pp. 983–993.

Wlosinski, J.H. and Collins, C.D., 1985: Confirmation of the Water Quality Model CE-QUAL-R1 Using Data from Eau Galle Reservoir, Wisconsin, US Army Engineer Waterways Experiment Station Vicksburg, MS. Environmental Lab Final Report. Technical Report E-85-11, 72 p.

Wagner, R.J., Mattraw, H.C., Ritz, G.F., and Smith, B.A., 2000, Guidelines and standard procedures for continuous water quality monitors-site selection, field operation, calibration, record computation, and reporting: U.S. Geological Survey Water-Resources Investigations Report 00-4252, 53 p.

7. CASE STUDIES

John M. Higgins

George Krallis

The case studies in this chapter are intended to illustrate the diversity of modeling applications and how a modeler might use the material presented in the previous chapters to facilitate management decisions. Implicit to each case study are key steps that are common to most modeling analyses (e.g. problem definition, data acquisition, limiting assumptions, model selection and application, interpretation of results, and management implications). The case studies also illustrate the fact that models do not usually provide a definitive solution to the problem, but rather yield information and insight that is combined with other factors to evaluate alternative courses of action.

Six case studies are presented. The first involves a large river system with multiple reservoirs and river segments. Water routing and water quality models are linked to assess the potential cumulative effects of proposed projects, power plants, and reservoir operations on beneficial uses of the river system. Large volumes of model output for each management alternative must be evaluated and presented in a manner consistent with water quality objectives and stream uses. Examples are given for comparing alternatives and relating model output to potential beneficial use impacts.

The second case study summarizes 30 years of water quality modeling of a western reservoir to guide management and operational decisions. Key issues include selective withdrawal to maintain suitable temperatures for downstream fisheries, nutrient management to control algal productivity, and the scheduling of reservoir releases and power generation to provide adequate tailwater flows. The model application was complicated by chemical stratification of the reservoir and internal nutrient recycling. Model results provided valuable insight into reservoir processes and assisted in evaluating management alternatives.

The third case study focuses on the details of model application and calibration. It involves water quality modeling of Old Hickory Reservoir in

Tennessee for assessing nutrient load and steam plant effects. Specific methods are discussed for converting available field data into input files that are representative of reservoir bathymetry and inflowing water quality. Procedures and examples are given for model calibration and the inclusion of cooling water intakes. The potential effect of reservoir geometry on hydraulic travel time is also evaluated.

The fourth case study demonstrates the use of two different models in determining whether the cooling water discharge from a Nebraska power plant would potentially exceed the NPDES permit limit for instantaneous maximum temperature. The study had two primary objectives: 1) to examine the frequency and duration of potential NPDES violations and 2) to examine cooling pond dynamics and suggest modifications to increase the cooling pond efficiency. The first objective was addressed by assessing the overall cooling system performance over many years using fully-mixed computations (a "systems model") that incorporated various water sources, station thermal loads, and varying meteorological conditions at an hourly time scale. The second objective was addressed by simulating the cooling pond dynamics using a 3-dimensional hydrodynamic and transport model at a 30 meter horizontal and 0.5 meter vertical spatial scale.

The fifth case study uses a 3-dimensional hydrodynamic and transport model to examine the effectiveness of a submerged weir on reservoir dissolved oxygen (DO) concentrations. State regulations require a minimum reservoir release of 4 mg/L. A submerged weir upstream of the dam would increase the DO concentration of the outflow by maintaining a well mixed forebay consisting mostly of high DO water skimmed by the weir. An added modeling complexity is groundwater leakage from sinkholes in the forebay area. If the weir is unable to maintain a downstream DO of at least 4 mg/L, a more costly alternative involving direct oxygen injection would be required.

The final case study involves near-field hydrothermal modeling to determine the configuration and operation of cooling towers for a steam electric power plant. The plant is located on a large reservoir, withdraws intake cooling water from one tributary stream embayment, and discharges the effluent into a second tributary embayment. Modeling was used to determine the optimal location of the cooling tower discharge; details of the discharge structure depth and exit velocity to maintain reservoir stratification; and cooling tower flows and temperatures to maintain zones of fish refuge and passage. As part of NPDES permitting, model verification of actual cooling tower operation was required (i.e. comparing model predicted temperatures with field data collected during tower operations).

7.1. Water Quality Modeling of the Tennessee River System to Support Beneficial Water Uses

John M. Higgins

Katherine F. Lindquist

Tennessee Valley Authority

Introduction

Water quality models usually focus on site specific issues involving a single river or reservoir. The purpose of this case study is to illustrate model application in the management of a river system, containing multiple reservoirs and river segments. Reservoir operations require an understanding of downstream impacts, cumulative effects, and seasonal changes over a large area. In the past a large geographic area usually meant a more simplistic modeling approach. This situation is changing, however, as increased computing capabilities allow more complex models and linking individual models.

The Tennessee Valley Authority (TVA) uses such an approach to assess the water quality implications of proposed projects and reservoir operating decisions in the Tennessee River system. The ultimate purpose is to operate the river system for the greatest benefit to the region. This includes the integrated management of 48 dams and reservoirs to provide navigation, flood control, power generation, recreation, water quality, water supply, and other benefits. To assist in this effort, individual river and reservoir models analyze site specific issues. Individual models are also linked together to simulate water quality throughout the system. The models provide input for a variety of management decisions (e.g., the potential water quality impact of proposed projects, the timing of reservoir releases to provide downstream benefits, and the assessment of reservoir operating alternatives, minimum flows, and target lake levels). This case study focuses on system-wide modeling by TVA to assess the potential effects of changes in reservoir operations.

Site Description

The Tennessee Valley encompasses 108,000 km^2 in the southeastern United States. Approximately four and one half million people live in the basin. The Tennessee River extends 1,050 km from its confluence with the Ohio River near Paducah, Kentucky to Knoxville, Tennessee (Figures 7.1.1 and 7.1.2). The mean annual discharge from the river is 1,750 m^3/s.

Fourteen of the 48 TVA dams are small, largely self regulating projects. The remaining 34 projects that make up the water control system fall into four groups: 9 multipurpose projects on the main Tennessee River that provide seasonal flood protection, maintain minimum pool levels for commercial navigation, and generate electric power; 12 multipurpose tributary projects that provide seasonal stream flow regulation for flood control, navigation, and hydroelectric power generation; 7 single-purpose power tributary projects that generate hydroelectric power; and 6 tributary multipurpose power projects that provide flood damage reduction, water supply, water quality, and recreation. Table 7.1.1 summarizes key characteristics of the 34 reservoirs. TVA also owns and operates 6 fossil and 3 nuclear power plants within the Tennessee River Basin and 5 fossil plants in adjacent basins.

Approach to Water Quality Assessment

TVA periodically reviews reservoir operations to ensure that the multi-purpose system is providing the greatest possible benefit to the region and its water users. A formal review, completed in 1990, resulted in minimum reservoir releases, improved oxygen concentrations, and higher lake levels (Tennessee Valley Authority, 1990; Higgins and Brock, 1999). The most recent review was completed in February 2004 (Tennessee Valley Authority, 2004). As part of this effort, river and reservoir models were used to simulate water quality under current operations (base case) and the potential effects of changes aimed at increasing overall system benefits (operating alternatives). The following paragraphs describe the role of water quality modeling in the assessment process. For simplicity, the examples of model output are based on hypothetical, rather than actual, operating conditions.

Each reservoir and river (or tailwater) model generates a time series of output for a variety of hydrodynamic and water quality parameters (e.g., flow, temperature, dissolved oxygen, nutrients). The output is summarized in a smaller set of **metrics** that relate more directly to beneficial water uses. The metrics serve as a basis for assessing the potential impact of expected changes in water quality. For example, an expected decrease in the dissolved oxygen (DO) concentration of water released from a dam may be expressed in the number of pounds of oxygen that must added be to continue meeting the DO target (i.e., the minimum DO concentration that must be maintained through aeration). This metric can then be used to estimate the increased annual cost of aerating releases from the reservoir.

Metrics are used to evaluate potential impacts for five categories of beneficial use.

- Water Quality Criteria—While not a specific beneficial use, water quality criteria represent conditions that must exist for a specific beneficial use to be maintained. As such, each criterion reflects a balance of political, economic, and technical factors surrounding a beneficial use impact that society has chosen to avoid. Water quality criteria also represent thresholds that trigger regulatory actions that usually involve additional costs.

- Aquatic Life—Water quality can impact aquatic life through habitat suitability, toxicity, or stress. Unsuitable habitat can limit the livable environment for various species.

- Recreation—Water contact recreation, fishing, and aesthetic enjoyment can be impacted by adverse water quality.

- Water Supply—Poor water quality can potentially increase water treatment requirements and cause taste and odor problems.

- Economic Growth—Wastewater discharges to streams with impaired water quality are not allowed to add additional amounts of the pollutant causing the impaired condition. Reduced flows, lower DO concentrations, or higher temperatures in waters that are not classified as impaired can also reduce the capacity to accept additional waste loads. Thus, economic growth may be limited along stream segments that are classified as impaired or that have their assimilative capacity significantly reduced.

The following metrics are designed to translate output from the water quality models into measures that can be used by the media specialist in assessing potential impacts (e.g., environmental engineers, aquatic biologists, and recreation specialists).

- Compliance with water quality criteria—Changes in water quality that either increase or decrease the spatial or temporal extent of compliance with established criteria. For example, the state criteria for minimum DO concentration may be 5.0 mg/L when measured at a depth of 1.5 meters. Any significant change in the extent of this condition in time or space is identified. Potential changes in 303 (d) listed streams and reservoirs are also identified. The metrics focus primarily on DO and temperature changes and are expressed in terms such as the additional number of miles or additional number of days that the water quality criteria are not met.

- Aquatic habitat suitability—Ability of the impoundments and downstream sections to support indigenous fish and aquatic life. This measure relies on habitat suitability conditions defined by aquatic biologist. For example, fishery specialists define the combinations of dissolved oxygen and temperature considered suitable for a given species. The water quality model defines how this metric varies for each operational alternative. Fishery specialists use the results to assess the potential impact on the fish species. In addition to DO and temperature, other habitat suitability metrics include useable wetted area, flow, nutrients, and potentially toxic parameters.

- Reservoir release DO and temperature—Significant changes in the DO concentration and temperature of reservoir releases. For example, changes in DO are expressed in the number of days and length of stream impacted, and in the tons of oxygen deviation from the base case.

- Waste assimilative capacity—Significant water quality changes that may not violate a water quality criteria, but which worsen or improve the ability of the receiving water to accept additional waste loads. Examples include significantly expanding an area of warm water or increasing the volume of low DO water in a reservoir. Metrics focus primarily on water volume, DO, and temperature measures expressed in terms of tons of oxygen and calories of heat.

- Additional metrics—Other metrics are used when the model suggests a potential for significant impacts to a beneficial use. Examples include suspended sediment and algal concentrations with a potential for impacting aquatic habitat or water supplies.

Hydrodynamic and Water Quality Models

TVA uses a variety of hydrodynamic and water quality models to evaluate water management issues. The most frequently used two-dimensional reservoir models are CE-QUAL-W2 (Cole and Buchak, 1995) and BETTER (Bender et al., 1990). TVA's River Modeling System (RMS) is used for one dimensional analyses of river segments (Hauser and Schohl, 2002). This case study focuses on applications of CE-QUAL-W2 and RMS as part of a broader effort to improve river system operations in the Tennessee Valley.

CE-QUAL-W2 is a laterally-averaged model that simulates the vertical and longitudinal distribution of thermal energy and water quality constituents through time. The model assumes lateral homogeneity and is best suited for relatively long and narrow water bodies with longitudinal and vertical water quality gradients. The hydrodynamics are based on standard equations with a finite difference numerical solution technique. The primary model inputs include reservoir geometry, water inflows and outflows, inflow water quality, meteorology, and initial conditions. TVA modifications to the model include new subroutines for selective withdrawal, surface evaporation, and reservoir aeration. Model results are examined using a graphical post-processor that allows a variety of plots and animations of the modeled parameters (Parsly and Hauser, 2001).

RMS is a longitudinal, dynamic model of river system hydrodynamics, water quality, and fish growth. The hydrodynamic component solves the one-dimensional, longitudinal, equations for conservation of mass and momentum (St. Venant equations) using a four-point implicit finite difference scheme with weighted spatial derivatives. Major inputs to the hydrodynamic component include channel geometry, roughness coefficients, upstream and lateral inflows, boundary rating curves, and initial water surface elevations and discharges throughout the modeled reach. The water quality component solves the mass transport equation with a choice of two different numerical schemes (4-point implicit, Holly-Priessmann). Major inputs include meteorology, inflow water quality, and initial conditions. The bioenergetics fish component simulates fish growth as a function of food availability, temperature, and dissolved oxygen. Inputs to the fish component include temperature and dissolved oxygen over time throughout the modeled segment from the water quality component, fish characteristics such as food consumption and respiration parameters, and food availability assumptions.

Other models provide input that is needed for simulating the potential water quality effects of changes in reservoir operations (e.g., water flows and reservoir elevations). Two system routing models provide the input information used in this case study: the Weekly Routing Model (Miller et al., 1989) and RiverWare (Zagona and Shane, 1996). These models translate reservoir operations into the hourly and daily input required for water quality simulation, based on system geometry, historic flow records, and specific operating procedures.

Model Application

Twenty eight reservoir models and 10 tailwater (river segments downstream of a dam) models are used to simulate water quality in the flow-regulated portion of the Tennessee Valley. Five reservoirs listed in Table 7.1.1 are not modeled due to their small size and limited operational flexibility (i.e., Wilbur, Normandy, and the three Ocoee projects). Hiwassee and Apalachia are combined due to a pump back capability. Only 10 tailwater models are included because many of the dams discharge to a downstream reservoir without a significant intervening river segment. The modeled tailwaters are downstream of Watauga, South Holston, Fort Patrick Henry, Cherokee, Douglas, Norris, Chatuge, Nottely, Apalachia, and Normandy dams.

To facilitate the simulation process 17 reservoir and 6 tailwater models are linked together to run as a single simulation. This linkage eliminates running each component separately and individually converting upstream output to downstream input for the next component. The linked models cover the four largest sub-basins in the Valley (the mainstem Tennessee River, Clinch River, Holston River, and French Broad River). Time constraints and computing capabilities prevented linking all of the models together. The remaining models are run as a watershed unit and the output used as input for the linked simulation.

An eight year simulation period (1987-1994) is used to identify potential water quality effects associated with changes in reservoir operations. This period was selected because suitable data are available and because it represents a range of hydrologic and meteorological conditions. Input data for the simulation period includes hourly discharge flow and surface elevation for each reservoir, hourly meteorology from the nearest weather station, and unregulated inflows and water quality for the headwater (most upstream) reservoirs. Due to the limited availability of water quality data entering the headwater reservoirs, the same data set at a given reservoir is generally used for each year of the simulation period.

Examples of Model Output and Evaluation Metrics

Figures 7.1. 3 through 7.1.10 depict model output for a hypothetical reservoir under three operating scenarios. The figures illustrate not only the model results, but also, metrics that can be used in assessing potential impacts of the operating alternatives.

The three alternatives reflect a key consideration in reservoir operations: the timing of the fall drawdown to provide storage space for the spring flood season. Lake users and adjacent property owners would like for the lake to remain at the higher summer level for as long as possible for recreational and scenic reasons. Downstream water users desire a gradual lake drawdown beginning earlier in the summer. The earlier drawdown allows higher downstream flows that support downstream recreation, navigation, and waste assimilation.

Figure 7.1.3 illustrates three drawdown rates: early drawdown (June 15), late drawdown (October 1), and intermediate drawdown. Each drawdown alternative is at the same elevation on June 15 and on December 31. The early drawdown occurs primarily in the first three months after June 15. The late drawdown maintains the summer lake elevation until October 1. The intermediate drawdown maintains lake levels mid way between the early and late drawdown alternatives.

Figure 7.1.4 shows the reservoir profile for temperature and dissolved oxygen for each alternative on September 15. The profiles reflect the higher lake levels and greater water volumes as the drawdown is delayed from June 15 to October 1. They also indicate that delaying the drawdown preserves cold water at the lower elevations and increases the zone of low oxygen concentrations. Figure 7.1.5 shows similar temperature and dissolved oxygen effects at one location (30 km upstream of the dam) over the entire simulation period.

Figure 7.1.6 plots the DO concentration in the reservoir forebay at a depth of 1.5 meters. For a state criterion requiring at least 5 mg/L, the model predicts that extending the summer pool elevation from June 15 until October 1 would increase the potential violation from 4 to 18 days. Similar analyses can be conducted for temperature (Figure 7.1.7) and to define the spatial extent of potential violations.

Figure 7.1.8 illustrates an extreme condition of adverse fish habitat suitability. Suitable fish habitat for this example is defined as a water temperature less than 29 °C and a DO concentration above 4 mg/L. On July 14 the modeled surface water temperatures and low dissolved oxygen concentrations in the lower portion of the reservoir restricted the suitable habitat to a narrow band of water just below the surface (i.e., the blue area shown in Figure 7.1.8). The profiles also suggest that the area of suitable conditions decreases with delayed drawdown. The suitable habitat can be quantified in terms of reservoir volume, but model animation indicates whether suitable zones are cutoff and depleted, forming dead ends that could lead to fish kills.

Figure 7.1.9 shows the projected temperature and DO concentration of reservoir releases (from the low level turbine intakes). Temperature is important to downstream fisheries. The model indicates cooler releases with delayed drawdown, as the colder water is preserved and released later in the year. The release DO is lower with delayed drawdown. The amount of oxygen that must be added to increase the release to 5 mg/L is less with delayed drawdown, however, because less water is discharged during the lowest DO period.

Figure 7.1.10 provides an overall view of dissolved oxygen in the reservoir relative to the reservoir's ability to assimilate oxygen demanding material (e.g., natural organic matter or manmade waste). If the state water quality criterion is 5 mg/L, reservoir zones (or model cells) with concentrations exceeding this level could be considered to have assimilative capacity. The amount of assimilative capacity can be expressed as the tons of oxygen in the water above 5 mg/L (i.e., the cell volume times the DO concentration above 5 mg/L). The upper left graph in Figure 7.1.10 shows this measure of reservoir assimilative capacity for each drawdown alternative. Similarly, the upper right graph shows the reservoir oxygen deficit (i.e., the amount of oxygen required to increase the reservoir zones with depressed DO concentrations to 5 mg/L). The lower graph shows the net assimilative capacity obtained by adding the capacity to assimilative oxygen demand and the oxygen deficit. In general, Figure 7.1.10 indicates that a delayed drawdown would result in a greater assimilative capacity in the fall (due primarily to increased reservoir volume) and a greater oxygen deficit in the summer (due to a larger zone of low DO at lower reservoir elevations).

Conclusions

These model results and examples illustrate potential analytical approaches and methods for displaying model output. No specific conclusion regarding reservoir operations are intended or reasonable based on the hypothetical simulation. The effects of reservoir drawdown will vary depending on site specific conditions. The examples do suggest, however, that water quality models can play a useful role in evaluating the potential effects of changes in reservoir operating procedures. Improved computer capabilities are making it easier for the modeler to relate model outputs to water management objectives in a way that facilitates the assessment of alternatives.

REFERENCES

Bender, M. D., G. E. Hauser, M. C. Shiao, and W. D. Proctor (1990). "BETTER: A Two-Dimensional Reservoir Water Quality Model, Technical Reference Manual and User's Guide," Engineering Laboratory, Report No. WR28-2-590-152, Norris, TN, October.

Cole, T. M. and E. M. Buchak (1995). "CE-QUAL-W2: A two-Dimensional, Laterally Averaged, Hydrodynamic and Water Quality Model, Version 2.0," Instruction Report EL-95-1, U.D. Army Corps of Engineers, Waterways Experiment Station, Vicksburg, MS, June.

Hauser, G. E. and G. A. Schohl (2002). "River Modeling System v4 – User Guide and Technical Reference," Report No. WR28-1-164, TVA River System Operations and Environment, Norris , TN, May.

Higgins, J. M. and W. G. Brock (1999). "Overview of Reservoir Release Improvements at 20 TVA Dams," ASCE, Journal of Energy Engineering, April.

Lindquist, K. F., W. D. Proctor, W. L. Harper, M. W. Lee, M. McCall, C.R. Montgomery, J. A. Parsly, and M. C. Shiao (1999). "Reservoir Level Policy Effects for Water Quality – Approach," TVA River System Operations & Environment, River Operations, River Scheduling, WR99-3-590-176, November.

Miller, B. A., P. A. Parsly, H. M. Goranflo, Jr., and W. G. Brock (1989). "Weekly Scheduling Model Analysis of Lake Level and Reservoir Release Alternatives, Flat Recreation Target Levels," Tennessee Valley Authority, WR28-1-500-166, August.

Parsly, J. A., and G. E. Hauser (2001). "Animation and Graphics Portfolio Manager," Loginetics, Inc. Knoxville, TN

Tennessee Valley Authority (1990). "Tennessee River and Reservoir System Operation and Planning Review, Final Environmental Impact Statement," TVA/RDG/EQS-91/1, Knoxville, TN, December.

Tennessee Valley Authority (2004). "Final Programmatic Environmental Impact Statement--Tennessee Valley Authority River Operations Study," February.

Zagona, E. A. and R. M. Shane (1996). "PRYSM: A Modeling Framework for Water Resource Operations Management," Proceedings of the Fifth Water Resources Operations Management Workshop: An International Review of Decision Support Systems in River Basin Operation, ASCE, Arlington, VA, March.

Table 7.1.1. Physical Characteristics of 34 Major TVA Reservoirs

Reservoir	River Basin	Drainage Area (sq km)	Mean Annual Flow (m^3/s)	Full Pool Area (ha)	Full Pool Volume (10^6 m^3)	Mean Depth[1] (m)	Hydraulic Retention Time[1] (days)
Mainstem							
Fort Loudoun	Tennessee	24,730	463	5,909	448	7.6	10
Watts Bar	Tennessee	44,830	778	15,783	1,246	7.9	17
Chickamauga	Tennessee	53,850	962	14,326	775	5.4	8
Nickajack	Tennessee	56,640	998	4,197	297	7.1	3
Guntersville	Tennessee	63,330	1,172	27,479	1,256	4.6	12
Wheeler	Tennessee	76,640	1,432	27,143	1,295	4.8	9
Wilson	Tennessee	79,640	1,489	6,273	782	12.5	6
Pickwick	Tennessee	85,000	1,573	17,443	1,140	6.5	8
Kentucky	Tennessee	104,120	1,754	64,873	3,502	5.4	19
Tributary							
Watauga	Watauga	1,210	20	2,602	702	27.0	325
Wilbur	Watauga	1,220	21	29	1	3.0	0
South Holston	Holston	1,820	27	3,068	811	26.4	262
Boone	Holston	4,770	70	1,744	233	13.4	30
Fort Patrick Henry	Holston	4,930	73	353	33	9.4	5
Cherokee	Holston	8,880	127	12,262	1,827	14.9	92
Douglas	French Broad	11,760	190	12,303	1,737	14.1	49
Fontana	Little Tennessee	4,070	112	4,306	1,752	40.7	124
Tellico	Little Tennessee	6,800	169	6,678	511	7.7	31
Norris	Clinch	7,540	115	13,841	2,517	18.2	169
Melton Hill	Clinch	8,660	137	2,303	148	6.4	11
Blue Ridge	Toccoa/Ocoee	600	17	1,331	238	17.9	117
Ocoee 1	Toccoa/Ocoee	1,540	39	765	105	13.7	28
Ocoee 2	Toccoa/Ocoee	1,330	36	0	0	0	0
Ocoee 3	Toccoa/Ocoee	1,270	32	194	4	1.8	1
Nottely	Hiwassee	550	11	1,692	210	12.4	134
Chatuge	Hiwassee	490	12	2,853	288	10.1	199
Hiwassee	Hiwassee	2,510	60	2,465	521	21.1	67
Apalachia	Hiwassee	2,640	60	445	71	16.0	13
Normandy	Duck	510	10	1,307	144	11.0	141
Tims Ford	Elk	1,370	27	4,836	654	13.5	240
Upper Bear Creek	Bear Creek	280	6	749	46	6.2	75
Bear Creek	Bear Creek	600	13	279	12	4.2	9
Little Bear Creek	Bear Creek	160	3	631	56	8.9	158
Cedar Creek	Bear Creek	460	9	1,700	116	6.8	113

[1] Mean depthand hydraulic retention time are based on average, rather than full pool, area and volume

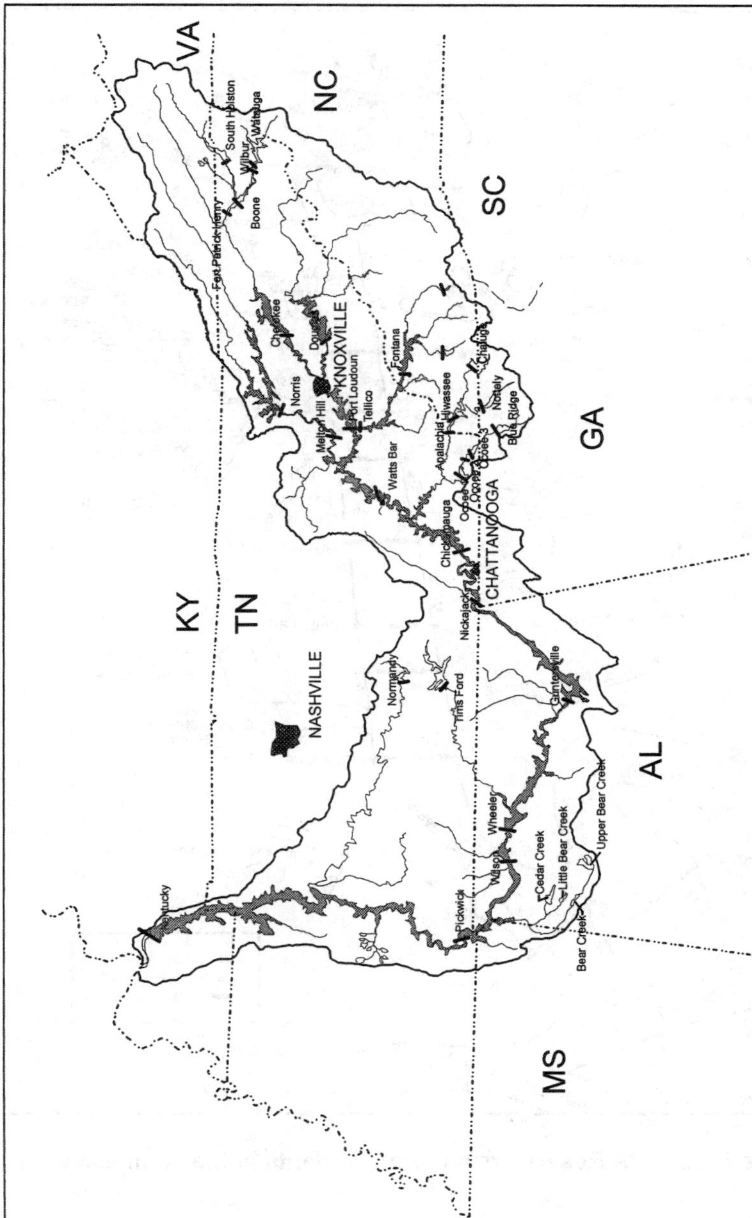

Figure 7.1.1. Tennessee Valley Region

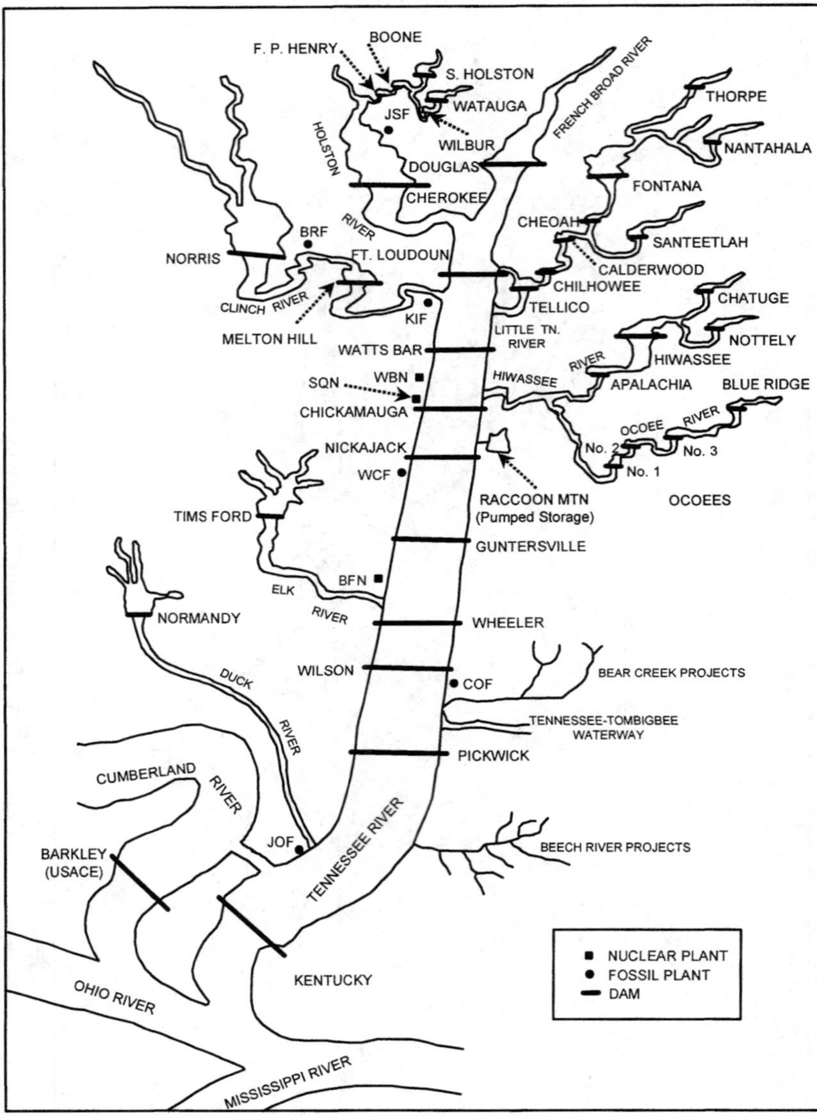

Figure 7.1.2. TVA Reservoirs and Power Plants in the Tennessee Valley

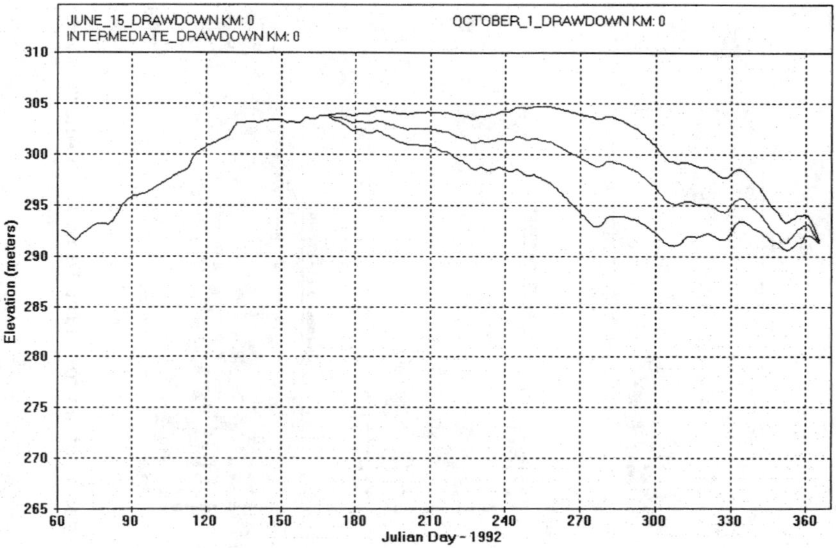

Figure 7.1. 3. Surface Water Elevation for Three Drawdown Alternatives

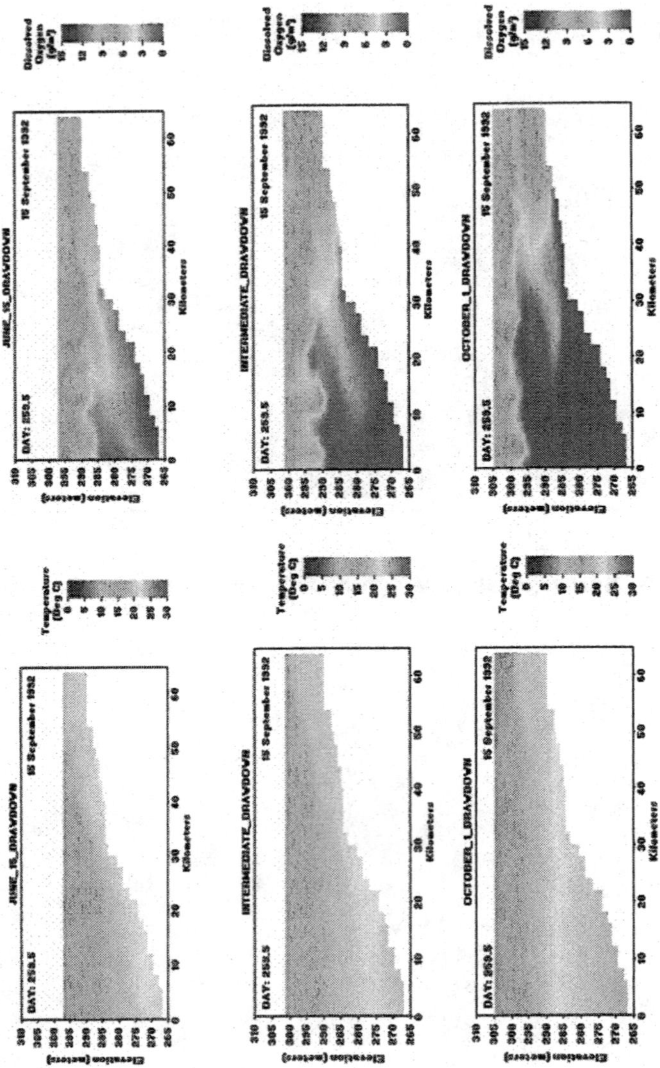

Figure 7.1.4. Temperature and Dissolved Oxygen Profiles for Three Drawdown Alternatives

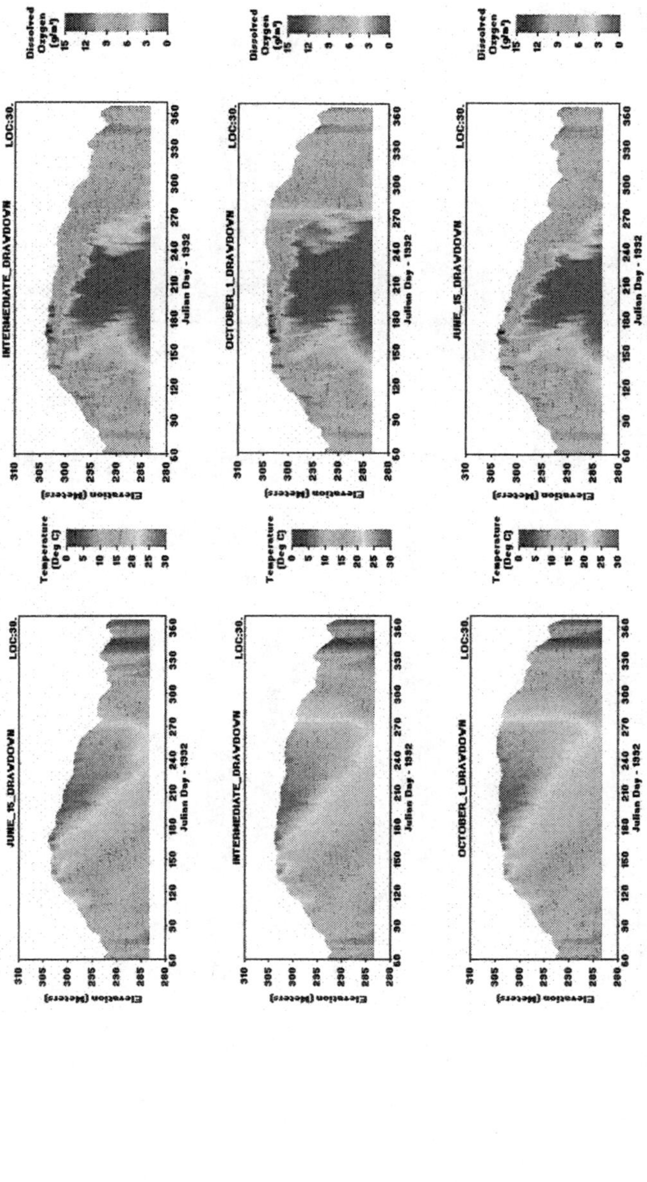

Figure 7.1.5. Time-Depth Plots of Temperature and Dissolved Oxygen for Three Drawdown Alternatives

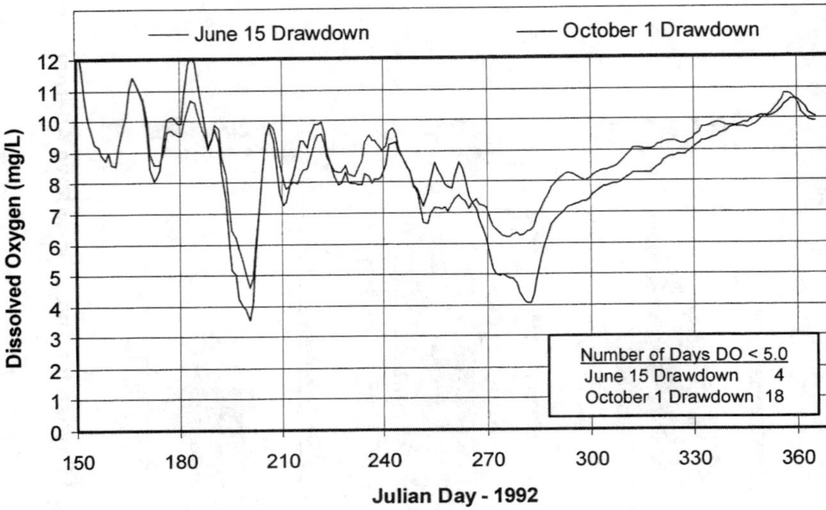

Figure 7.1.6. Dissolved Oxygen Concentration in the Forebay at 5-Foot Depth

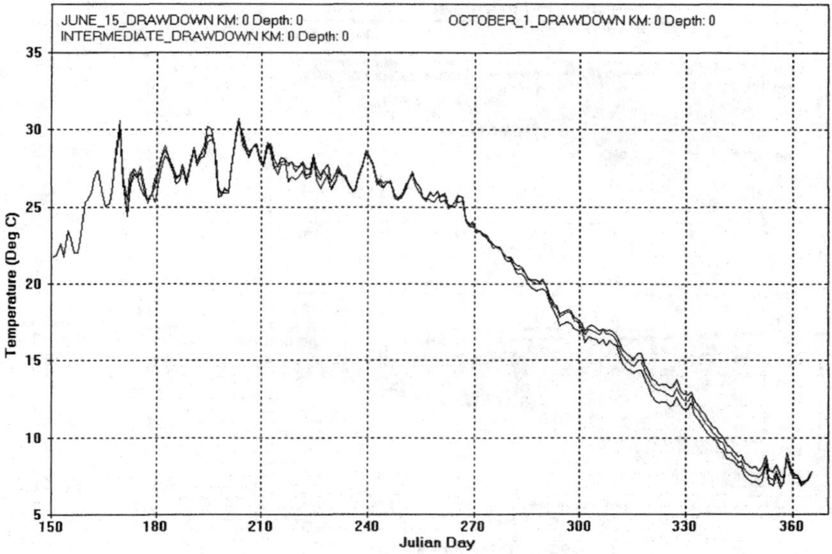

Figure 7.1.7. Water Temperature in the Forebay at the Surface

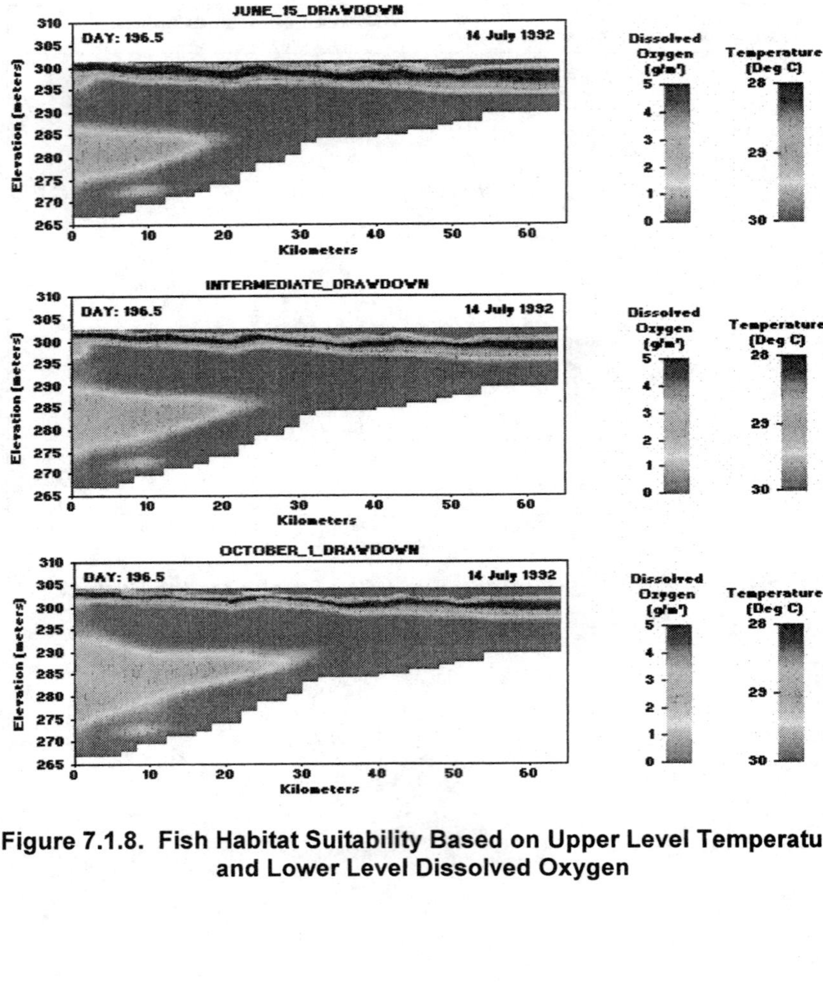

Figure 7.1.8. Fish Habitat Suitability Based on Upper Level Temperature and Lower Level Dissolved Oxygen

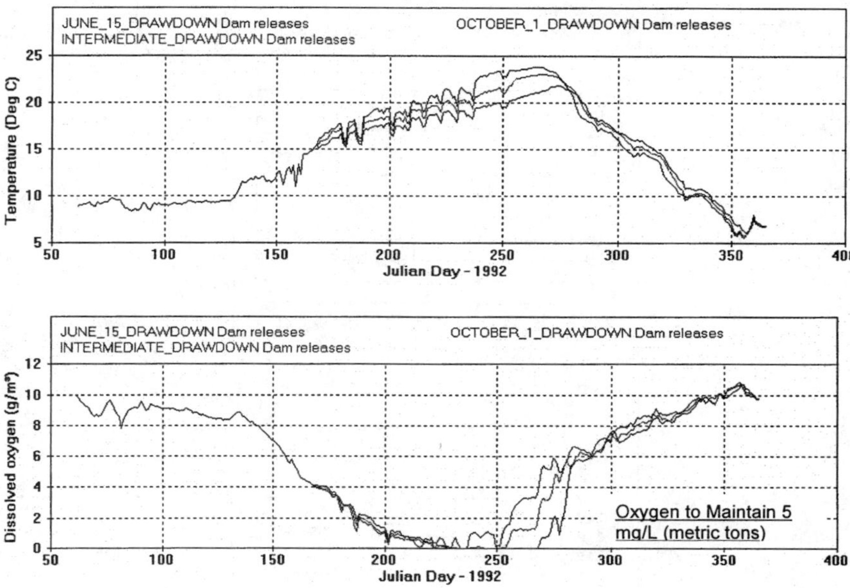

Figure 7.1.9. Temperature and Dissolved Oxygen of Reservoir Releases for Three Drawdown Alternatives

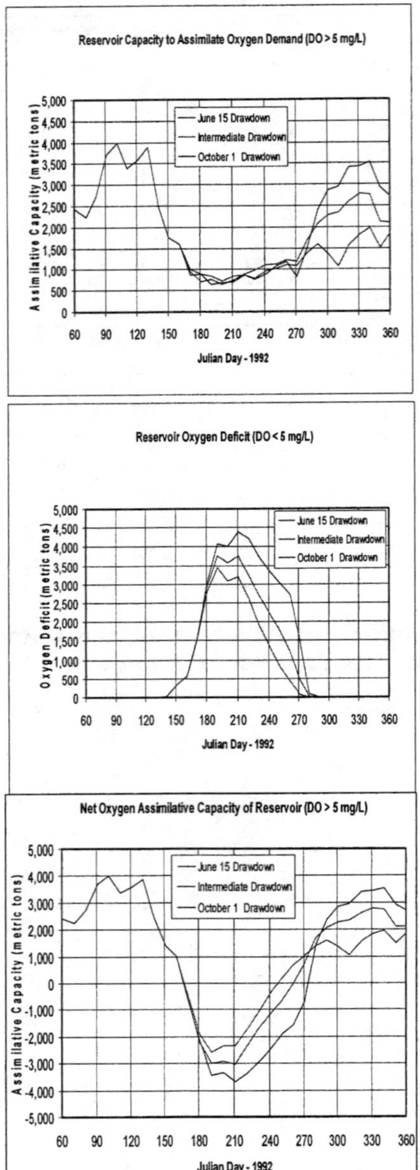

Figure 7.1.10. Time Series of Reservoir Oxygen Assimilative Capacity and Deficit Variations from 5 mg/L

7.2. Limnology and Water Quality in a Multi-Zone Reservoir: Flaming Gorge Reservoir

J.B. Miller

S.G. Campbell

Flaming Gorge Dam is a major feature of the Colorado River Storage Project along with Glen Canyon Dam (Lake Powell), and Boulder Dam (Lake Mead). Flaming Gorge reservoir straddles the state line between Wyoming and Utah, USA. Power production and long term water storage to allow the upper Colorado River Basin States to meet lower basin state water delivery commitments were the primary purposes of building Flaming Gorge Dam. Therefore, annual delivery of local water supplies is not a major operating feature. This is unique among arid region western reservoirs.

Description of Project

Flaming Gorge Reservoir impounds the Green River in southwestern Wyoming and northeastern Utah. The headwaters for the Green River are the Wind River Range in Wyoming. The Green River is the largest tributary of the Colorado River and contributes nearly 50% of the discharge volume at their confluence. The Green River basin encompasses a large region of high desert plateau and mountainous terrain and has a total length of near 744 miles (1200 km). Air temperature ranges from sub-zero in winter to near 100°F (38°C) in summer. Night time temperatures drop dramatically and cooler inflows can begin by late August.

Flaming Gorge Dam construction started in 1956 and was completed in 1964. Initial filling of the reservoir took several years, 1963-1971, with full operation of the reservoir with power production beginning in 1967 (Smith and Green, 1991). The reservoir surface area is 42,000 acres and is 91 miles (147 km) long. The reservoir live storage capacity is 3.75 million acre feet (MAF) with an annual mean inflow volume of approximately 1.7 MAF. The full pool elevation is 6040 feet-msl (1841 m). The average hydraulic retention time with full pool is 2.2 years, but hydrology is highly variable. Initially, the reservoir was drawn down 20-30 feet annually and over 50 feet during multi-year droughts. Historical operation of the reservoir has been to maintain near full pool to maximize head and power production. Releases were adjusted hourly, daily, weekly, and seasonally to maximize peaking power revenue power until 1992. Flow could range from 800-4200 cfs daily based on power demand.

Tests of seasonally adjusted flows began in 1987 and were formalized in 1992. Since 1992, a snowmelt flood event high release to coincide with high flows of downstream tributaries has been used to increase flooded river bottom habitat for fish. Unfortunately, some of the non-native fish may be more adept and competitive at utilizing this habitat than non-natives, so long term continued monitoring, adaptive management, and control of non-natives is still necessary if endangered native fish are to actually realize some recovery. These seasonally adjusted flows included a low flow during the summer and fall. This resulted in very little reservoir drawdown in the July-October period when sediment resuspension and resulting internal phosphorus loading drove huge blue-green algal blooms in the riverine/transitional zone above the confluence in Wyoming. Since 1987 the longitudinal extent and overall magnitude of blue-green algal blooms has been greatly reduced with this change in operation.

Geologic Setting-Paleolimnolgy

The geologic setting of Flaming Gorge contains a great deal of lake paleolimnology. The Uintah Mountains are the longest East-West range of mountains in North America. They rose and essentially eroded away, and rose again over millions of years. The Green River at Flaming Gorge Canyon essentially cuts across the East flank of the mountain. At various times the mountain rose thousands of feet faster than the river could cut through. The Green River Lakes at times covered much of Southwestern Wyoming and Northeastern Utah. The building of Flaming Gorge Dam was simply a miniature re-enactment of this periodic lake formation over geologic time scales. The inflow area of Flaming Gorge is on Green River Lake deposits which have been the subject of paleolimnological studies. Trona (sodium bicarbonate) deposits formed in the receding Green River Lakes are the primary source of sodium bicarbonate production world wide. Sodium and sulfate sources from this area produce salt leaching from bank storage during reservoir drawdown. Oil shale deposits are a good indication that eutrophication processes were also important, since sufficient plankton were present to form the oil rich lake deposit shales.

Longitudinal zonation

Flaming Gorge, WY and UT, is an excellent example of a reservoir having multi-zonal characteristics (Figure 7.2.1). The riverine zone in Flaming Gorge Reservoir originates in two rivers inflowing into the impoundment, the Green and Black's Fork Rivers. The Green River supplies an average of 81% of the inflow water to the reservoir (Madison and Waddell 1973).

Longitudinally, the Flaming Gorge riverine zone exhibits both seasonal and annual fluctuations because it is a function of 1) inflow discharge, 2) changing water surface elevation, and 3) the relative temperature of the inflow water compared to impoundment water temperature (Miller, 1984). The riverine zone is an area of higher velocities, deposition and resuspension of sediment, seasonal phosphorus loading with sediment head cutting, and light limitation. The transition zone in Flaming Gorge fluctuates longitudinally over these same temporal periods, but in addition, can fluctuate vertically as a function of water velocity, temperature, and salinity. These seasonal fluctuations in water density produce density currents. They also "fingerprint" the water so that it can be traced through the reservoir. The transition zone is the beginning of density current formations, and its upper end is marked in the autumn by a sharp plunge line that separates sediment laden water from algal laden water. The transition zone is an area of large noxious blue-green algal blooms and anoxia in late summer and fall. The lacustrine zone in the reservoir exhibits less longitudinal fluctuation and a greater degree of vertical fluctuation in water quality. The lacustrine zone is highly phosphorus limited and usually has chlorophyll-a concentrations less than 2 ug/L. It contains a sharp vertical chemical stratification. A chemocline formed by the sharp increase in salinity below the penstock resulted in an anoxic monimolimnion from early filling until the spring of 1982 (see Figure 7.2.1).

Issues involved

The issues at Flaming Gorge include six major topics:

1. Hypolimnion withdrawal, tailwater low temperatures, river ecology, and management of native versus non-native fish;
2. Design, construction, and operation of a variable selective level withdrawal structure to control tailwater temperature;
3. Analysis of the environmental impacts of operating the selective level withdrawal on reservoir water quality;
4. Analysis of the environmental impacts of changing river operation from peaking power to a more seasonally natural flow with resulting seasonal changes in reservoir fluctuation pattern, and defining the environmental impacts to the river and reservoir;
5. Determining the relative importance of internal versus external phosphorus loading to summer and fall algal blooms in the transition zone with seasonal changes in reservoir elevation fluctuation, general lack of primary productivity reservoir wide, and the relationships to dam operation and seasonal reservoir elevation fluctuations; and
6. Developing and utilizing reservoir water quality models as decision making tools for all of the above.

The issues in dam operation included optimizing peaking power over other ecological and recreation benefits. The loss in power revenues was not insignificant. The original design had a deep hypolimnion outlet from which summer-time tailwater temperatures ranged only from about 4-8°C with a full pool. This was too cold to sustain even a cold water sport fishery, let alone the warmer water native fish. Reservoir fluctuation caused increased salt leaching, and salinity is a major concern in the arid Colorado River Basin. Reservoir summer/fall drawdown also caused increased sediment resuspension in the riverine zone which released up to 400 ug/L bioavailable phosphorus and produced large blue-green algal blooms.

Seasonally, the inflow salinity (as total dissolved solids - TDS) varied from near 300 mg/L to over 1000 mg/L. A few ungauged small tributaries had TDS measured in the 3000-5000 mg/L range. Lower TDS inflows came during spring/summer snowmelt events. Since the river warms faster than the reservoir prior to thermal stratification, the low TDS inflow became an overflow density current. In the fall and winter the more saline inflows cooled faster than the large stratified reservoir; these inflows become a high TDS bottom hugging underflow density current. The chemical stratification and destratification is illustrated in Figure 7.2.2. The reservoir became chemically stratified during the first few years of initial filling, and a chemocline formed leaving a saline and anoxic zone (monimolimnion) in the deep hypolimnion immediately beneath the penstock elevation. This monolimnion persisted from initial filling until about February of 1982. The selective level withdrawal structure was installed and began operation with summertime epilimnion withdrawals in August of 1978. It is significant that the chemocline broke down and partial mixing occurred in 1981-82, essentially a drought cycle. In 1983, a very wet year resulted in large spills from the surface spillway, penstock, and deeper by-pass tubes. The mixing and dilution of very high flows in 1983-85 resulted in nearly complete chemical destratification of Flaming Gorge as can be seen in Figure 7.2.2. Chemical stratification of TDS has returned since 1985, but not to the degree of re-establishing a monolimnion that prevented vertical mixing and partial reaeration of the entire hypolimnion.

Sediment/phosphorus diagenesis in the riverine/transitional zone

Reservoir fluctuation caused scouring of the sediment delta formed during filling. This sediment scouring (head cutting) resulted in the release of large phosphorus loads that drove large blue-green algae blooms in the first 10-15 miles of the reservoir in the riverine/ transitional zone. The phosphorus/ sediment diagenesis in the riverine/transitional zones of Flaming Gorge is unique in that resuspended sediment yielded as high as 400 ug/L bioavailable phosphorus, which dwarfed the inflow budget. This meant that a watershed phosphorus wasteload allocation or Total Maximum Daily Loading (TMDL) would have been ineffective at reducing algal blooms in the transition zone. It is interesting that the paleolimnology in this same area is oil shales formed in eutrophic stages of the ancient Green River Lake. Figure 7.2.3 shows the extent of the algal bloom associated with a near 50 foot drawdown in 1978. Since this was just after initial operation of the selective level outlet, Wyoming residence asked, "Is this what we get for the warmer water below the dam?" In the years with reduced summer drawdown following 1987-92, it is interesting to note that these algal blooms did not extend below the confluence.

Anoxia developed quickly in the algal blooms in the transition zone after thermal stratification set up in June. This resulted in seasonal movement of fish away from the transitional zone. The noxious and potentially toxic producing blue-green algae reduced recreation and fishing opportunities in the portion of the reservoir nearest to the population centers in Wyoming.

Operational changes to Flaming Gorge Dam that optimize more than power revenue resulted in seasonally adjusted releases to replace habitat.

Studies and Modeling

Scientists became very interested in Flaming Gorge Reservoir both from the standpoint of limnology, salinity, and fisheries management very early in its history. The tailwater became an outstanding cold water (trout) sport fishery during the first few years of filling; however, as the reservoir continued to fill, the large cold hypolimnion hydraulic retention time quickly exceeded summer releases. The releases became very cold with summertime temperature only reaching a maximum of 8C. The trout fishery disappeared by 1974. Scientists of the U.S. Geological Survey began studies of Flaming Gorge Reservoir limnology and particularly salinity issues from 1968-1975 (Madison and Waddell, 1973; and Bolke and Waddell, 1975. In 1975, EPA included Flaming Gorge in the National Eutrophication Survey (NES, 1975); the Southwest Wyoming Council of Governments received a section 208 Clean Water Act Grant and began studies of Flaming Gorge's algal bloom problems and documented nutrient loading sources from the watershed.

Initially, the models tested resulted in complete turnover of the reservoir during each fall turnover. The models had numerical instabilities which over-mixed the reservoir at the on-set of fall turnover. Since mixing of bottom nutrients back up to the euphotic zone of the reservoir is an important nutrient source to fall algae blooms, then accurate simulation of complex hydrodynamic functions is very important. Similarly, understanding salt retention and routing is important to understanding salinity variations throughout the Colorado River Basin. Subsequently, the 2-dimensional BETTER and CE-QUAL-W2 models both proved to be capable of multi-year simulations that maintain vertical salinity gradients within the accuracy of the input data. The ability to chemically fingerprint seasonal inflows into Flaming Gorge and track the hydrodynamics of TDS as a tracer makes Flaming Gorge an outstanding reservoir to test reservoir hydrodynamics.

The WRE one-dimensional model was utilized to help evaluate the eutrophication problem and was utilized again in 1977-78 by the U.S. Bureau of Reclamation to examine the feasibility and options of adding a selective level outlet works to control temperature of downstream releases. The LARM model was applied in the early 1980's by the Bureau of Reclamation), and the TVA BETTER model was tested.., The CE-QUAL-W2 (Cutler and Miller, 2002) is still under development. All of these models had strengths and weaknesses, and gave pieces of predictive capabilities which helped resolve the issues most pressing at the time. Each model did an adequate job of predicting temperature and water quality of the releases at the dam, but only BETTER and CE-QUAL-W2 could adequately evaluate all of the issues, particularly eutrophication in the transition zone. BETTER and CE-QUAL-W2 both showed that the large blue-green algal blooms could not be produced by the inflowing nutrient budgets of the Green and Blacks Fork Rivers.

While the models demonstrated that the watershed nutrient budgets could not generate the algal biomass present in the reservoir, they did not simulate the sediment/phosphorus diagenesis that was responsible for the algae. However, knowing that internal nutrient recycling might be more significant than external loading should lead to research on phosphorus/sediment diagenesis in the riverine zone. Discovery that drawdown and sediment head cutting resulted in the release of 400 ug/L bioavailable phosphorus right in the inflow riverine zone allowed the modelers to include the additional phosphorus concentration in the inflow during drawdown. Increasing the inflow phosphorus concentration during drawdown until the algal biomass and resulting oxygen demand could be reproduced demonstrated that internal phosphorus recycling was more significant than the watershed budget. Furthermore, the blue-green algae are nitrogen fixers, and allowing the model to go nitrogen-limited also resulted in low algal biomass. Therefore, calibrating the model to produce the algal biomass also required increasing the inflow nitrogen budget so that nitrogen could not become limiting. It requires years of experience and some water quality art to know when, how, and why to do appropriate sensitivity studies. However, when predicting if a watershed wasteload allocation will produce the desired end-points in the reservoir, these tools and evaluation become invaluable. Inappropriate use of these models could result in large expenditures that will not produce the appropriate desired end-points.

Selective Level Outlet works

In 1978, a multi-level outlet was installed at Flaming Gorge Dam. Prior to this, the reservoir operation was through a hypolimnetic release and the only constraint was a minimum 800 cfs release to maintain a trout fishery in the Green River below Flaming Gorge. Peaking power demand determined reservoir releases both hourly and seasonally until 1992, when a biological opinion for endangered fish resulted in mimicking a seasonally adjusted natural flow pattern. An interim flow agreement was put in place to protect nursery habitats for endangered fish (Colorado pikeminnow, razorback sucker, humpback chub and bonytail chub). The temperatures of the release range from 12-15 C for as long as possible each summer to provide for a blue-ribbon trout fishery in the first 45 km downstream and to provide warmer water for native fish thereafter. Both flow and temperature targets have been developed for the protection of endangered fish species in the Green River (Muth, *et al.* 2000). The changes in reservoir operation not only affected downstream fish communities but also the physical-chemical limnology and biota in the reservoir as well. The changes also had a considerable impact on revenues from peaking power generation.

Summary

Flaming Gorge Reservoir has a long history of water quality concerns (Madison and Waddell 1973, Bolke and Waddell 1975, Miller 1984, Verdin 1984). Initial studies focused on the concentration of dissolved solids (Madison and Waddell 1973) in the reservoir, but soon a chemocline in the main basin near the dam that was completely anoxic developed and persisted for many years (Bolke and Waddell 1975). Changes in thermal regime and water quality in the Green River below the Dam were also identified and in September, 1975, the first documentation of an extensive algal bloom in the upper reservoir including the Blacks Fork and Green River Arms of the inflow zone was reported in 1979. Bureau of Reclamation began monitoring Flaming Gorge Reservoir water quality and productivity in 1981 and was an active participant in the development of a remote sensing application for trophic status estimation in the reservoir (Verdin 1984). Figure 7.2.3 depicts the surface extent of algae in 1978.

Dam and reservoir operation has also affected the water quality and productivity of the impoundment in a variety of ways. The biota has experienced many changes, partly because the reservoir fishery now consists of introduced sport and forage fishes. Recreational use of the reservoir has increased along with the overall population growth and increased opportunity for access as recreation facilities continue to improve. The annual fall algal blooms persist, but are decreased in magnitude, duration, and longitudinal pervasiveness as a result of the altered operational scheme.

REFERENCES

Bolke, E.L., and K.M. Waddell. 1975. Chemical quality and temperature of water in Flaming Gorge Reservoir, Wyoming and Utah, and the effect of the reservoir on the Green River. Geological Survey Water-Supply Paper 2039-A. U.S. Government Printing Office, Washington, DC. 26 pp.

Madison, R.J., and K.M. Waddell. 1973. Chemical quality of surface water in the Flaming Gorge Reservoir area, Wyoming and Utah. Geological Survey Water-Supply Paper 2009-C. U.S. Government Printing Office, Washington, DC. 18 pp.

Miller, J. 1984. Intermountain West reservoir limnology and management options.

Muth, R.T., L.W. Crist, K.E. LaGory, J.W. Hayse, K.R. Bestgen, T.P. Ryan, J.K. Lyons,and R.A. Valdez. 2000. Flow and temperature recommendations for endangered fishes in the Green River downstream of Flaming Gorge Dam. Final Report. Upper Colorado River Endangered Fish Recovery Program Project FG-53. 241 pp.

Verdin, J.P. 1984. Limnological applications of multi-spectral remote sensing, USBR Report No. REC-ERC-85-2, U.S. Bureau of Reclamation.

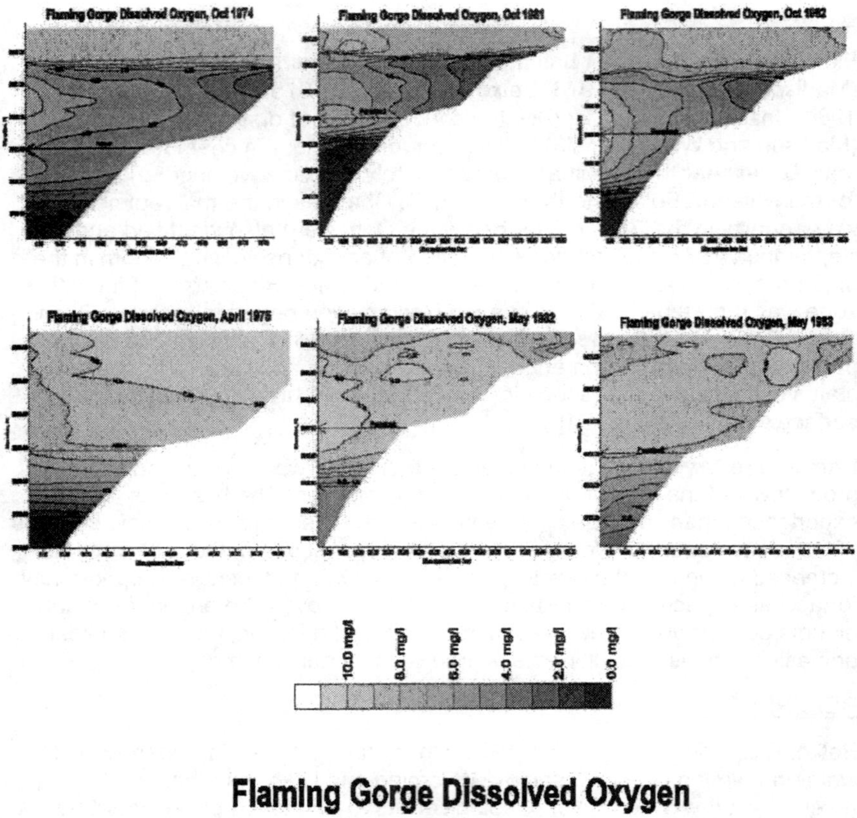

Figure 7.2.1 Flaming Gorge DO profiles

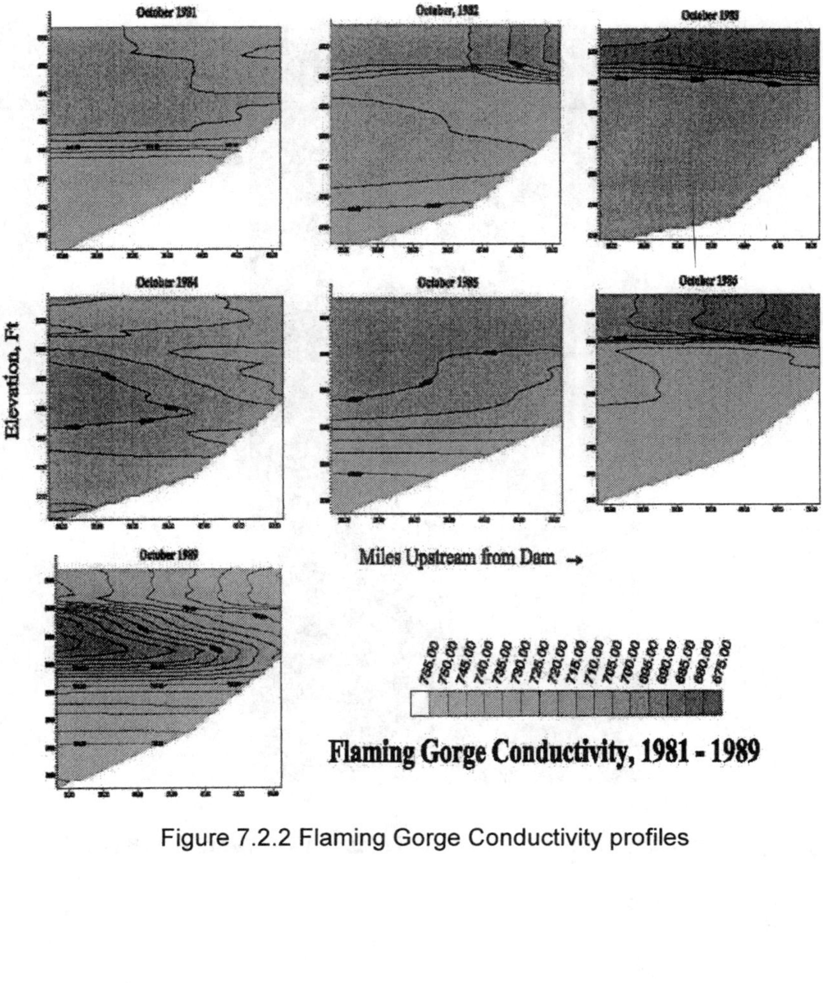

Figure 7.2.2 Flaming Gorge Conductivity profiles

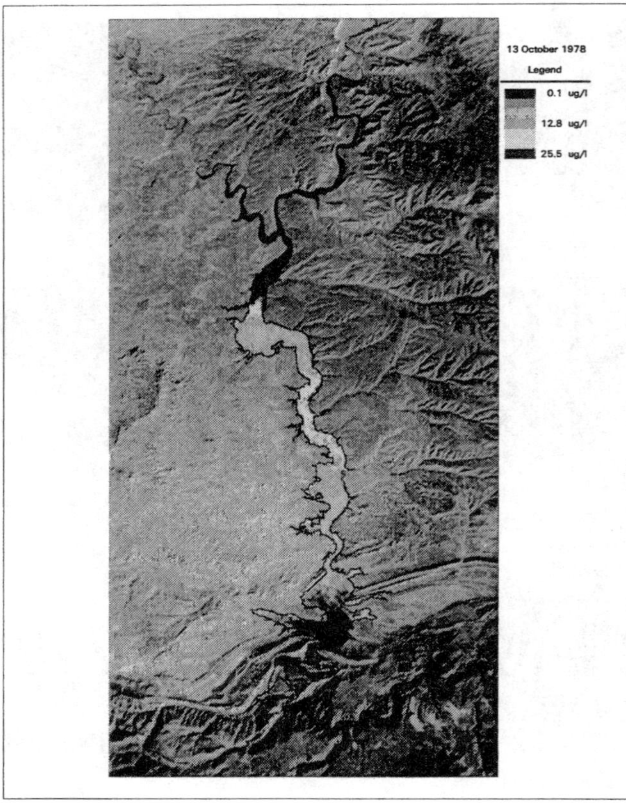

Figure 7.2.3 Flaming Gorge algal bloom 1978 (Verdin 1984)

7.3. Old Hickory Reservoir

Edward M. Buchak

Rajeev Jain

G. Lynn Jarrett

This summary is based on a study and report entitled "Application of CE-QUAL-W2 to Old Hickory Lake, Comprehensive Report" (11 September 2002) and submitted to the Nashville District, USACE, Nashville, Tennessee, under Delivery Order 1 of Contract DACW62-01-D-0014 and by subcontract to Northwest Hydraulic Consultants. The report authors were Rajeev Jain, Edward M. Buchak, and G. Lynn Jarrett. In general, the work was supervised, the model was set up and calibrated for temperature by E. M. Buchak; Rajeev Jain calibrated the model for water quality; and G. Lynn Jarrett performed the watershed quantity and quality modeling. All three authors shared in the writing.

Old Hickory Lake (OHL) is located just east of Nashville, one of a series of Kentucky River multi-purpose reservoirs operated by the USACE. The dam that forms OHL includes both hydroelectric and locking facilities. OHL is also used for water supply, flood control, and recreational purposes. OHL receives inflows from Cordell Hull dam, itself a multi-purposes structure that includes hydroelectric and locking facilities, and from Caney Fork, a large tributary whose flow rate and temperature are also controlled by an upstream dam. OHL dam replaced an earlier series of low elevation lock and dam structures that were inundated but not demolished. Gallatin Steam Plant is located about two-thirds of the way down the Lake and uses once-through cooling.

The primary objective of this study was to provide the Nashville District of the USACE with a calibrated CE-QUAL-W2 model application capable of estimating changes in Old Hickory Lake water quality due to changes in watershed nutrient loads and due to changes in the operations of Gallatin Steam Plant. An additional objective of the Nashville District was to develop a technique for estimating time-of-travel more accurately than typical CE-QUAL-W2 applications do. Besides improving longitudinal temperature and water quality gradient resolution, success in this objective would allow use of the model for real time predictive purposes in the case, for example, of computing arrival times at drinking water intakes downstream of a toxic substance spill.

Careful model setup is critical to the success of CE-QUAL-W2 applications. Important components of the setup are accurate bathymetric representation of the waterbody and incorporation of the above-mentioned geometric and structural features. These were relatively easy to do with CE-QUAL-W2's comprehensive boundary condition capabilities when combined with GEMSS

(Generalized Environmental Modeling System for Surfacewaters) software that provides GIS capabilities and allows these features to be accurately located on a common geographic reference system.

The time-varying data sets necessary to make simulations with the model were similarly detailed, but are not discussed in the following summary. The reader should note that the available datasets included the years 1987-1988 and 1997-1999

Bathymetry

The CE-QUAL-W2 model grid (Figure 7.3.1) was prepared from information supplied by Nashville District staff, chiefly bathymetric surveys conducted in 1994 and 2001, but also including various reports, maps, and notes obtained during a reconnaissance in late October 2001. The bathymetric surveys were in the form of transects ("sediment ranges") carried out during 1994 and 2001; these were pooled for development of the CE-QUAL-W2 grid. The grid was prepared by processing the bathymetric data using a Fortran program. The program created a grid with a single CE-QUAL-W2 segment for each of the transects and a layer thickness of 0.5 m. This grid therefore corresponds exactly in longitudinal resolution to the bathymetric surveys.

A feature of the grid development for OHL is specialized treatment of the cross-sections. The main stem below the OHL riverine-lacustrine dividing line at River Mile (RM) 254 was divided into two parallel branches: a conveyance branch and a storage branch. The conveyance branch represents the main channel. The storage branch represents the volume in shallow embayments and overbank areas. For each segment in the conveyance branch there is a corresponding segment in the storage branch. There are 40 such pairs. Each segment in the conveyance branch is connected to the corresponding segment in the storage branch using a lateral (submerged) weir boundary condition. To illustrate this approach, Figure 7.3.2 and 7.3.3 show the individual and combined components for the segment nearest OHL dam.

Using CE-QUAL-W2's lateral weir to model the flow between adjoining storage and conveyance segments results in flows that are uni-directional for any particular time step, i.e., flows are computed either from the storage segment to the conveyance segment or the vice versa. The exchanged flows are modeled as CE-QUAL-W2 "withdrawals" from the conveyance segment and "tributaries" from the storage segment. The direction of flow is controlled by the head difference. The vertical extent of the flow is controlled by the SELECT algorithm on the withdrawal side and by control file parameters on the tributary side.

The purpose of this approach is to address a problem inherent in CE-QUAL-W2 applications, that is, its inability to distinguish hydraulically between conveyance and storage volumes. The traditional approach results in decreased velocities in the main channel, essentially proportional to Q/A, where Q is the flow and A the combined conveyance and storage areas. The specialized treatment of cross-sections in the present approach allows increased velocities proportional to Q/A', where A' is the conveyance area. These increased velocities result in decreased, and presumably more accurate, time of travel estimates. Not only does the present approach allow for more accurate longitudinal gradients, it is especially important for event-based simulations, for example, computing arrival times at drinking water intakes subsequent to a toxic spill.

Time-of-travel tests

The time of travel tests were done by comparing simulations made with the combined conveyance and storage and simulations made with the separated conveyance and storage cross-section level grids. The test setup included a steady inflow rate of 100 $m^3\ s^{-1}$ at Cordell Hull that was numerically dyed to a value of 100 mg l^{-1}. Branches 2 through 9 had inflow rates of 10 $m^3\ s^{-1}$ and no dye. The Branch 1 outflow rate was 180 $m^3\ s^{-1}$, so that there was no change in reservoir volume. The equilibrium, fully-mixed value of the dye would be 55.6 mg l^{-1} (= 100 $m^3\ s^{-1}$ x 100 mg l^{-1} / 180 $m^3\ s^{-1}$). The flows selected for the experiment are equivalent to very low flows through OHL. Inflow temperatures were held constant and surface heat exchange was turned off. Other features of the time-of-travel tests can be determined from an examination of the control file in corresponding directory.

Results for the two grids are shown in 7.3.4, which shows that 90% of the tracer arrives at the dam 14 days sooner when the storage volume was separated from the conveyance. This decrease in arrival time represents a substantial increase in longitudinal velocity. Based on the success of this and other experiments, the separated storage and conveyance grid was used to develop the hydrodynamic grid and therefore was used in all subsequent simulations.

Calibration

Model calibration was done in three phases: water balance, temperature, and water quality, each phase depending on the successful completion of the prior phase. In all phases time-series observations, when available, took precedence over synoptic measurements (i.e., vertical profiles at a particular time). This approach is consistent with the goal of modeling the overall system properties and behavior. Furthermore, 1988 has significantly more data than the other years, so it became the priority. The 1998 and 1999 comparisons were made without additional coefficient tuning.

The water balance was calibrated by programmatically comparing computed and observed water surface elevations and adding or subtracting flow (as tributary inflows or as additional outflows at the dam, respectively) so that the observed elevation time series was reproduced. This approach was necessary because OHL, like many lakes and reservoirs, does not have a complete or accurate water balance. For the 1988 simulation, the mean value of the flow correction is -21.5 $m^3 s^{-1}$ while the mean value of the correction for the 1998-1998 simulation was -15.7 $m^3 s^{-1}$. The negative values indicate either the inflow derived independently from data sources was too large or that the outflow similarly derived was too small.

The water temperature profiles and time series show very good agreement using standard CE-QUAL-W2 parameters, and therefore, additional calibration simulations for temperature calibration were not made.

Data limitations strongly constrain the calibration strategy for a water quality model application. Comparisons of computed and observed time series are most suitable for calibration of water quality constituents as they display the seasonal and diurnal dynamics well. However, time series were not considered appropriate in this case, as there were too many distinct location and depth combinations, and very few observations for each combination. Similar difficulties occurred when considering water quality profile data. Seven stations measuring TP, NH3, TKN, NO2-NO3, Chla, and TSS were available for calibration purposes, but because data were not available systematically at all dates, profile comparisons plots had little value. The water quality calibration relied instead on obtaining a overview of seasonal and spatial magnitudes and trends from the two datasets (time series and profile data) and reproducing in a general way the observed magnitudes and trends.

Water quality calibration was achieved using several iterations of improvement in load estimates and model coefficients. The calibration strategy was to review key predictions from model "snapshot" output against in-reservoir and boundary data. The calibration was not staged in that it did not focus first on inorganic solids, or on constituents in any particular order, and it retained all model parameters for calibration. However, the early simulations generally focused on getting the seasonal dynamics of the various constituents approximately reproduced, and then later simulations focused more on the profiles.

Although the water quality calibration achieved was satisfactory for the purposes of this study (estimating changes in Old Hickory Lake water quality due to changes nutrient loads and due to changes in the operations of Gallatin Steam Plant), there were several features of the water quality field data that proved difficult to capture: DO supersaturation, the DO minimum at intermediate depth, and high ammonia and nitrate concentrations in the lake, together with Chla values that remain below 25 ug/l. The reasons for these are likely interrelated. The difficulty appears to arise from the inability of the model to capture supersaturation of DO at the surface, which is in turn related to PO4 depletion in the surface waters. Light limitation was encountered in some simulations at the surface cells, because the model uses a light inhibition function that suppresses photosynthesis at light levels higher than saturation. In most of the simulations there was PO4 limitation, and this is due to depletion of P due to algal uptake as well as locking up of the P in various organic matter forms. Even without rapid sinking, a large pool of organic matter and algae can reduce PO4 enough to prevent the rapid photosynthesis that is required for supersaturation. The re-aeration equation was found to make a difference to the surface DO, but did not completely explain the lack of DO supersaturation.

CE-QUAL-W2 may be structurally unable to model a supersaturation at the surface. This can be due to the use of a light inhibition function. It may also be related to the fixed stoichiometry used in the model for algal and other dead organic matter. Stoichiometry was a very important determinant of algal dynamics. Because the choice for this parameter needs to be consistent with that used in the boundary condition development, it was difficult to experiment with this parameter. Still, a limited number of runs were made using two sets of boundary conditions developed using two reasonable values for this parameter. These runs suggest that stoichiometric ratios are an important parameter in the calibration.

Other limitations of calibration (e.g., the metalimnetic DO minimum) appear to be consequences of the smaller productivity experienced in the model at the surface. Metalimnetic DO is usually explained by one or more of several mechanisms, one of which is the outflow structure location. The organic matter fixed in the surface layers settles to the metalimnetic layers where its decay leads to a DO minimum. Below the metalimnion, the DO increases again because of the relatively young water flowing in from upstream sections to a near bottom deep outflow structure location at the dam.

The metalimnetic DO minimum was captured in many simulations, but never as strongly as in the field data. Following considerations of transients and seasonality of inputs as discussed in the boundary condition development, a question could be raised that the DO minima observed was specific to the two dates sampled. Because the boundary nutrient load was input as an annual average, such transients would not be modeled. However, because the minima occurred on two separate dates, and was observed at upstream locations as well with a steadily decreasing strength, it is likely a real and seasonal time scale feature that should be captured in a future upgrade to the model.

To address this several features not reproduced in the model, the following source-level code changes were implemented

- Coding of a new light limitation function that does not show inhibition at high light levels

- Decoupling algal excretion from light limitation
- Second order algal mortality, to prevent large blooms in the absence of light inhibition
- Stoichiometric ratio of organic matter and algae

The new light limitation function helped substantially to capture the additional DO production at the surface. However, Chla values were increased well beyond the 25 ug/l or so that was indicated by the limited amount of field data available. Various strategies were tried to allow for large photosynthesis at the surface, and yet maintain a small algal population. This was found to be not possible.

The fact that a large primary production flux occurs at the surface by a relatively small algal population in the field data, but not in the model, is likely due to the accompanying changes in the stoichiometry. Phosphorus is mineralized much faster than nitrogen in nature, but in the model it must mineralize at the same rate as organic carbon and nitrogen. The organic matter may also in reality get depleted in phosphorus thus releasing more phosphorus for uptake and incorporation into algal matter. Experiments with organic matter and algal stoichiometry suggested a strong sensitivity to this parameter. Note that the stoichiometry for all organic matter compartments in the model must be maintained the same, else the model does not conserve nutrients.

Gallatin Steam Plant thermal and water quality impacts

The effects of removing the observed Gallatin thermal discharge load (plant flow and condenser temperature rise) on OHL water quality were examined. For the no load case, temperatures were reduced significantly within the first two segments upstream and downstream of the Gallatin, as expected. There were insignificant effects on dissolved oxygen, and even these were not evident beyond the two segments upstream and downstream.

A feature of the Gallatin simulations is noticeable recirculation of heat from the discharge to the intake. Although known to occur in once through cooling systems, the specific amount of recirculation may be related to the coarseness of the OHL grid. In the event that Gallatin effects become important for regulatory reasons, the OHL grid should be refined in the vicinity of Gallatin, a relatively simple task.

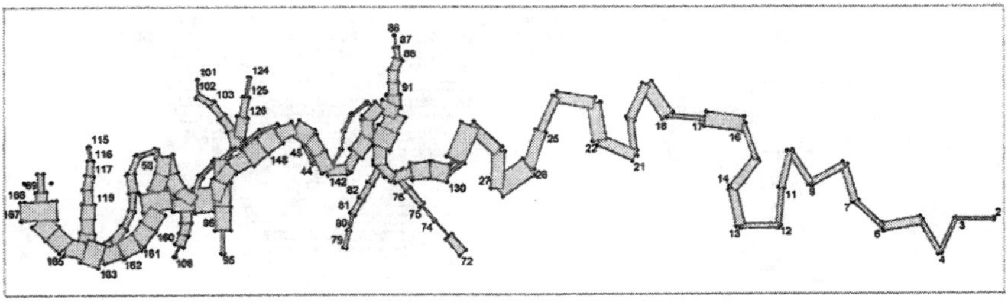

Figure 7.3.1 CE-QUAL-W2 grid showing segments and orientations

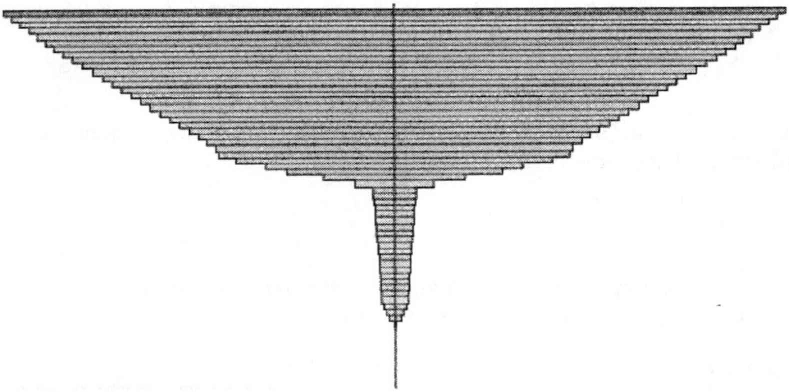

Figure 7.3.2 CE-QUAL-W2 cross-section at OHL dam showing combined storage and conveyance (no scale)

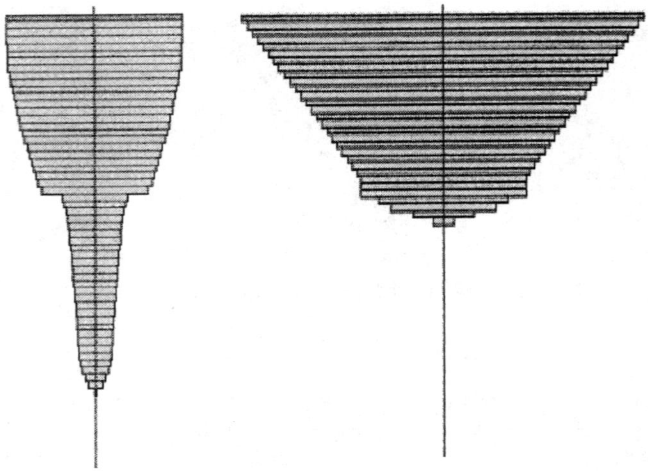

Figure 7.3.3 CE-QUAL-W2 cross-section at OHL dam showing conveyance component (left) and storage component (right)

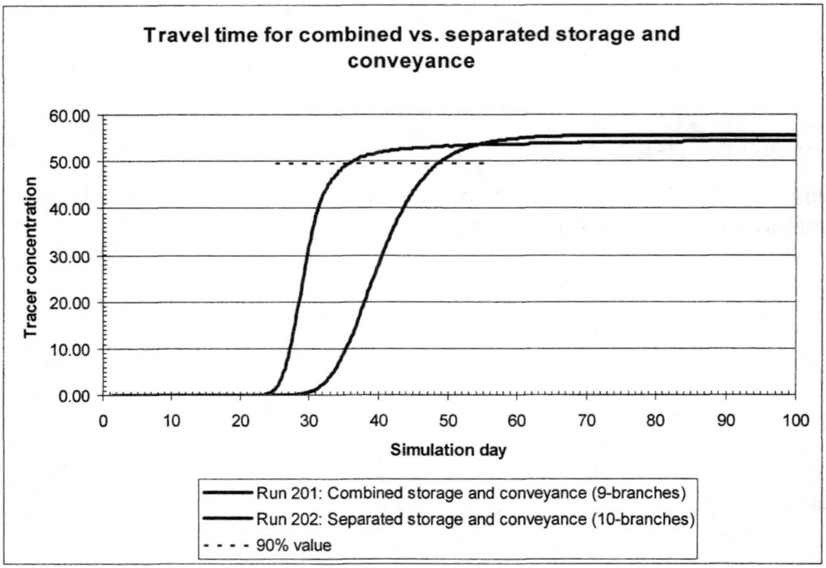

Figure 7.3.4. Travel time for dye introduced at the upstream boundary

7.4. Use of Systems Models in Conjunction with 3-D Models: Gerald Gentlemen Station, Nebraska

Edward M. Buchak

George A. Krallis

Nebraska Public Power District's Gerald Gentlemen Station (GGS), a fossil fuel steam electric station located near the junction of the North and South Platte Rivers, utilizes several interconnected and seasonally available water sources for once-through cooling and evaporative makeup water. Normal operations circulate North Platte River, South Platte River, and Sutherland Reservoir water to the GGS intake, through the condensers and into a 2.9 mi cooling canal, a 220 ac cooling pond, and finally to the Sutherland Reservoir (7.4.1). Selection of the possible source water depends on the season and water availability.

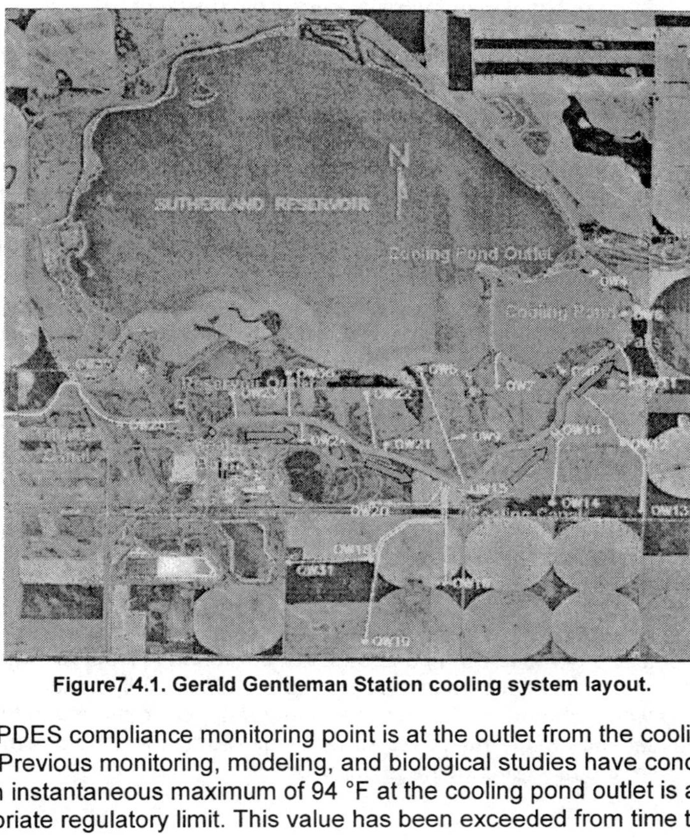

Figure7.4.1. Gerald Gentleman Station cooling system layout.

The NPDES compliance monitoring point is at the outlet from the cooling pond. Previous monitoring, modeling, and biological studies have concluded that an instantaneous maximum of 94 °F at the cooling pond outlet is an appropriate regulatory limit. This value has been exceeded from time to time in recent summers due to higher utilization of GGS and warmer natural conditions that affect both the temperature of the source water and the cooling capacity of the canal, pond and reservoir.

The Nebraska Public Power District (NPPD) requested that a model be constructed (1) to assess the frequency with which the limit of 94 °F is exceeded and (2) to examine cooling pond dynamics and suggest modifications that will increase its efficiency. The first objective was addressed by assessing overall cooling system performance over many years using fully-mixed computations (a "systems model") that incorporated the various water sources, the varying station thermal loads and pumping rates, and the varying meteorological conditions at an hourly time scale that allowed computation of surface heat exchange rates in the canal, pond, and reservoir. The second objective was addressed by simulating the cooling pond dynamics using GEMSS®, (Generalized Environmental Modeling System for Surfacewaters), a 3-D hydrodynamic and transport model. The pond was model at a 30 m horizontal and 0.5 m vertical space scale.

The systems model and the cooling pond model are related because cooling pond performance as determined with the 3-D model was used in the systems model to assess the changes in the frequency and duration of the discharge temperatures under the proposed modifications. Mapping of 3-D model results to the system model relies on the fact that reservoir temperatures in the presence of heated discharges can be separated into two components: natural temperatures and waste heat-induced temperature rises. The temperature rise component was computed using GEMSS. The long term, natural temperature component was developed using the systems model. Although the 3-D model has the capability to compute all of the components incorporated into the systems model directly, it is impractical to run the 3-D model for the very long time periods needed to obtain reliable frequency-duration estimates, as it is best suited to simulations on the order of 1 year. The systems model, in contrast, can simulate decades in a few minutes.

The 3-D model was calibrated against dye and temperature surveys; the systems model was calibrated against the 6-year record of cooling pond outlet temperatures (1995-2000).

<u>Systems Model</u>

The temperature at the cooling pond outlet to Sutherland Reservoir can be computed by assuming the canal and pond are a single, fully-mixed waterbody with heat added at the condenser outlet, heat removed at the cooling outlet and heat removed (or added under some meteorological conditions) at the air-water interface as follows:

$$V\frac{dT}{dt} = QT_{Condenser\ outlet} - QT + \frac{R_{net}A}{\rho c_p} \qquad \text{Equation 7.4.1}$$

where

V = GGS canal and cooling pond volume, m^3

T = GGS canal and cooling pond temperature at the current time, C

t = time, s

Q = discharge into and out of GGS canal and cooling pond, $m^3\ s^{-1}$

$T_{Condenser\ outlet}$ = condenser outlet temperature, C

R_{net} = net rate of surface heat exchange, W m^{-2}

A = GGS canal and cooling pond surface area, m^2

ρ = density of water, 1000 kg m^{-3}

c_p = specific heat of water, 4186 J kg^{-1} °C^{-1}

Note that because of the fully-mixed assumption, the cooling pond temperature applies to the entire cooling canal and pond in general as well as the cooling pond outlet in particular.

Rearranging Equation 7.4.1 to solve for the canal and cooling pond temperature using finite differences,

$$T' - T = \frac{\Delta t Q}{AD} T_{Condenser\ outlet} - \frac{\Delta t Q}{AD} T' + \frac{\Delta t R_{net}}{D \rho c_p} \qquad \text{Equation 7.4.2}$$

where

T' = GGS cooling pond temperature at the next time step, C

recognizing that

$$D = \frac{V}{A} \qquad \text{Equation 7.4.3}$$

where

D = GGS canal and cooling pond average depth, m

and placing in implicit form

$$T'\left(1 + \frac{\Delta t Q}{AD}\right) = T + \frac{\Delta t Q}{AD} T_{Condenser\ outlet} + \frac{\Delta t R_{net}}{D \rho c_p} \qquad \text{Equation 7.4.4}$$

A final rearrangement of Equation 7.4.4 allows for direct solution

$$T' = \frac{T + \dfrac{\Delta tQ}{AD}T + \dfrac{\Delta tR_{net}}{D\rho c_p}}{1 + \dfrac{\Delta tQ}{AD}} \qquad \text{Equation 7.4.5}$$

Equation 7.4.5 is iterated through time at an hourly timestep.

The temperature of the water that enters the cooling canal and pond, $T_{Condenser\ outlet}$, is computed from the flow-weighted average of the Unit 1 and 2 outlet temperatures as follows:

$$T_{Condenser\ outlet} = \frac{Q_{Unit\,1} \cdot T_{Out\ Unit\,1} + Q_{Unit\,2} \cdot T_{Out\ Unit\,2}}{Q_{Unit\,1} + Q_{Unit\,2}} \qquad \text{Equation 7.4.6}$$

where the variables on the right hand side of the equation are hourly, observed values:

$Q_{Unit\,1}$ = Unit 1 cooling water flow

$Q_{Unit\,2}$ = Unit 2 cooling water flow

$T_{Out\ Unit\,1}$ = Unit 1 cooling water outlet temperature

$T_{Out\ Unit\,2}$ = Unit 2 cooling water outlet temperature

Equation 7.4.6 was used for the calibration simulations. For the full load simulations, $T_{Condenser\ outlet}$ was computed as follows

$$T_{Condenser\ outlet} = T_{Condenser\ inlet} + \Delta T \qquad \text{Equation 7.4.7}$$

where

$$T_{Condenser\ inlet} = \frac{Q_{Unit\,1} \cdot T_{In\ Unit\,1} + Q_{Unit\,2} \cdot T_{In\ Unit\,2}}{Q_{Unit\,1} + Q_{Unit\,2}} \qquad \text{Equation 7.4.8}$$

with

$T_{In\ Unit\,1}$ = Unit 1 cooling water inlet temperature

$T_{In\ Unit\,2}$ = Unit 2 cooling water inlet temperature

For the full load simulations, the condenser temperature rise, ΔT, is computed as the flow weighted average of the individual unit full-load condenser temperature rises.

Calibration procedure and results

The calibration procedure was based on the following comparisons of computed and observed cooling pond outlet temperatures, listed in order of priority:

- the mean difference of all computed and observed pairs above 90 F
- time series plots of the peak temperature month for each of the six study years, the summers of 1995 through 2000
- the mean difference of all computed and observed pairs
- the mean difference of pairs of computed and observed daily maximum temperatures
- the mean difference of pairs of computed and observed daily minimum temperatures
- the mean difference of pairs of daily amplitudes, defined as one-half the daily computed and observed maximum minus the daily computed and observed minimum

The calibration procedure showed that computed pond temperatures are most sensitive to wind speed (affecting mostly the mean temperature through the evaporative heat loss term in the head budget equation) and the cooling pond depth (affecting mostly the diurnal amplitude). A range for the depth, D, was derivable from the known volumes and surface areas of the cooling pond and cooling canal. Several experiments showed that varying D changed the diurnal amplitude of the computed cooling pond outlet temperature. An acceptable value was of the depth was 7.5 ft, halfway between the average depth for the cooling pond (10 ft) and the average depth of the cooling canal (5 ft).

The error analysis on the hourly pairs (computed minus observed) at the cooling pond outlet is shown in the table 7.4.1.

Note that positive values in Table 7.4.1. indicate that the systems model is computing values that are higher than those observed. The model over predicts water temperatures by 0.03°F when temperatures are above 90 F and by 0.32°F when water temperatures are between 85 F and 90 F.

The error analysis on the daily maximum, minimum and amplitude pairs (computed minus observed) at the cooling pond outlet is shown in the table 7.4.2. There are 720 data pairs because only the four summer months of the six-year analysis were included. These statistics show that the systems model also slightly over predicts daily maximum and minimum temperatures, but within typical model error ranges.

Table 7.4.1 Calibration statistics for the systems model

Cooling pond temperature range, F	Number of pairs	Mean error, °F	Variance, °F	Standard deviation, °F	Mean absolute error, °F	Standard error, °F	RMS error, °F
<65	43	-2.31	2.71	1.65	2.31	0.25	2.82
65-70	276	-0.50	10.91	3.30	2.77	0.20	3.33
70-75	1614	0.47	6.15	2.48	1.95	0.06	2.52
75-80	3913	0.74	4.21	2.05	1.72	0.03	2.18
80-85	6201	0.49	3.68	1.92	1.56	0.02	1.98
85-90	3844	0.32	1.66	1.29	1.03	0.02	1.33
>90	245	0.03	2.32	1.52	1.14	0.10	1.52
All pairs	16136	0.48	3.74	1.93	1.53	0.02	1.99

Table 7.4.2. Additional calibration statistics for the systems model

Daily cooling pond value	Number of pairs (daily for 6 months)	Mean error, °F	Variance, °F	Standard deviation, °F	Mean absolute error, °F	Standard error, °F	RMS error, °F
Max	720	0.30	2.60	1.61	1.23	0.06	1.64
Min	720	0.59	3.52	1.88	1.55	0.07	1.97
Amplitude	720	-0.14	0.49	0.70	0.54	0.03	0.71

Frequency of exceedance statistics

Model results are presented as frequency of exceedance statistics that allow NPPD to assess the risk of exceeding the 94 F limit at the cooling pond outlet.

From the 6-year record of computed temperatures, frequency of exceedance statistics were developed to obtain the normal (2-year), the 3-year, the 5-year, the 10-year, and the 20-year return period temperatures using the Gumbel distribution of extreme values. The three steps in the analysis are

- determine for each year the number of hours that a given temperature is equaled or exceeded
- determine for each year the temperature which is equaled or exceeded for a specified number of hours (the duration)
- determine from the statistics of these temperatures the temperatures that are exceeded for specified annual frequencies and hourly durations.

The Gumbel extreme value distribution (Haan, 1977, p. 110) was formulated as

$$T = a(-\ln(-\ln(1-\frac{1}{R}))) + b \qquad \text{Equation 7.4.9}$$

$$a = \frac{S_d}{1.283} \quad \text{Equation 7.4.10}$$

$$b = M - 0.45 S_d \quad \text{Equation 7.4.11}$$

where

T	=	temperature for the selected return period
R	=	annual return period, number of years
S_d	=	standard deviation at each daily duration
M	=	mean temperature at each daily duration

Figurer 7.4.2 shows the number of hours that the cooling pond outlet temperature would be exceeded for return periods of 2, 3, 5, 10, and 20 years using historic operations at GGS. Figure 7.4.3 is similar, except that input for the systems model is full-load operating conditions. Each diagram is essentially the risk or probability of exceeding temperatures.

3-D Model

The temperature at the cooling pond outlet to Sutherland Reservoir can also be computed by using a 3-D hydrodynamic model, in this case the Generalized Environmental Modeling System for Surfacewaters (GEMSS). The GEMSS model of the GGS Cooling Pond was developed using bathymetric data from a 1995 survey, meteorological data from Lee Bird Field

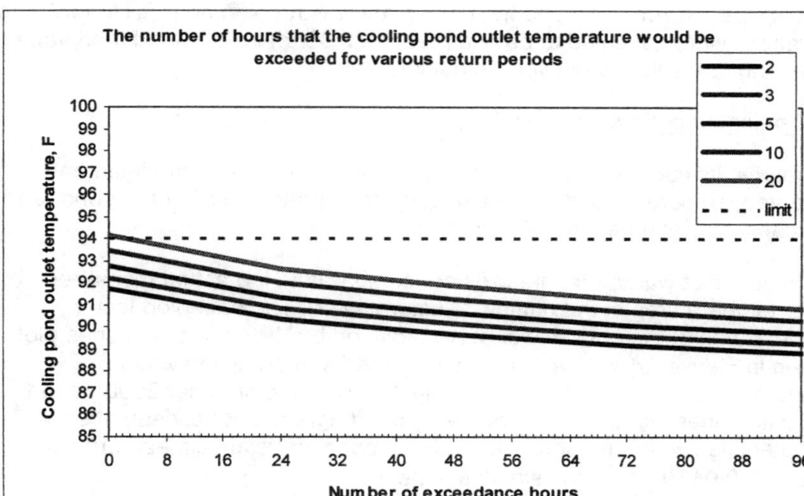

Figure 7.4.2. Frequency of exceedance by return period for historic 1995-2000 operations.

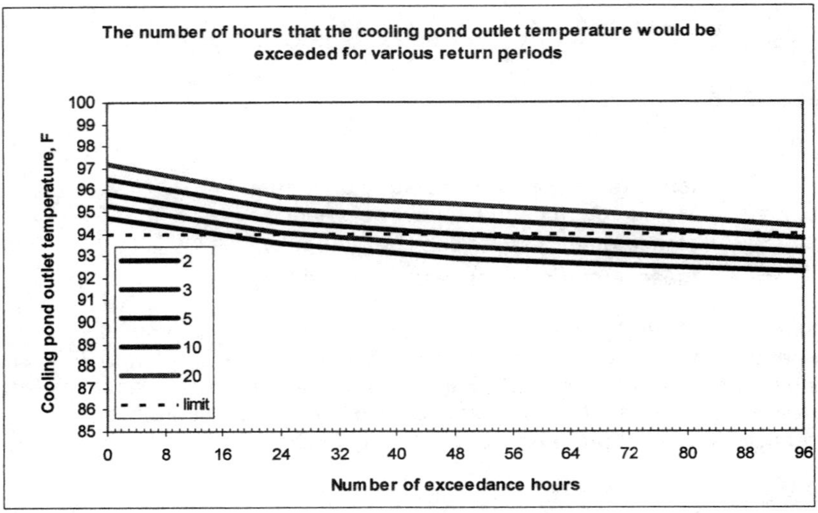

Figure 7.4.3. Hours of exceedance for the full load case.

(NWS station), and flow rates and water temperatures into the cooling pond from GGS provided by NPPD. The simulation period is for the summer months, June 1 to September 30, 2000. In consultation with NPPD, this simulation period was chosen because it contained the most complete input data, it was a warm year, and the GGS plant was running near full load. Pond elevations were assumed to be 3061 feet (933.32 m) initially, and then varied according to the flow rate over the outflow weir.

Calibration procedure and results

Model results compare well to observed data in three ways: modeled and observed temperatures at the outlet, circulation patterns within the pond, and vertical stratification at discrete locations.

Modeled outlet water temperatures are compared to the actual measured water temperatures at the outflow structure (weir). The observed water temperatures at the outflow weir were provided by NPPD. A time series plot, shown in Figure 7.4.4, illustrates the observed and predicted water temperatures at the outlet at 2-hour intervals over the summer 2000 period. The time series comparison shows very good agreement between observed and predicted water temperatures, with an absolute relative mean error of 0.025°C (0.045°F) over the simulation period.

The daily maximum outlet temperatures were specifically analyzed to focus on the ability of the model to capture the high end of the dataset. Figure 7.4.5 shows the difference between each modeled and observed daily maximum

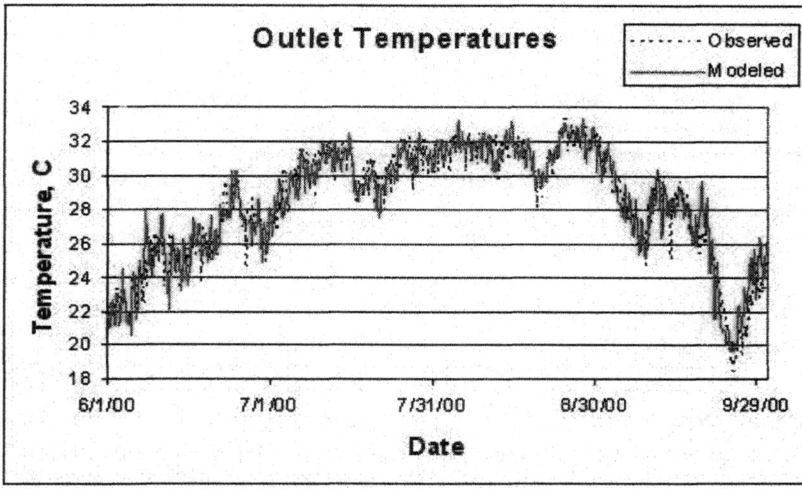

Figure 7.4.4. Modeled surface layer water temperatures.

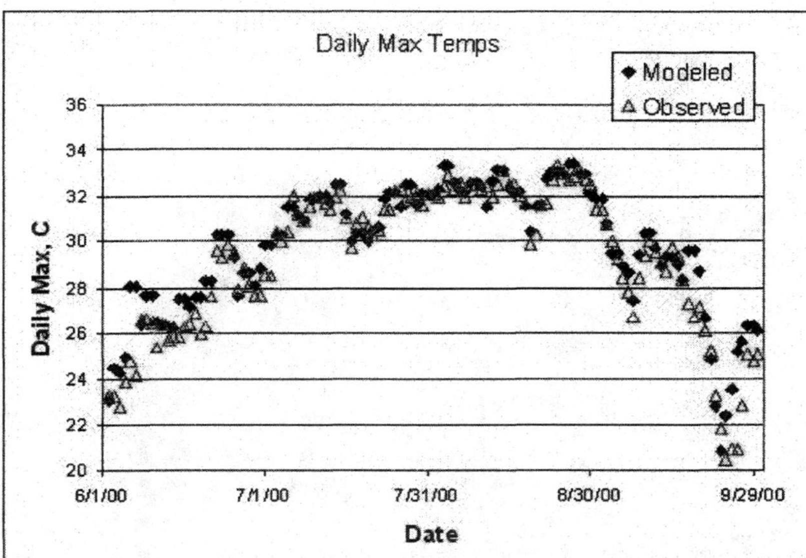

Figure 7.4.5. Modeled and observed daily maximum temperatures.

temperature. The average of the modeled daily maximum temperatures is 0.54°C (0.97°F) higher than the average of the observed daily maximum temperature during the June to September 2000 period. The standard deviation of the difference is 0.895°C (1.61°F). The modeled temperatures during the warmest period, July and August are 0.30°C (0.54°F) higher than the observed temperatures. The highest observed cooling pond outlet temperature during the year 2000 was 33.3C (91.94 F) on August 24, 2000. The model predicted a maximum temperature of 33.06C (91.51 °F) for that day, which is 0.24°C (0.43°F) lower than the observed temperature.

A comparison of the observed and modeled daily maximum temperatures suggests that the model-predicted temperatures are too high, but the maximum of the maximum day shows otherwise. Therefore, the initial parameter selection was unchanged and further calibration was not needed.

For the second type of verification, a contour plot is provided, Figure 7.4.6, to show the general circulation at the surface of the cooling pond. The day selected for Figure 7.4.6 is the day with the observed maximum discharge

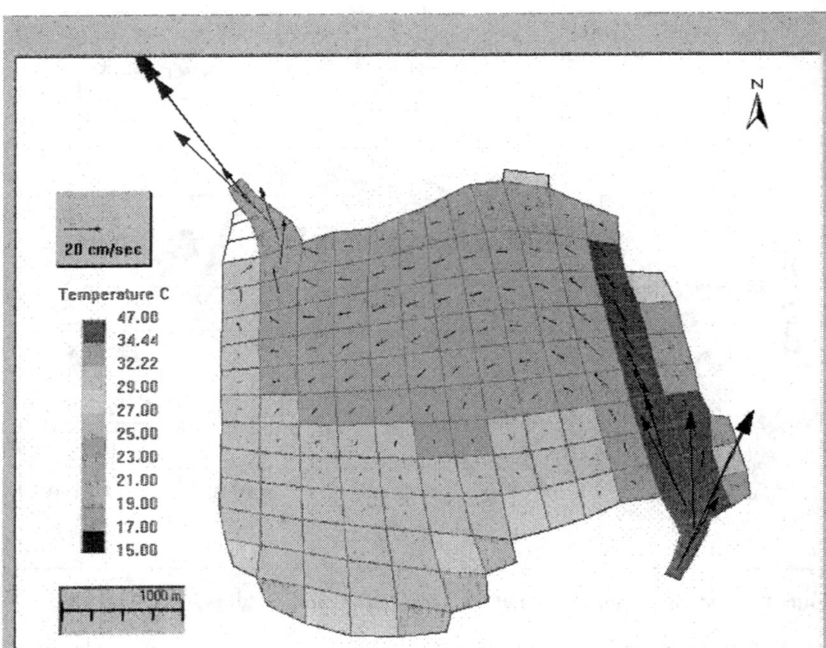

Figure 7.4.6. Modeled GGS cooling pond surface temperature and velocity for August 24, 2000.

temperature, August 24, 2000. Circulation patterns show good agreement with flow patterns inferred from earlier dye studies.

The third type of model verification is vertical temperature profiles. The data collected by NPPD show that the GGS cooling pond had a pond-wide average ΔT from top to bottom of 0.71°C (1.28°F). The model prediction for the same day produces a similar degree of temperature stratification.

Simulated Pond Modifications

The model was reconfigured and model simulations were conducted to simulate the placement of a curtain wall as a modification to the existing cooling pond. Three curtain wall alternatives were initially simulated: 1) 1000-feet long, full depth from water surface to pond bottom, extending southward at the cooling pond outlet; 2) 1000-feet long, full depth, extending westward near the inlet; and 3) a submerged curtain wall enclosing the inlet and a floating curtain wall enclosing the pond outlet. Other physical modifications to the cooling pond such as splitting the inflow and allowing it to enter the pond at two locations, dredging within the existing pond, extending the pond surface area to the south, and a combination of these with the curtain walls were also considered.The purpose of the placing a curtain wall at selected locations within the cooling pond is to increase the flow path length, to reduce short-circuiting in the pond, and to allow increased cooling of the water, thereby decreasing the outlet temperatures. The model was run for the June to September 2000 period.

Predicted outlet water temperatures with and without the 1000' North-South curtain wall show the effects of the curtain wall on water temperatures. It appears that the curtain wall provides minor relief (i.e., decreased average outlet temperatures). The mean temperature of the simulation period decreases from 28.41°C (83.14 °F) to 28.27°C (82.89 °F), a drop of only 0.14°C (0.25°F). The maximum modeled temperature of 33.36°C (92.05 °F) occurs on 8/27/00 at 4:00 pm, which does not coincide with the maximum observed temperature. In comparing the modeled alternatives, we make reference to the maximum modeled temperature. The maximum modeled water temperature shows a decrease of only 0.18°C (0.32°F) with the curtain wall installed. At other times within the modeled temperature record, the curtain wall produces differences in temperatures larger than 0.18°C (0.32°F), but they do not occur when the peak temperatures occur. The other two alternatives produce similarly small reductions in outlet temperatures.

Circulation flow patterns are affected, mostly near the curtain wall for all three alternatives. A time delay in the peaks is caused by the curtain wall and varies according to the pond modification. The magnitude of the delay is approximately 2 hours.

Additional cooling pond modifications simulated were the addition of 40 acres to the southern end of the existing pond, and combining the addition with diverted flow into the pond. This additional area represents an increase of approximately 18% of the existing surface area. The bed elevation of the addition is 3058 feet. The additional area alone produced an average temperature of 28.27°C (82.89 F) or a decrease of 0.14°C (0.25°F). Maximum temperature increased 0.25°C (0.45°F) and minimum temperature decreased 0.48°C (0.86°F). The additional pond area alone does not provide much additional cooling because it is out of the flow path between the inlet and the outlet, and it does not delay maximum temperatures from reaching the outlet. When combined with diverted flow into the additional area, the new scenario average temperature is 28.06°C (82.51 °F) or a decrease of 0.35°C (0.63°F). Maximum temperature increased 0.35°C (0.63°F) and minimum temperature decreased 0.48°C (0.86°F).

Several alternative pond modifications were simulated to illustrate the use of the model and the feasibility of such modifications. The results of the completed simulations show that minor temperature reduction at the outlet is possible by installing a curtain wall at various locations within the pond. The small temperature reduction is due to a small increase in detention time, which shifts the peaks to later in the day. Detention time can only be slightly increased because of the small pond volume relative to the flow rates through it. Also, the pond is essentially fully mixed vertically during peak temperature periods, which precludes the use of cooler hypolimnetic water. Other physical modifications, such as dredging portions of the pond, diverting flow into different parts of the pond, and increasing the pond surface area and volume showed limited effect on outlet temperatures.

Conclusion

Results of the frequency-duration modeling with the systems model indicated an unacceptably high probability of exceeding the 94 °F limit at the outlet, and the results of the detailed 3-D modeling of the cooling pond showed that modifications would not provide the additional cooling required. NPPD subsequently investigated the construction of supplemental cooling ponds, auxiliary heat dissipation devises (cooling towers and spray modules), and developed a well field capable of providing 50-60 °F dilution water during high temperature events. The systems model developed here has been converted to a predictive code that uses forecast meteorological parameters to provide 7-day forecasts of cooling pond outlet temperature trends so that the wells can be used only when required.

REFERENCES

"Statistical Methods in Hydrology" by Charles T Haan, Iowa State University Press, Ames, 1977.

7.5. Effectiveness of an Internal Weir to Improve Water Quality in a Hydroelectric Reservoir

George A. Krallis, Ph.D.

The dissolved oxygen content of water released from the Logan Martin Reservoir (LMR) on the Coosa River in Alabama frequently falls below the regulatory limit of 4 mg/l. Alabama Power Company (APC) is considering two engineered solutions to this problem: construction of a submerged weir, and installation of a hypolimnetic oxygenation system. Since surface waters are higher in dissolved oxygen (DO) than bottom waters, it is expected that installation of a submerged weir upstream of the dam will improve the DO content of the outflow by maintaining a well-mixed forebay consisting mostly of high DO water skimmed by the weir. If DO improvement due to the weir is not sufficient to meet the 4 mg/l limit, direct injection of oxygen will be used to make up the remaining deficit. An additional consideration is that the flow downstream of LMR consists of power generation releases as well as groundwater leakage from sinkholes in the forebay area. APC requires an examination of the effects of the proposed weir on forebay stratification and on the DO content of the leakage flow.

The DO content of forebay water is controlled by several in-reservoir processes. Atmospheric exchange adds DO at the surface, and sediment oxygen demand (SOD) removes it at the bottom. In the bulk of the water column, DO is produced due to algal photosynthesis and is consumed due to oxidation of organic matter produced by photosynthetic activity. When considering the submerged weir, the effect of these processes on the DO of the water released further depends on contact time and effective area as well as the corresponding kinetic rate of DO depletion or addition from each process.

In addition to the water quality effects mentioned, mixing processes related to selective withdrawal and weir height are also important. When the weir is too near the surface, high velocities induce mixing which decreases surface water DO. When the weir is too low, too little of the anticipated skimming may occur.

The Hydrodynamic and Water Quality Model

The model used in this study is called GEMSS (Kolluru, 1999). It is an integrated system of 3-D hydrodynamic and transport models embedded in a geographic information and environmental data system (GIS), field data viewer, grid generator and editor, control file generator, 2-D and 3-D post processing viewers. Additional tools include a meteorological data processor, time varying data generator and editor, and a USGS flow data processor to support 3-D modeling. Customization of the suite of hydrodynamic, transport, thermal and different water quality models is achievable through the use of a modular design reflecting the needs of each user's application.

GEMSS uses GLLVHT (Generalized Longitudinal Lateral Vertical Hydrodynamic and Transport Model) which computes time-varying velocities, water surface elevations, and water quality constituent concentrations. The theoretical basis of the three-dimensional GLLVHT model was first presented in Edinger and Buchak (1980). The GLLVHT model has been peer reviewed and published (Edinger and Buchak, 1985 and 1995; Kolluru et al.1999; Edinger and Kolluru, 1999). The model is built to accept a large number of transport constituents and constituent relationships depending on the water quality model, sediment transport model, oil and chemicals model, and toxic model being used. The properties of the GLLVHT model are given in Kolluru et al. (1999). The water quality model module incorporated in GEMSS includes the water quality components and processes affecting dissolved oxygen.

Input Conditions

The grid used in the 3-D model developed for the Logan Martin Reservoir is shown in Figure 7.5.1, with the existing bathymetric contours shown in the background. The grid extends approximately 3 miles upstream from the dam. The grid consists of 591 grid cells. Each cell is approximately 1.77 feet thick, and horizontal dimensions vary from 130 to 600 feet.

Sinkholes detected at the floor of the reservoir in the forebay area are included in the model. The model removes water from those cells at a constant rate and contributes it as leakage to the downstream flow. APC determined that sinkhole leakage rates are a steady value of 620 cfs, as used in the simulations. The DO concentration of the sinkhole leakage is the value determined for the cell from which it was removed.

Temperature and DO profiles at the dam and at locations 3-miles upstream of the dam are available intermittently from May 1992 to February 2002. Several simulation periods are chosen to represent each of the four seasons, and critical times of high water temperature and low DO levels. The initial dissolved oxygen and temperature profile data required for model input are available for the simulation time periods selected. This discussion presents only the results from the July 1993 simulation period.

The water quality model includes the pertinent DO processes, and in this case sediment oxygen demand. Observed sediment oxygen demand (SOD) rates were provided by APC for reservoirs throughout Alabama, Florida, Georgia, Tennessee, and Kentucky. The average of SOD values at Weiss Reservoir, 2.86 $g/m^2/day$, was used in the model. Inflow water quality data for other processes affecting dissolved oxygen such as ammonia decay to nitrate, BOD decay, nutrient cycling, and algal growth including photosynthesis and respiration did not exist in sufficient detail through the chosen modeling periods to be included. However, initial studies with and without SOD showed that SOD was sufficient alone to account for most of the dissolved oxygen changes through the short reach of reservoir being studied.

The turbine release schedule flow rates during the middle of the simulation period are shown in Figure 7.5.2. Actual release rates were used for the existing condition in July 1993, and proposed release rates were used for the proposed weir configuration simulations. Both observed and proposed release rates exhibit a pattern of high peaking flow rates followed by periods of low or no flow.

Three simulations for the July 1993 period are presented here: 1) the existing condition, 2) the proposed pumping schedule, and 3) the proposed pumping schedule with the new weir. The existing condition simulation, consisting of the existing release rates and no new weir, demonstrates the veracity of the model. Additional simulations not shown here utilize the proposed release rates and a new weir at three different elevations (low, medium, and high) to optimize weir height.

<u>Model Simulation Results</u>

The simulation of existing conditions shows qualitative agreement with the observed data according to the DO profile data at the dam shown in Figure 7.5.3. Within the forebay area, modeled and observed DO concentrations for July 1993 are above 6 mg/l at the surface and below 0.5 mg/l at the bottom, as observed. Measured tailrace DO data could not be used for detailed model verification because it was not sampled frequently enough through the study period. Measured tailrace DO concentrations from the entire record, however, are in the range of values provided by the model.

The DO concentrations modeled under existing July 1993 conditions are illustrated in a vertical-longitudinal slice plot in Figure 7.5.4. The location of the slice (Figure 7.5.4) is longitudinally down the thalweg of the old riverbed. The date of the plot is the end of the simulation period (7/30/93) to show the extent of the DO field after 30 days of pumping. The Logan Martin dam is on the left side of the figure. Figure 7.5.5 shows the turbine and sinkhole dissolved oxygen levels and the combined DO downstream of the dam under existing conditions at its lowest DO levels within the simulation period. During no-release periods, downstream DO is lower than during release periods because of the high percentage of flow emanating from the sinkholes.

The longitudinal thalweg slice, Figure 7.5.6, shows the DO concentrations in the reservoir with the proposed release rates and no new weir installed. New pumping schedules alter the DO in the reservoir at specific points in time, but at the time shown in the slice figures, there appears to be little change from Figure 7.5.4 in the DO field. Figure 7.5.7 shows the turbine and sinkhole dissolved oxygen levels, and the combined DO downstream of the dam under proposed pumping schedule with no new weir. This figure shows the effects of the proposed pumping schedule alone. The less frequent – higher intensity proposed pumping schedule increased DO concentrations during the pumping periods, but DO declined steadily until pumping resumed. DO at the sinkholes remained at zero throughout this period.

The longitudinal, thalweg slice, Figure 7.5.8, shows the DO concentrations in the reservoir with the proposed release rates and the new weir installed. Flows over the proposed weir draw high DO water deeper into the forebay profile. The weir also retains lower DO water further upstream form the turbines and the sinkholes. Figure 7.5.9 shows the turbine and sinkhole dissolved oxygen levels, and the combined DO downstream of the dam under proposed weir conditions. The less frequent – higher intensity proposed pumping schedule increased DO concentrations during the pumping periods. As in the previous case, DO declined until pumping resumed, but at a rate less than in the previous case. DO at the sinkholes rose above zero daily during this period.

Conclusions

The proposed pumping schedule alone altered the downstream DO levels, but did not increase the overall DO concentration in the downstream reach or at the sinkholes. The proposed pumping schedule in conjunction with the new weir provided an increase in DO at the sinkholes, but only a minimal increase the overall DO in the downstream reach. Installation of the weir provides a more stable DO environment between pumping periods in which to operate a hypolimnetic oxygenation system.

REFERENCES

Edinger, J. E., and E. M. Buchak. 1980. *Numerical hydrodynamics of estuaries.* In Estuarine and Wetland Processes with Emphasis on Modeling, edited by P. Hamilton and K. B. Macdonald. New York: Plenum Press: 115 – 146.

Edinger, J. E., and E. M. Buchak. 1985. Numerical waterbody dynamics and small computers. In *Proceedings of ASCE 1985 hydraulic division specialty conference on hydraulics and hydrology in the small computer age*. Reston, VA: American Society of Civil Engineers.

Edinger, J. E., and E. M. Buchak. 1995. Numerical intermediate and far field dilution modeling. *J. Water Air Soil Pollution* 83:147–160.

Edinger, J. E., and V. S. Kolluru. 1999. Implementation of vertical acceleration and dispersion terms in an otherwise hydrostatically approximated three-dimensional model. In *ASCE estuarine and coastal modeling: Proceedings of the 6th international conference, November 3–5,*

1999, New Orleans, Louisiana. Reston, VA: American Society of Civil Engineers.

Kolluru, V. S., 1999. GEMSS – Technical Document on Generalized Environmental Modeling System for Surface waters. J. E. Edinger Associates, Wayne, Pennsylvania.

Kolluru, V. S., E. M. Buchak, J. Wu, 1999. Use of Membrane Boundaries to Simulate Fixed and Floating Structures in GLLVHT. In Spaulding, M.L, H.L. Butler (eds.). *Proceedings of the 6th International Conference on Estuarine and Coastal Modeling*. American Society of Civil Engineers, Reston, VA. pp. 485 - 500.

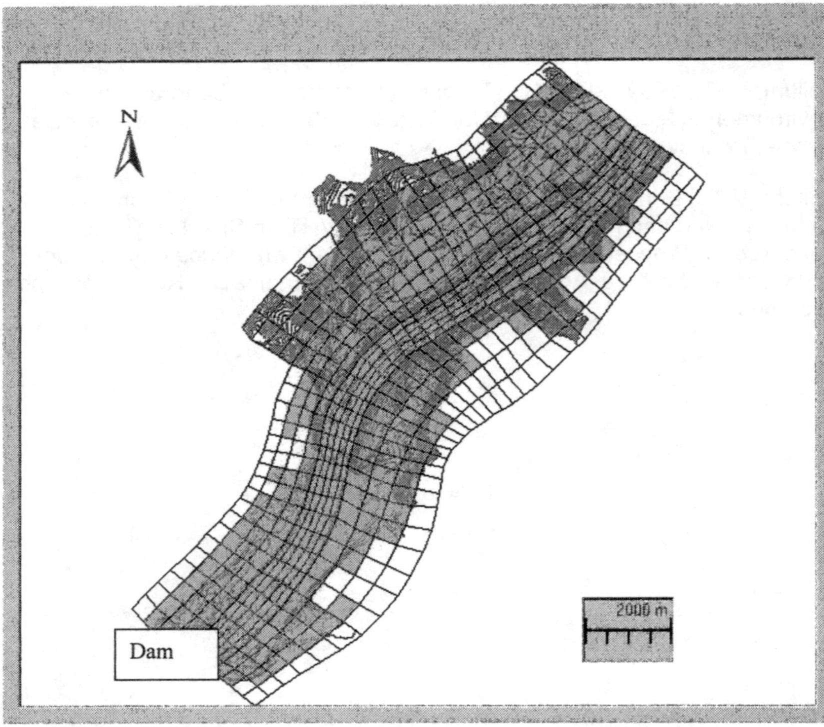

Figure 7.5.1 - Reservoir bathymetry near the dam, and model grid

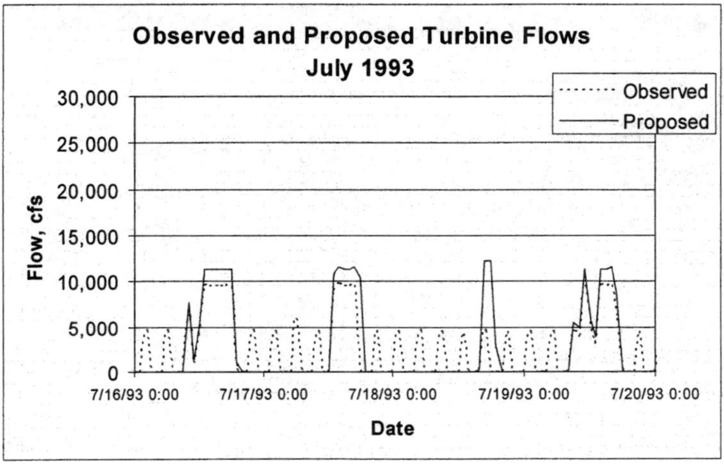

Figure 7.5.2 – Portion of observed and proposed turbine release schedule used in July 1993 simulations.

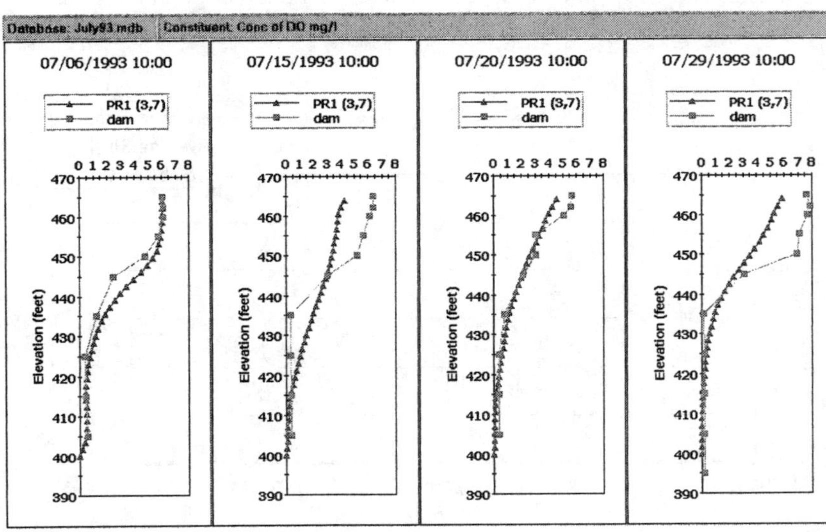

Figure 7.5.3 - Observed and modeled DO profiles at the dam for existing July 1993 conditions. The triangles represent the model results and the squares represent the observed data.

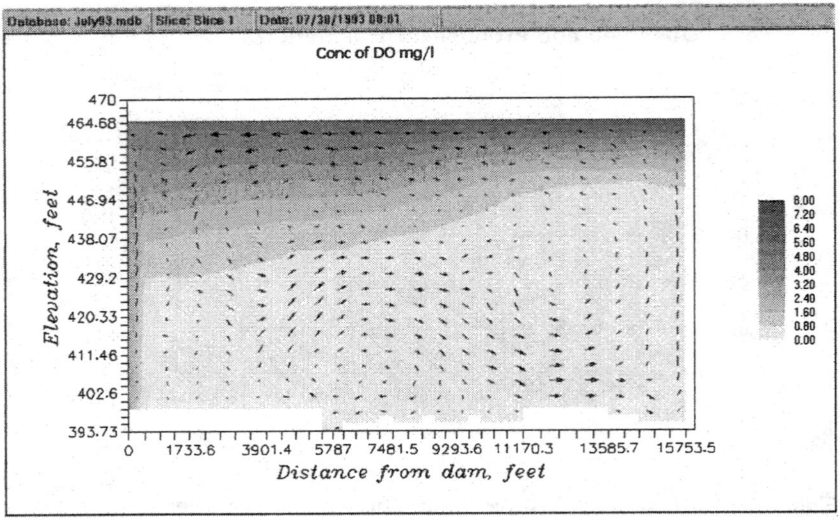

Figure 7.5.4 - North-South slice through reservoir thalweg with existing conditions. The Logan Martin dam is located on the left edge of the figure.

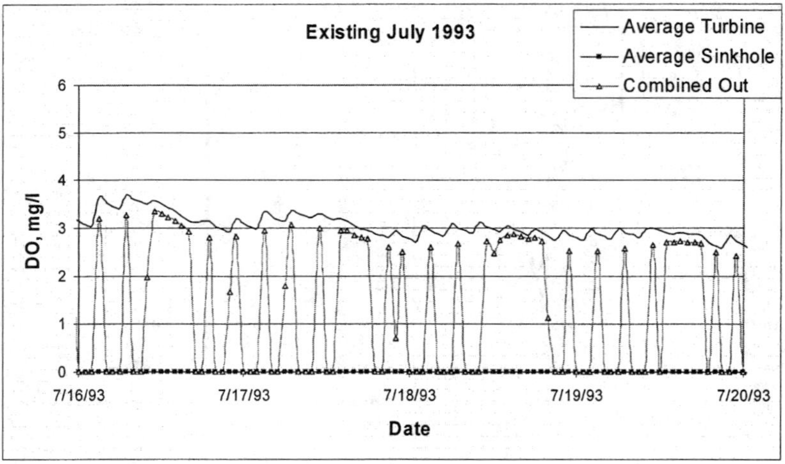

Figure 7.5.5 - Average turbine and sinkhole dissolved oxygen levels, and combined DO downstream of dam with existing conditions.

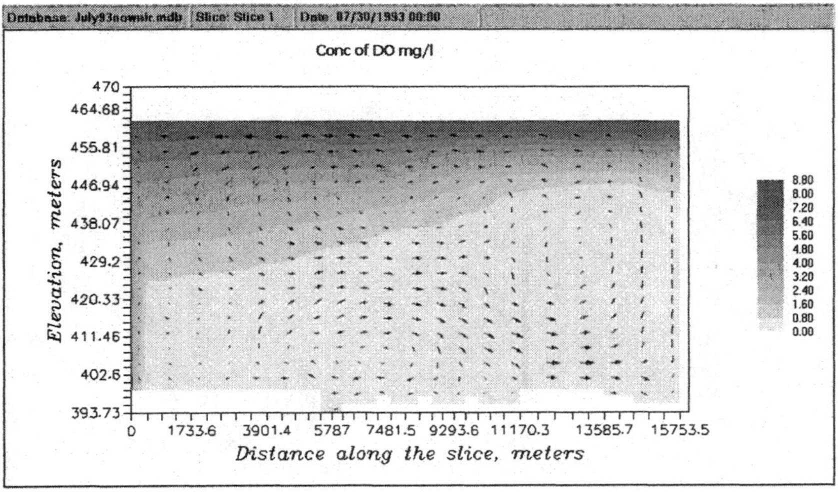

Figure 7.5.6 - North-South slice through reservoir thalweg with new pumping rate and no proposed weir condition.

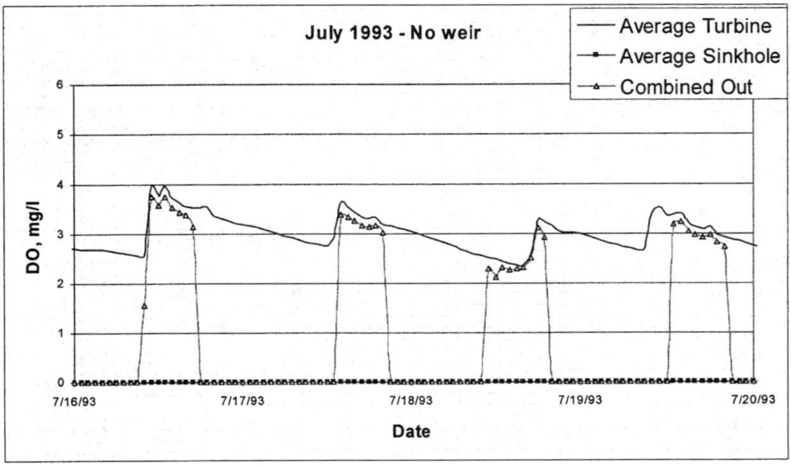

Figure 7.5.7 - Average turbine and sinkhole dissolved oxygen levels, and combined DO downstream of dam for new pumping schedule and no proposed weir condition.

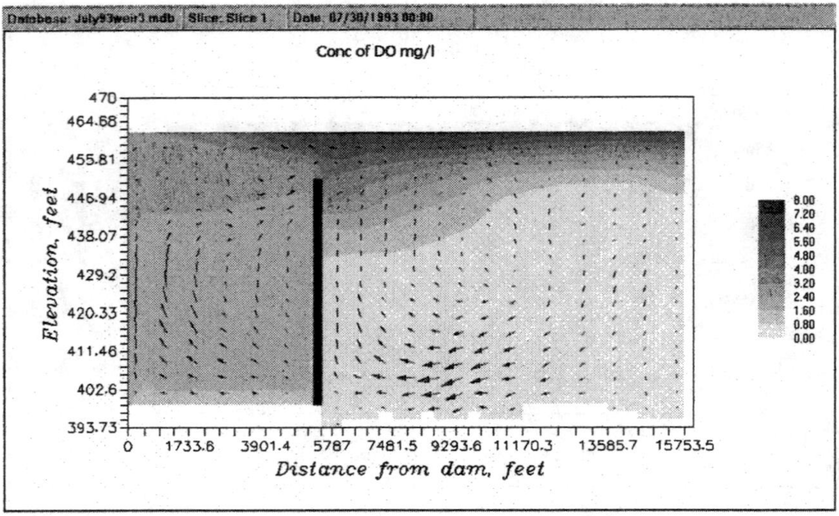

Figure 7.5.8 - North-South slice through reservoir thalweg with proposed pumping schedule and new weir conditions. The proposed weir is located approximately 5200 feet from the dam.

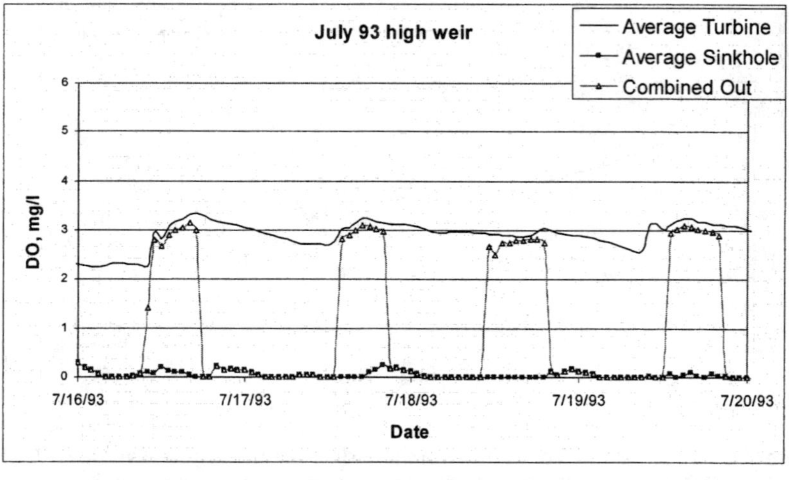

Figure 7.5.9 - Average turbine and sinkhole dissolved oxygen levels, and combined DO downstream of dam with proposed pumping schedule and proposed new weir condition.

7.6. Hydrothermal Modeling Studies of Cooling Tower Alternatives

John Eric Edinger, Ph.D.

A tributary creek of a large reservoir receives the cooling water discharge from a steam electric power plant and its auxiliary cooling towers. The map in Figure 7.6.1 shows the location of the plant discharge and cooling tower discharge on the receiving creek. The plant intake is located on another creek that joins the reservoir south of the receiving creek. The receiving creek has a length of 3,000 meters and a maximum depth at its mouth of 14 meters. The bottom slopes upstream to a depth of 6 meters where it passes under a bridge. The creek is 1200 meters long between its headwater and the bridge.

Hydrothermal modeling was undertaken to determine cooling tower configurations and operations that would maintain a refuge of water and zone of passage through the bridge opening of less than 33.9 C (93 F). Since the plant has a condenser temperature rise of 10 deg C (18 deg F), tower operations was designed to begin when the plant intake temperature exceeds 24 C (75 F). A portion of the plant discharge water is transferred to the cooling towers to reduce its temperature to approximately 30 C (86 F).

During the development of the cooling tower configurations and operations hydrothermal modeling was used to determine:

- The optimal location of the cooling tower discharge to inject cooler bottom water into upper receiving creek for the zone of refuge and passage.
- The details of the cooling water discharge structure depth and exit velocity to maintain the cold water layer.
- The details of the plant discharge structure to place the higher temperature water on the surface and maintain stratification through upper receiving creek and the bridge opening.
- The cooling tower flows and temperatures required during different times of year to maintain the zone of refuge and passage.

As part of the NPDES permitting of the cooling water discharge facilities to the receiving creek it was required to perform a hydrothermal modeling verification study with the cooling towers in operation. The verification study consisted of collecting temperature profile data throughout the receiving creek during operations, running the model to predict temperatures for that day, and making a comparison of the predicted and computed temperatures. A later part of the study required examining relationship between the cooling water discharge and the distribution of dissolved oxygen in the receiving creek.

This case study presents the results of the verification study to illustrate the kinds of temperature and dissolved oxygen distributions produced by the flow fields computed by the model for tower operations, and the results of the field data and predicted temperature and dissolved oxygen comparisons.

Model Description and Input Data

The model used in this study is the three-dimensional generalized longitudinal-lateral-vertical-hydrodynamic and transport model that is described in Edinger, 2002 and Kolluru and Edinger, 2002. It is a finite difference hydrodynamic model derived from the primitive equations of fluid motion and uses a heat budget routine to account for surface heat exchange. It uses a complete dissolved and particulate organic carbon, nitrogen and phosphorous algal nutrient cycling model with dissolved oxygen and sediment exchange, and zooplankton density dependent grazing (Edinger, et al. 2002).

The GLLVHT model requires for this application the detailed digital bathymetry and the plant and cooling tower flows. The detailed digitized model bathymetry is given in Table 7.6.1, and a mapping of the bathymetry data is presented in Figure 7.6.2. The GLLVHT model was set up using a rectangular grid with surface cells 60 meters long by 20 meters wide, and with vertical layer thickness of 0.5 meters. As shown in the model bathymetry plot, the route 441 bridge is located 1200 meters from the head of the creek.

For application to the cooling tower studies the model requires for input; the plant pumping rate, the plant intake temperature, the plant discharge temperature, the tower flow rate, and the tower discharge temperature. These data for the one of the verification survey are as follows:

Parameter/Survey	Aug 28, 02
Plant Condenser Flow rate, thds gpm	791.1
Plant intake temperature, C	30.6
Plant discharge temperature, C	40.9
Tower flow rate, thds gpm	315.0
Tower discharge temperature,	29.9

The model was initialized using a temperature profile measured at the downstream junction of receiving creek and the reservoir. This temperature profile was also used for the open boundary temperature profile at the same location.

The hydrodynamic and transport routine (GLLVHT) are modules within the Generalized Environmental Modeling System (GEMSS; Kolluru, 1999) which manages the modeling data input, simulation conditions, modeling results output routines, and different statistical methods for comparison of field data to model results applicable to the receiving creek study.

Model Temperature Results

Figure 7.6.3 is a longitudinal-vertical slice down the thalweg of the receiving creek giving the contours of computed temperature, and the horizontal velocity vectors. It shows a lens of cooler water along the bottom in the vicinity of the cooling tower discharge. The upper receiving creek is stratified with the warmer surface layer resulting from the warmer discharge from the plant and the cooler bottom water resulting from the tower discharge. Within the upper receiving creek, there is a surface flow upstream from the vicinity of the discharge to the head of the creek, and a bottom cooler density flow along the bottom toward the bridge.

At the bridge, the temperature is still stratified and the flow is all in the down stream direction. Downstream of the bridge, the warmer water stratifies on the surface, and cooler bottom water intrudes up the creek, mixes upward near the bridge and flows out with the surface flow.

Figure 7.6.4 is a longitudinal-vertical slice of the excess temperature, or increase in temperature over background, due to the combined plant and tower discharge. It shows how much of the stratification through and below the bridge is maintained by the combined discharges, and how much mixing is induced by the buoyant stratified discharge between the bridge and the mouth of the receiving creek.

Figure 7.6.5 is a lateral-vertical slice at a cross-section between the discharge and the bridge. The model results show temperature stratification with a thick cooler layer extending across the bottom of the whole cross-section.

Verification of Temperature Results from Survey Data

The verification studies required that temperature profiles be measured on each of seven cross-sections located up and down the receiving creek. The location of the cross-sections is given in Table 7.6.2 as the distance from the upper end of the receiving creek for reference to Figure 7.6.2 and Figure 7.6.3. Three temperature profiles were taken on each cross-section at half-meter depths. There were approximately 150 observations to compare to predicted temperatures on each survey.

The computed and modeled receiving creek thalwag temperatures for all five surveys were combined for a statistical analysis. A comparison of the computed versus observed thalweg temperatures over all surveys is given in Figure 7.6.6 along with the statistics of the comparison. The overall slope of the comparison is 1.012. It is compared to an ideal slope of unity using a t-test of slopes as described in Neville and Kennedy (1966). The t-test value of the comparison indicates that there is no significant difference between the slope of the comparison and the ideal slope of unity at the 5% rejection level.

When comparing model and data results the errors, or differences between computed and observed results, should be normally distributed to be true errors. Any deviation from normally distributed errors is a result of modeling bias and/or data. The Chi^2 test can be used to determine if the errors are normally distributed. The Chi^2 of the errors can be computed as (Kirkman,1996):

$$Chi^2 = \sum_{i=1}^{N} (xi - \mu)^2/\sigma^2$$

xi = the error between the observed and simulated value

μ = the mean of the error over the N observations

σ = the standard deviation of the error

The errors are normally distributed if Chi^2 is very close to N or the total number of observations. The computed and observed temperatures and the computation of Chi^2 for the thalweg temperatures of a survey day are given in Table 7.6.3. Table 7.6.3 shows that the value of Chi^2 for this day is within 2% of the number of observations indicating that the errors are normally distributed and are truly errors.

The overall RMS error and the AME is often used as a comparative statistic. Figure 7.6.6 shows that the receiving creek thalweg data are 1.11 and 0.067 respectively. One way to determine if these represent a good verification is to compare to modeling results performed by others on a different water body. Martin and Cole (2000) report an RMS error ranging from 0.16 to 2.47 and an AME error ranging from 0.14 to 2.06 for a similar number of observed versus computed profile comparisons over three years. The values given in Figure 7.6.6 are well within the range reported for their case.

The mean and standard deviation of error statistics are the only ones that have a true theoretical basis for comparison (Bayes Theorem as described in Neville and Kennedy, 1966). The mean and standard deviation statistics of the errors over each cross-section and for all five surveys are given in Table 7.6.2. They show that the statistics are consistent from survey to survey at each cross-section and thalweg. The individual cross-section values are within the overall value for the survey and the values across all surveys showing the comparisons are internally consistent. Their acceptability can be further judged by comparison to similar statistics in other studies. Below is a summary of mean and standard deviation error statistics reported in GPC (1998) for a previous pre-tower study and by the COE for studies of J. Percy Priest Reservoir of Lake Norman (Martin and Cole, 2000) for comparison to the results shown in Table 7.6.4:

Source/Location	Mean Error, deg C	Standard deviation, deg C
Previous study w/o towers	0.55 – 0.88	0.56-1.09
COE Lake Norman	0.44 - 1.13	0.53-1.32
J. Percy Priest	0.84- 1.09	0.44- 0.56

The Beaverdam Creek overall statistics for each survey and across all surveys are within the summary results for the other independently evaluated cases.

Dissolved Oxygen Simulations and Results

The August, 2002 HMV thermal modeling was extended to compute the distribution of dissolved oxygen based on the concentration of dissolved oxygen in the cooling water discharge, the dissolved oxygen in the remaining plant flow over the weir, the background oxygen demand in the water column as represented by biochemical oxygen demand (BOD), surface reaeration, and bottom sediment oxygen demand. All process rates are temperature dependent as described in Edinger (2002). The resulting longitudinal-vertical dissolved oxygen profile for these conditions is given in Figure 7.6.7 along with a comparison of the observed and computed dissolved oxygen values at the different profile stations in BDC.

Figure 7.6.7 shows that the higher dissolved oxygen water produced by the combined cooling tower and plant weir discharge upstream of the Rt. 441 bridge is stratified and remains stratified at the bridge. The surface DO decreases through and downstream of the Rt. 441 bridge due to mixing with the lower DO water from the deeper portions of BDC. Figure 7.6.7 shows that the BDC water that is mixed into the surface layers is replaced by the intrusion deeper of low DO water from the main portions of Lake Sinclair that moves up BDC toward and through the Rt. 441 bridge. The DO of the intruding water is further depleted by the BOD and sediment oxygen demand in BDC.

The comparison of computed and observed dissolved oxygen concentrations for the center stations in BDC is quite good indicating that the processes included in the DO modeling to explain its longitudinal and vertical distribution are correct. The analysis of the errors between computed and observed dissolved oxygen is given in Table 7.6.5. The Table 7.6.5 Chi^2 test shows that the errors are normally distributed within 2.2% representing true errors with no bias.

The cooling tower discharges coupled with the plant induced circulation in the receiving creek shown in both Figure 7.6.3 and Figure 7.6.4 for temperature and excess temperature, and in Figure 7.6.7 for dissolved oxygen is very important are very important in maintaining the high dissolved oxygen levels in the portion of the receiving creek upstream of the bridge. If the plant and tower discharge were not there the dissolved oxygen would drop to very low values as found during previous surveys in GPC (1998).

Conclusion

The three dimensional GLLVHT hydrodynamic and transport model used to examine the design alternatives for the steam plant cooling towers showed that a cooling tower discharge could maintain the necessary stratified condition for a zone of refuge upstream of and through the Bridge across the receiving creek. The modeling techniques provided accurate representation of the distribution of temperatures and dissolved oxygen in comparison to independent observed conditions using rigorous statistical tests.

REFERENCES

Edinger, J.E. 2002. Waterbody Hydrodynamic and Water Quality Modeling. Three dimensional hydrodynamic and transport modeling. A workbook with 3-D model CD ROM. 215 pgs. ASCE Press, Reston, VA. ISBN 0-7844-0550-6.

Edinger, J.E. , S.Dierks and V.S. Kolluru. 2002. Density Dependent Grazing in Estuarine Water Quality Models. Accepted for publication in Water, Air and Soil Pollution, Kluwer Academic Publishers, December 2002.

GPC .1998. Documentation and Verification of the GLLVHT Hydrodynamic and Transport Modeling of Beaverdam Creek and the Oconee River for Plant Branch on Lake Sinclair, GA. Prepared for GPC Environmental Affairs, Atlanta, GA and accepted by EPA Region IV and EPA Headquarters. Prepared by J.E.Edinger and V.S.Kolluru, J.E.Edinger Associates, Inc., Wayne, PA, September, 1998.

Kirkman, T.W. 1996. Statistists to Use. http://www.physics.csbsju.edu/stats/ (January 23, 2005)

Kolluru,V.S. 1999. The Beaverdam Creek overall statistics for each survey and across all surveys are within the summary results for the other cases.. GEMSS-Technical Document on Generalized Environmental Modeling System for Surface Waters. Available from J.E.Edinger Associates, Wayne, PA (www.jeeai.com).

Kolluru, V.S. and J.E.Edinger. 2002. Combined Hydrodynamic and Water Quality Modeling. ASCE Proceedings of Engineering Mechanics Division Conference on Environmental Fluid Mechanics. June 2002.

Martin, J.L. and T.M.Cole. 2000. Water Quality Modeling of J.Percy Priest Reservoir Water Quality Modeling Using CE-Qual-W2. U.S. Army Corps of Engineers, Engineering Research and Development Center. Environmental Laboratory Report ERDC/EL SR-00-9. August 2000.

Neville, A.M. and J.B. Kennedy (1966). Basic Statistical Methods for Engineers and Scientists. International Textbook Company. Scranton, PA.

Table 7.6.1 - Digitized receiving creek bathymetry used in GLLVHT modeling. Depths are to the nearest 0.5 meters. Cells are 60 meters long and 20 meters wide. The bridge is located at I=20. The creek thalweg runs along J=10.

I/J	2	3	4	5	6	7	8	9	10	11	12	13	14	15	16	17	18
18	0	0	0	0	0	0	0	0	2	3	2	0	0	0	0	0	0
17	0	0	0	0	0	0	0	0	0	2	3	2	0	0	0	0	0
16	0	0	0	0	0	0	0	0	0	0	2	3	2	0	0	0	2
15	0	0	0	0	1	1	2	0	0	2	0	3	3	3	3	3	3
14	0	0	0	0	1	1	2.5	2	2.5	3	3	3	4	4	3.5	3.5	3.5
13	0	0	0	0	1.5	1.5	3	3	3	3.5	3.5	3.5	4	4	4	4	4
12	0	0	2	2	1.5	1.5	3	4	4	4	4	4	4	4	4.5	4.5	5
11	1	1.5	2	2	2	2	3	4	4	4.5	4.5	4.5	4.5	4.5	5	5	6
10	1	1.5	2	2.5	2.5	2.5	3	4	4	5	5	5	5	5	5	6	7
9	1	2	2	3	3	3	3	4	4.5	5.5	5.5	5.5	5.5	5.5	6	6	6
8	1	1.5	2	3	3.5	3.5	3.5	4	4	5	5	5	6	6	6	6	6
7	1	1.5	2	2.5	3	3	4	4	4	4	4	4	5.5	5.5	6	6	6
6	0	1.5	2	2.5	3	3	3.5	4	4	3	3	3	5	5	5	5	5
5	0	1.5	1.5	2.5	3	3	3	3.5	3.5	3	3	3	4.5	4.5	4.5	4.5	4.5
4	0	0	0	2	2	2	3	3	3	3	3	3	4	4	4	4	4
3	0	0	0	0	1	1	3	3	2	2	2	2	3.5	3.5	4	4	4
2	0	0	0	0	0	0	2	2	2	2	2	2	3	3	3	3	3

I/J	19	20	21	22	23	24	25	26	27	28	29	30	31	32	33	34	35
18	0	0	0	2	2	2	2	2	2	2	2	2	2	2	3	3	3
17	0	0	2	2	2	2	2	2.5	2.5	3	3	3	3	3	4	4	4
16	3	0	2	2	2	2	2.5	3	3	3	3	3	3.5	4	5	5	5
15	4	0	3	3	3	3	3.5	4	4	5	5	5.5	6	6	7	7	7
14	4	0	3.5	3.5	3.5	3.5	4	4.5	4.5	5	5	5.5	6	6	7	7	7
13	4	0	4	4	4	4	4.5	5	5	6	6	6	6.5	6.5	8	8	8
12	5	0	5	5	5	5	5.5	6	6	7	7	7	7.5	7.5	8	8	8
11	5	0	6	6	6	6	6.5	7	7	8	8	8	9	9	9	9	9
10	7	7	7	7	7.5	8	8	9	9	9	9.5	9.5	9.5	9.5	9.5	9.5	9.5
9	6	6	6	6	8	8	8	8	9	9	9	10	10	10	10	10	10
8	6	0	6	6	7	7	7.5	8	8	8	8	8	9	9	9.5	9.5	9.5
7	6	0	6	6	6	6	6.5	7	6	6	6	6	7	7	9	9	9
6	5	0	6	6	6	6	6	6	6	6	6	6	7	7	9	9	9
5	5	0	4	4	4	4	4	4.5	4.5	5	5	5	6	6	9	9	9
4	5	0	3	3	3	3	3	3.5	3.5	4	4	4	5	5	7	7	7
3	5	0	2	2	2	2	2	3	3	3	3	3	3	3	6	6	6
2	5	0	2	2	2	2	2	2	2	2	2	2	2	2	3	3	3

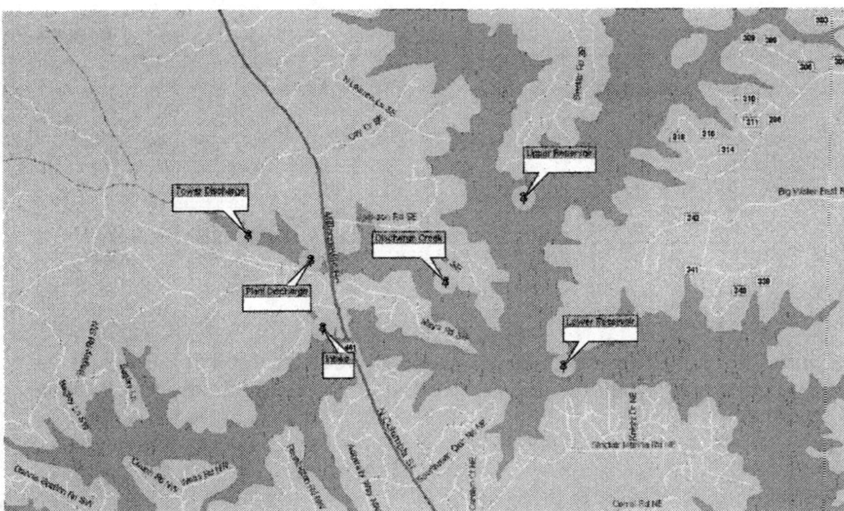

Figure 7.6.1 – Location of Steam Plant and Receiving creek Study Site

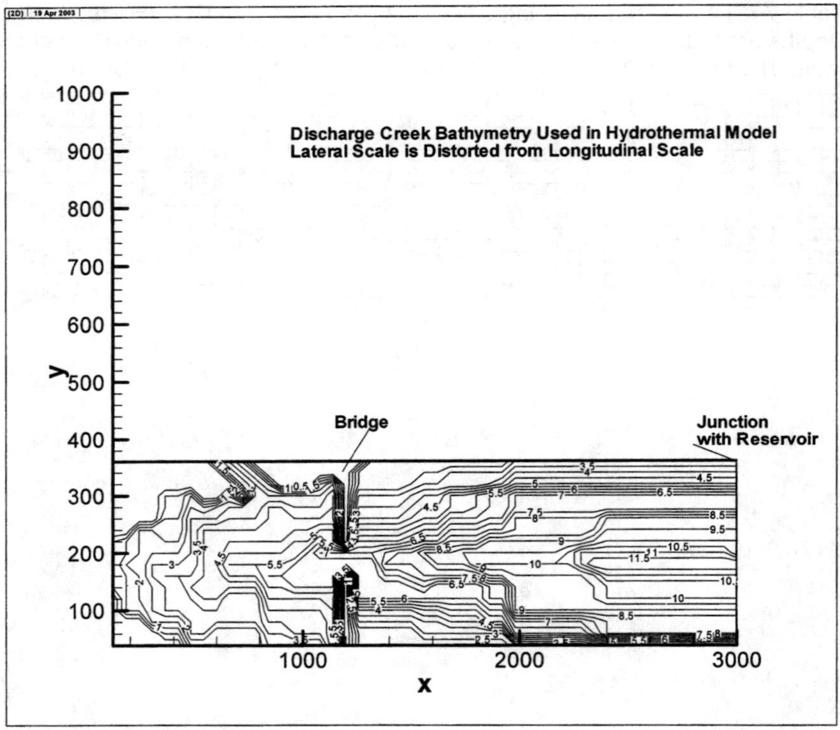

Figure 7.6.2 - Bathymetric map of receiving creek as used in Introgllvht. All dimensions are in meters. The width scale is different than the length scale.

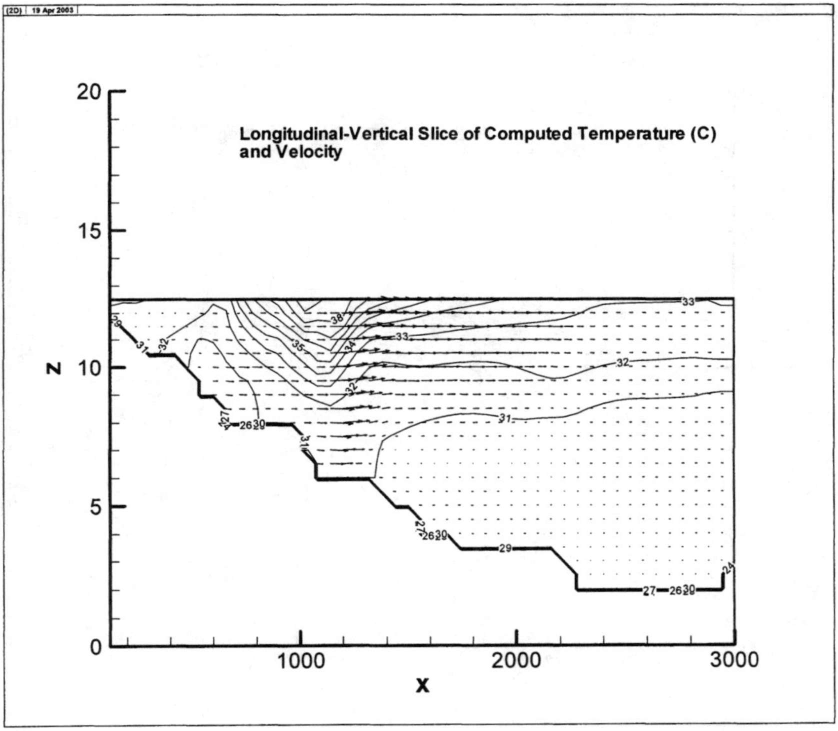

Figure 7.6.3 – Receiving creek Predicted Longitudinal-Vertical Temperatures and Velocity Predictions for Facility Operating Conditions During a Field Survey.

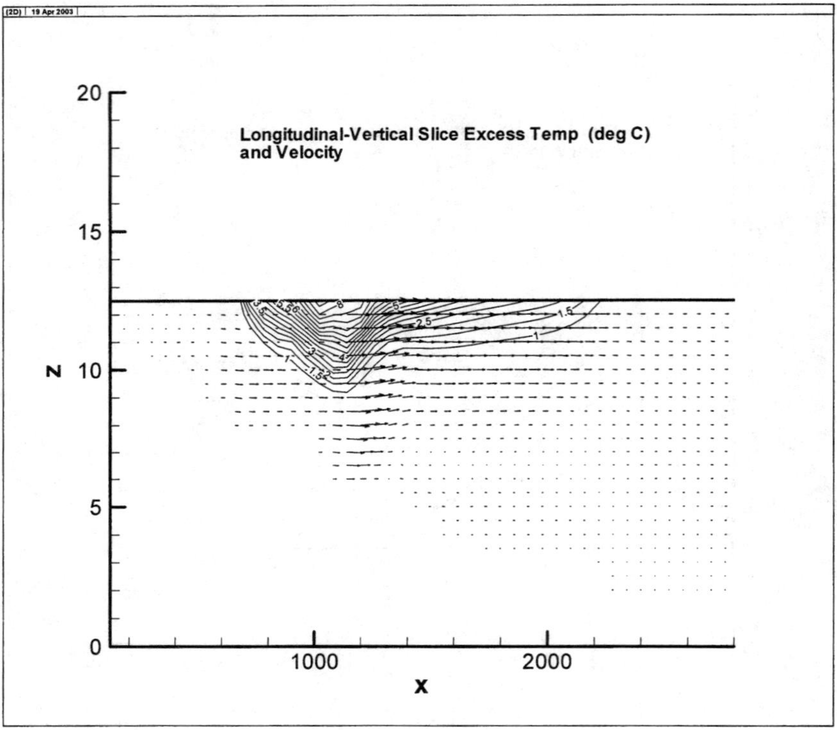

Figure 7.6.4 – Receiving creek Predicted Longitudinal-Vertical Excess Temperature and Velocity Predictions for Facility Operating Conditions During a Field Survey.

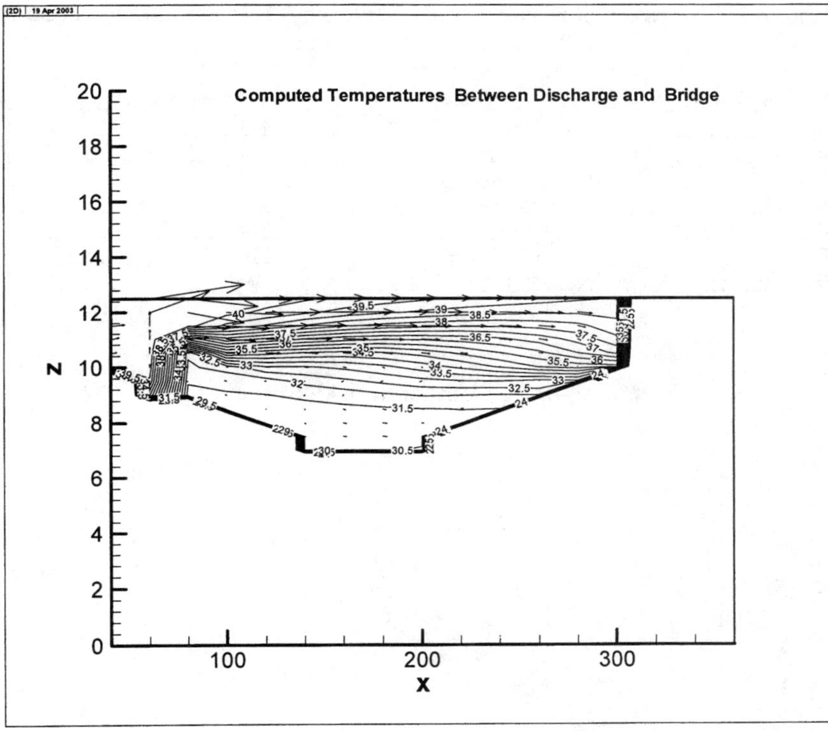

Figure 7.6.5 - Receiving creek Lateral-Vertical Temperatures and Velocities at Discharge Cross-section #D for Facility Operating Conditions during a survey. See Table 7.6.2 for location of Cross-section #D.

Table 7.6.2 - Location of Survey Cross-sections by Distance from the head of Receiving creek relative to Figure 7.6.2 and Figure 7.6.3.

Cross-Section #	Description	Distance from Head of Receiving creek, meters
A	Up str. Cool tower discharge	420
B	Cooling tower discharge	480
C	Half-way between B and D	980
D	Plant discharge	1090
E	Dn. Str of Bridge	1330
F	Half-way bridge to Mouth	2100
G	Mouth Beaverdam Creek	3000

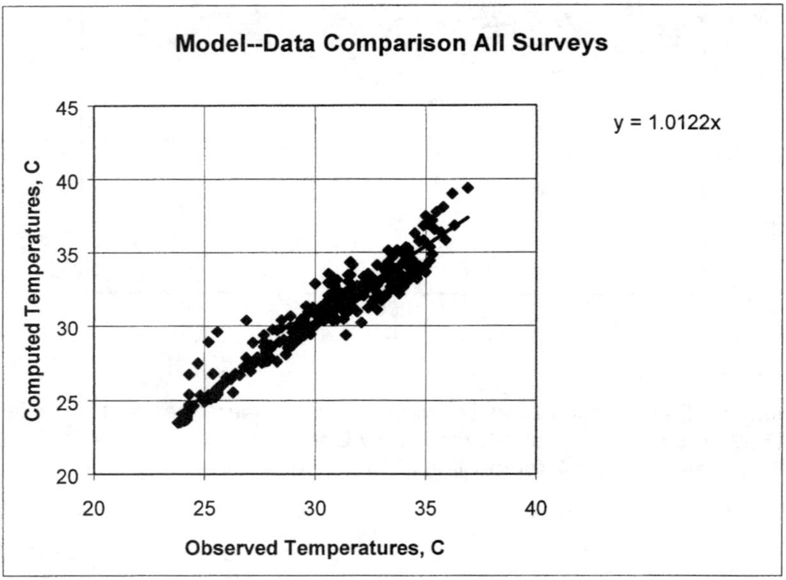

Overall Properties	
t-slope	1.089664
RMS	1.113815
Abs mn er	0.067765

Figure 7.7.6 - Comparison of Computed versus Observed Temperatures for All Surveys.

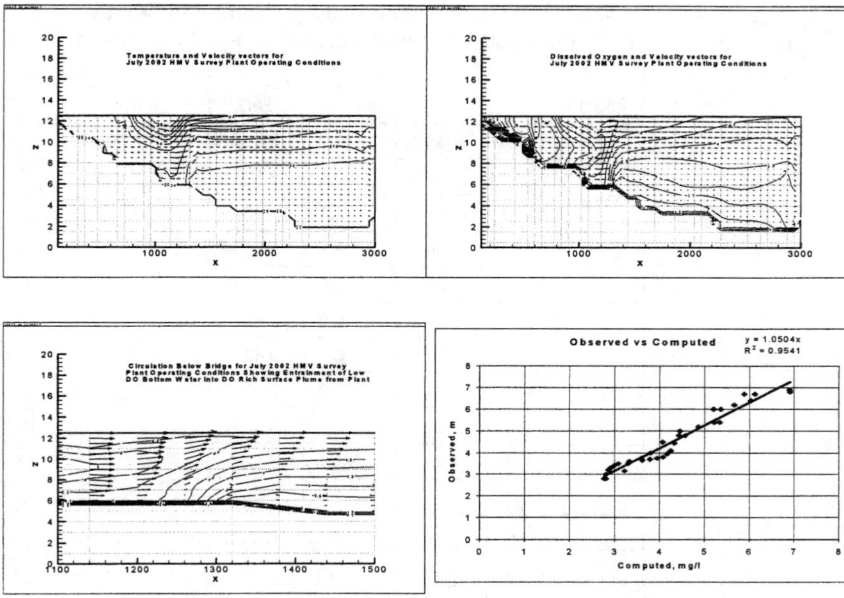

Figure 7.6.7 - Modeled longitudinal and vertical centerline profile of temperature, dissolved oxygen, near bridge circulation and comparison of observed and computed dissolved oxygen values.

Table 7.6.3 - Analysis of errors between computed and observed temperatures for August 2002 Thalweg survey using Chi^2 analysis.

Model	Obs	Cmp-obs	(X-u)2/s2
33.19	33.4	-0.2	2.9241
33.19	33.3	-0.1	2.5921
33.12	31.0	2.1	0.3844
32.88	30.0	2.9	1.9044
32.64	32.6	0.0	2.25
34.35	34.4	0.0	2.4025
33.95	33.1	0.9	0.4225
33.55	32.4	1.2	0.1225
33.02	31.5	1.5	0.0004
32.75	30.9	1.9	0.1225
32.65	30.9	1.8	0.0625
36.83	36.3	0.5	0.9409
36.35	35.7	0.6	0.7225
35.12	33.3	1.8	0.1024
34.35	31.6	2.8	1.5625
39.41	36.9	2.5	1.0201
39.02	36.2	2.8	1.7424
38.09	35.8	2.3	0.6241
36.57	35.4	1.2	0.1089
35.14	33.7	1.4	0.0036
34.15	31.7	2.5	0.9025
33.43	30.7	2.7	1.5129
Mean Error		1.5	22.4307
stand dev Error		1.0	22

Table 7.6.4 – Summary of mean and standard deviations of differences between computed and observed values at each cross-section for five surveys.

Cross-sectional Areas												
	5-13-02 HMV		6-04-02 HMV		6-26-02 HMV		7-10-02 HMV		8-28-02 HMV		Overall	
Sec	mean	std	mean	std	mean	std	mean	std	mean	std	mean	std
A	0.40	0.26	0.01	0.11	0.28	0.92	0.90	1.14	2.53	0.20	0.82	0.68
B	0.57	0.38	0.65	0.59	-0.34	1.20	0.91	1.22	-0.06	1.20	0.35	0.98
C	0.71	0.78	1.13	0.73	1.58	1.04	1.32	0.90	1.97	0.51	1.34	0.81
D	1.01	0.50	1.21	0.76	1.54	1.17	2.00	1.23	2.59	1.25	1.67	1.03
E	1.65	1.44	-0.72	0.67	-0.04	0.96	0.01	0.78	-0.27	0.69	0.13	0.95
F	-0.03	0.48	-0.87	0.84	-0.34	0.65	-0.65	0.52	-0.63	0.63	-0.50	0.64
G	0.54	0.88	0.09	0.89	0.64	0.85	0.06	0.70	0.74	5.35	0.41	2.50
Overall	0.64	0.96	0.01	1.08	0.39	1.16	0.36	1.22	0.64	2.97	0.41	1.66

Table 7.6.5 - Analysis of errors between computed and observed dissolved oxygen for August 2002 Thalweg survey using Chi^2 analysis.

Computed	**Observed**	**Computed minus Observed**	**Chi^2**
5.89	6.7	-0.81	3.5679
6.13	6.7	-0.57	1.0000
6.91	6.8	0.11	2.3058
6.90	6.9	0	1.2345
5.21	6.0	-0.79	3.2935
5.37	6.0	-0.63	1.4938
5.67	6.2	-0.53	0.7256
6.04	6.4	-0.36	0.0493
5.34	5.4	-0.06	0.7901
5.22	5.4	-0.18	0.1975
4.87	5.2	-0.33	0.0123
4.45	5.0	-0.55	0.8573
2.81	2.8	0.01	1.3182
3.79	3.7	0.09	2.0864
3.62	3.65	-0.03	1.0000
3.34	3.6	-0.26	0.0219
3.09	3.5	-0.41	0.1659
2.99	3.4	-0.41	0.1659
2.92	3.3	-0.38	0.0877
2.87	3.2	-0.33	0.0123
2.82	3.0	-0.18	0.1975
2.77	2.8	-0.03	1.0000
Mean Error		-0.30136	21.58
Standard Deviation		0.273727	22

INDEX